Srinivasa Ranga Reddy Vemuru

CIRCUIT ANALYSIS BY COMPUTER
from algorithms to package

To my family
Kathleen, Robert and Merin
ROBERT SPENCE

To my parents
Charles and Marjorie Burgess
JOHN P. BURGESS

CIRCUIT ANALYSIS BY COMPUTER
from algorithms to package

Robert Spence
Professor of Information Engineering
Imperial College, London

and

John P. Burgess
Professor of Electrical Engineering
University of New Brunswick, Canada

Prentice/Hall International

Englewood Cliffs N.J.   London   Mexico   New Delhi   Rio de Janeiro
Singapore   Sydney   Tokyo   Toronto   Wellington

Library of Congress Cataloging in Publication Data

Spence, Robert.
  Circuit analysis by computer.

  Bibliography: p.
  Includes index.
  1. Electric circuit analysis--Data processing.
I. Burgess, John P., 1938--   II. Title.
TK454.S66   1984   621.319′2′02854   84-16046
ISBN 0-13-134024-7
ISBN 0-13-134016-6 (pbk.)

British Library Cataloguing in Publication Data

Spence, Robert
  Circuit analysis by computer.
  1. Electronic circuit design—Data processing
  I. Title   II. Burgess, John
621.3815′3′02854   TK7867

ISBN 0-13-134024-7
ISBN 0-13-134016-6 Pbk

© **1986 Prentice-Hall International (UK) Ltd**

Prentice-Hall Inc., *Englewood Cliffs, New Jersey*
Prentice-Hall International (UK) Ltd, *London*
Prentice-Hall of Australia Pty Ltd, *Sydney*
Prentice-Hall Canada Inc., *Toronto*
Prentice-Hall Hispanoamericana S.A., *Mexico*
Prentice-Hall of India Private Ltd, *New Delhi*
Prentice-Hall of Japan Inc., *Tokyo*
Prentice-Hall of Southeast Asia Pte Ltd, *Singapore*
Editora Prentice-Hall do Brasil Ltda, *Rio de Janeiro*
Whitehall Books Ltd, *Wellington, New Zealand*

Printed and bound in Great Britain for
Prentice-Hall International (UK) Ltd,
66 Wood Lane End, Hemel Hempstead, Herts. HP2 4RG,
at the University Press, Cambridge.

1 2 3 4 5 90 89 88 87 86

ISBN 0-13-134024-7
ISBN 0-13-134016-6 PBK

Contents

Preface

The way in which electronic circuit simulation is taught can exhibit wide variation. It depends upon the interest of the teacher as well as the chosen text and other supporting materials and techniques. The present book has an individuality which reflects our own approach to the subject, and incorporates a number of features which distinguish it very clearly from other texts. These features include its *algorithmic* emphasis, the consideration of *package* design, the notation/programming *language* used in the exposition and the exercises, and an emphasis on student learning through extensive 'hands-on' experience at an *interactive computing* facility. We briefly examine each of these features in turn.

Conventional wisdom suggests – unfortunately – that a fundamental and exhaustive consideration of circuit theory should dominate a text concerned with computer-aided circuit design. In contrast, we take the view that although circuit theory is undoubtedly fundamental to the systematic calculation of circuit behavior, an adequate understanding of this topic is only one of many prerequisites to the creation, modification and use of a circuit analysis package. Thus a limited but adequate treatment of circuit theory is quickly used to formulate the *algorithms* which lie at the core of the systematic analysis of the d.c., a.c. and transient behavior of circuits.

By themselves, these algorithms are inadequate for realistic industrial use. They must appropriately be integrated into a software package, but with a clear understanding of their ultimate use. The design and implementation of such a package is very much an art, involving skill and insight as well as an appreciation of the needs of the industrial circuit designer who will use the package. The second half of the book therefore addresses itself to package design, and in so doing may make the student's understanding of conventional topics such as circuit theory and numerical methods far more meaningful.

The text also differs from others currently available in its pedagogical approach. Most teachers would maintain that, to gain insight, the student must spend a substantial amount of time exploring relations which are presented on the chalkboard or in the textbook. Unfortunately, whereas computers can facilitate such exploration, they typically impose the need to 'translate' a simple relation on the chalkboard into a complex program acceptable to the computer; as a consequence, the student's attention is diverted from the topic being studied. In this book the expression on the chalkboard is *identical* to what the student enters on the keyboard of the computer in order to execute the expression. Such a desirable situation is possible because the mathematical notation used is also an interactive programming language called APL (*A Programming Language*). After exploring an expression interactively – perhaps by choosing a variety of numerical examples – it is also a simple matter to embody the expression in a program for later use.

This book developed from our common interest in computer-aided circuit design, and had its origin during the 1976/77 sabbatical year which Burgess spent at Imperial College. At this College, Spence had been making full use of the APL notation and programming language in teaching a course on Computer Aided Circuit Design. Burgess had previously been investigating the advantages of interactive circuit analysis using APL at the University of New Brunswick, Canada, beginning with a preliminary circuit analysis package in 1973. Our discussions and sharing of ideas led to the concept of a useful new APL circuit analysis package. Our collaboration continued following Burgess' return to Canada, as he developed the package called BASECAP (Burgess and Spence Electronic Circuit Analysis Package), a package which incorporates and extends many of the algorithms originally developed by Spence for his course on Computer Aided Circuit Design. Thus, BASECAP was designed to function not only as a useful circuit solution tool, but also as an effective teaching tool, and is structured to aid the transition in the student's understanding from the nature of the individual algorithms to their integration in a well-structured package.

Later, we prepared a four-day course with the title 'Circuit Analysis by Computer – from Algorithms to Package'. This course was presented to predominantly industrial audiences in 1979 at Imperial College, London, and the Technical University, Eindhoven and, in 1980, at the Technische Hochschule, Zurich, and at George Washington University, Washington, DC. Presentations to university students in the UK and Canada followed. Our experience was that students having no previous knowledge of APL could successfully acquire an understanding of that language sufficient for them to emerge after four days with an understanding of d.c. and a.c. circuit analysis. This understanding was gained, not only from lectures, but also from extensive exercises carried out at an interactive terminal. Indeed, since no previous computer experience at all was called for, we concluded that such a course is well-suited to the mature engineer whose background is such that he still does not feel at home with a computer.

The BASECAP package itself deserves additional comment. As explained above, it was developed largely as a teaching tool and is discussed in our courses after the presentation of algorithms. Nevertheless, it has also been used, at the University of New Brunswick, as an integral part of courses on electronic circuit analysis and design. In view of its modular construction and clean interfacing, and as a result of its implementation in APL, it can easily be modified and extended by the user. The BASECAP package has also been put to the test – sometimes very stringently indeed – by those attending the course 'Circuit Analysis by Computer – from Algorithms to Package', and has clearly emerged as a package capable of realistic industrial use. Thus, in using this text, the student can be assured about the realism of the final result.

The history and philosophy of this text, as explained above, should enable the instructor to decide whether it will serve his or her purpose. It can be used for a senior option; it can be tackled by juniors having the necessary exposure to the fundamentals of circuit theory; it can also support courses in electronic circuit design; and experience has shown that it can form the basis of a valuable course for industrial personnel. To aid both the student and the instructor, sample solutions are provided for most of the exercises; solutions to the remaining exercises are to be found in the instructor's manual. In the intensive four-day course, a substantial proportion of the terminal sessions was devoted to repeating the model solutions, with variations, at an interactive APL terminal. More extensive problems, suitable for thesis or term projects, are outlined.

Readers can purchase both the BASECAP package described in Part 2, and the programs presented in Part 1 of this book on a single floppy diskette suitable for use with an IBM PC/XT/AT or compatible microcomputer. Initially these programs will be in versions for two different APL language implementations. Version 1 is for use with APL-PLUS/PC from STSC Inc., and Version 2 is for use with Sharp APL/PC from I. P. Sharp Associates Ltd. Enquiries should be directed to the publisher of this book at the following address:

Prentice Hall International (UK) Ltd.
66 Wood Lane End,
Hemel Hempstead, Herts. HP2 4RG.
(Attention: Glen Murray)

Many individuals and institutions have given the invaluable help that has enabled us to develop the approach presented in this text. One of us (Robert Spence) was fortunate enough to work with the creator of APL (Kenneth Iverson) at the IBM Philadelphia Scientific Center in 1971, and in so doing develop a text, *Resistive Circuit Theory*, which made exclusive use of the APL notation/programming language. In the early 1970's, financial support from the APL Foundation (Denmark) aided the development of teaching material, as did provision of interactive APL by I. P. Sharp Associates. Professors Jess and van Bokhoven of the Technical University of Eindhoven offered useful suggestions, and Professor George Moschytz and Herr Schlatter of the Technische Hochschle, Zurich, gave useful comments. Important contributions to the BASECAP package were made by several students, particularly at the University of New Brunswick. Valuable input has come also from the many students exposed to this material in both our university courses and in our industrial courses.

The evolution of our original manuscript into the present text was guided, with tact and diplomacy, by Glen Murray, and immensely helped by the suggestions of two referees. Its production was handled most professionally by Ron Decent who confronted and effectively overcame innumerable difficulties in dealing with the exacting requirements inherent in typesetting our APL functions. To these two gentlemen we are indeed most grateful. Finally, the long evolution of this book could hardly fail to have been noticed by our families to whom we are infinitely grateful for their understanding and support.

R.S.
J.P.B.

THE ALGORITHMS

CHAPTER

1

Linear Resistive Circuits

The design of an electronic circuit to satisfy a customer's specification is usually a process of 'trial-and-error', in which an initial design is gradually refined in the light of the performance it exhibits. The more experienced the designer, or the easier the design, the fewer tends to be the number of iterations necessary before the circuit design is judged to be satisfactory.

Before the widespread availability of computers, the performance of a designed circuit was typically *measured* by constructing a prototype. A comparison of measured with desired performance then enabled the designer to suggest a modification which, on the basis of his prior experience and his insight into circuit behavior, he would expect to lead to either a satisfactory design or a useful improvement.

The construction of a prototype is both expensive and time consuming. By contrast, it is easy and inexpensive to *simulate* circuit behavior on a computer by analyzing a model of the circuit. An additional advantage is the designer's ability to selectively modify a single parameter — for example, the current gain of a transistor — to study its individual effect. A disadvantage is that the performance of a *model* of the circuit is computed rather than that of the circuit itself: this is why an understanding of device modelling is so important.

There is, therefore, a need to be able to compute the quiescent (d.c.), frequency-domain (a.c.) and time-domain performance of electronic circuits as an aid to the circuit designer. That is the subject of this book.

The kernel of many algorithms for the d.c., frequency-domain and time-domain analysis of electronic circuits is — or closely parallels — the analysis of a linear resistive circuit. We shall therefore study this class of circuit first. For the moment, the circuits will be assumed to contain only two-terminal linear resistors. The reader wishing to consult other sources treating the material of this and some later chapters is referred to Spence (1979).

1.1 Component Relations

Let the circuit contain a number B of resistors. For a circuit containing five resistors, these may be as shown in Fig. 1.1 where, it should be noted, the eventual connections between the components are not shown. With each resistor is associated a current and a voltage known as the *branch current* and *branch voltage.* We let *IB* denote the vector of all branch currents. Thus, for the set of components in Fig. 1.1, *IB* is a 5-element vector and *IB*[2], for example, is the current flowing through the 1 ohm resistor. Similarly, we let *VB* denote the vector of branch voltages. Thus if,

    VB←1 3 5 2 10

we know that the voltage across the 2 S resistor, labelled component number 5, is 10 V.

The vectors *IB* and *VB* are related by the matrix equation

    IB←*GB* + .×*VB*   (1.1)

which is a collective statement of Ohm's law for all the resistors in the circuit. For the collection of resistors shown in Fig. 1.1.

$$GB = \begin{matrix} 4 & 0 & 0 & 0 & 0 \\ 0 & 1 & 0 & 0 & 0 \\ 0 & 0 & 3 & 0 & 0 \\ 0 & 0 & 0 & 5 & 0 \\ 0 & 0 & 0 & 0 & 2 \end{matrix}$$

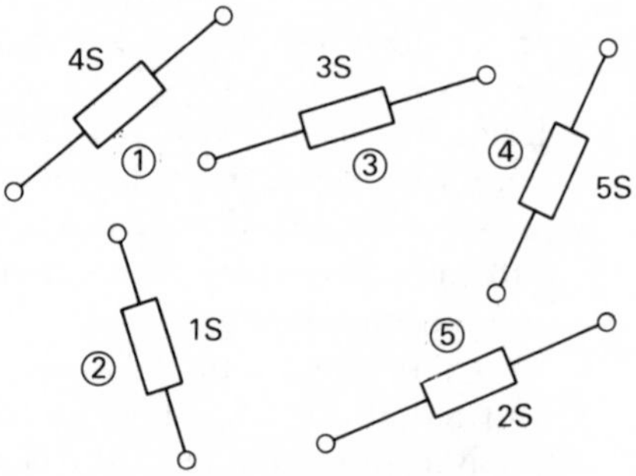

Fig. 1.1
A collection of resistors. Each resistor is numbered, and its conductance shown

Thus, if the branch voltages are described by

    VB←1 3 5 2 10

then the branch current vector is

            GB+ ×*VB*
     4   3   15   10   20

In other words, the current through component 1 is 4 A, that through component 2 is 3 A, and so on. Note that, up until now, we have not described the manner in which the components are connected together.

1.2   Component Interconnection

The connection of components imposes constraints on their voltages and currents. Kirchhoff's current law (KCL) requires that a sum of the currents flowing into a connection point is zero, and Kirchhoff's voltage law (KVL) states that the sum of the component voltages around any closed loop of components is zero.

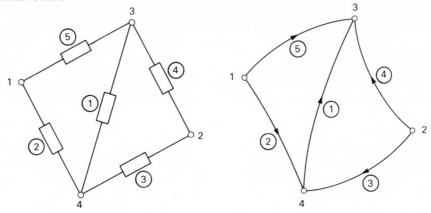

Fig. 1.2   **Fig. 1.3**
Circuit with node numbering   **Directed graph of circuit of Fig. 1.2**

To discuss the constraints imposed by the topology (i.e., the connectedness) of the circuit we need a model for each component which simply defines the current path it offers and the points between which a voltage can be measured. We use the *branch/node representation* illustrated, for the circuit of Fig. 1.2, by the *directed graph* of Fig. 1.3. For reference purposes the branches and nodes are identified by number, and arbitrary reference directions are assigned to branch currents. The reference polarity for branch voltages is implied by the reference current arrow as shown in Fig. 1.4.

A directed graph of B branches and N nodes can equivalently be described by an *incidence matrix* of size N,B – that is, having N rows and B columns. For the graph of Fig. 1.3 the incidence matrix is

$$A = \begin{matrix} 0 & 1 & 0 & 0 & 1 \\ 0 & 0 & 1 & 1 & 0 \\ \bar{1} & 0 & 0 & \bar{1} & \bar{1} \\ 1 & \bar{1} & \bar{1} & 0 & 0 \end{matrix}$$

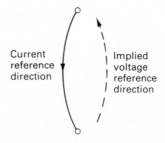

Current
reference
direction

Implied
voltage
reference
direction

where entries of 1 and $\bar{1}$ denote a branch reference current flowing *away from* and *into* a node respectively, and a 0 entry indicates that the branch is not connected to the node. Each column of A sums to zero

Fig. 1.4
Reference directions for current
(indicated) and voltage (implied)

$$+/[1]A$$
$$0 \quad 0 \quad 0 \quad 0 \quad 0$$

because the reference current associated with a given branch, and hence a column of A, must leave one node ($+1$) and enter another ($\bar{1}$). Thus, one row of A can be regarded as redundant.

1.2.1   Kirchhoff 's Current Law

For a circuit with no externally applied current sources (e.g., Fig. 1.2), Kirchhoff 's current law can be expressed as

$$(N\rho 0) = A +. \times IB$$

Thus for the circuit of Fig. 1.2, whose directed graph is shown in Fig. 1.3, and for which a set of branch currents obeying Kirchhoff's current law is

$$IB \leftarrow \bar{}3 \quad \bar{}2 \quad \bar{}1 \quad 1 \quad 2$$

we find that

$$A +. \times IB$$
$$0 \quad\quad 0 \quad\quad 0 \quad\quad 0$$

However, if currents are injected into the circuit from independent external sources as shown in Fig. 1.5, the product $A +. \times IB$ is modified.

To take an illustrative numerical example, suppose the externally injected nodal currents shown in Fig. 1.5 are denoted by the vector I , where

$$I \leftarrow \bar{}4 \quad \bar{}1 \quad 2 \quad 3$$

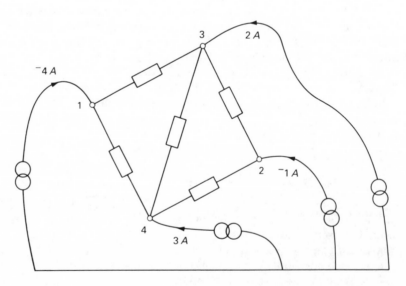

Fig. 1.5
The circuit of Fig. 1.2 with independent currents injected at the nodes

One set of branch currents IB which, in conjunction with the externally injected nodal currents, satisfies Kirchhoff 's current law, is

$$IB \leftarrow \bar{}1 \quad \bar{}6 \quad 2 \quad \bar{}3 \quad 2$$

The product

$$
\begin{array}{cccc}
 & & A+.\times IB & \\
\bar{4} & \bar{1} & 2 & 3
\end{array}
$$

is then seen to return the vector of injected nodal currents. That the expression

$$I = A+.\times IB \tag{1.2}$$

summarizes Kirchhoff's current law for the entire circuit can be verified by reference to the definition of the incidence matrix.

1.2.2 Kirchhoff's Voltage Law

The incidence matrix can also be used to express Kirchhoff's voltage law in the expression

$$VB = (\lozenge A)+.\times V \tag{1.3}$$

where V is the N-vector of nodal voltages. Thus, if we arbitrarily choose nodal voltages for the graph of Fig. 1.3,

$$V \leftarrow 3 \quad 1 \quad 2 \quad 5$$

then the corresponding branch voltages are

$$
\begin{array}{ccccc}
 & & (\lozenge A)+.\times V & & \\
3 & \bar{2} & \bar{3} & \bar{1} & 1
\end{array}
$$

a result which can easily be verified by inspection. Expression (1.3) can also be verified by reference to the definition of the incidence matrix: a branch voltage is the difference between two nodal voltages, and each row of A (with one 1 and one $\bar{1}$) selects the appropriate nodal voltages.

1.3 Circuit Description

Three relations describe a circuit: the component branch relations (1.1); Kirchhoff's current law (1.2); and Kirchhoff's voltage law (1.3). Substituting for IB from (1.1) into (1.2), and then for VB in the expression that results, we have

$$I \leftarrow (A+.\times (GB+.\times (\lozenge A)+.\times V)) \tag{1.4}$$

However, since matrix multiplication is associative, (1.4) can be rewritten as

$$I \leftarrow (A+.\times GB+.\times \lozenge A)+.\times V \tag{1.5}$$

Expression (1.5), representing what are known as the *nodal equations* of the circuit, is useful because it relates a circuit's nodal voltages to the currents flowing into the nodes from external sources. In other words, it is an *external* description of the circuit. It is often written as

$$I \leftarrow Y + . \times V \tag{1.6}$$

where Y is the complete nodal conductance matrix of the circuit. It is of size N, N and can be computed according to the relation

$$Y \leftarrow A + . \times GB + . \times \lozenge A \tag{1.7}$$

Example 1.1
For the circuit of Fig. 1.2 and the component values shown in Fig. 1.6,

$$GB = \begin{matrix} 4 & 0 & 0 & 0 & 0 \\ 0 & 1 & 0 & 0 & 0 \\ 0 & 0 & 3 & 0 & 0 \\ 0 & 0 & 0 & 5 & 0 \\ 0 & 0 & 0 & 0 & 2 \end{matrix}$$

$$\square \leftarrow Y \leftarrow A + . \times GB + . \times \lozenge A$$

$$\begin{matrix} 3 & 0 & \bar{2} & \bar{1} \\ 0 & 8 & \bar{5} & \bar{3} \\ \bar{2} & \bar{5} & 11 & \bar{4} \\ \bar{1} & \bar{3} & \bar{4} & 8 \end{matrix}$$

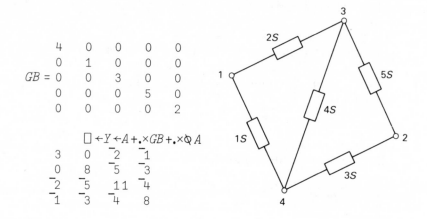

Fig. 1.6
A circuit comprising linear two-terminal conductances

Note that each row and column of Y sums to zero:

$$+/[1]Y$$
$$0 \quad 0 \quad 0 \quad 0$$

$$+/[2]Y$$
$$0 \quad 0 \quad 0 \quad 0$$

If the circuit is excited by voltage sources as shown in Fig. 1.7, then the vector V is known. For this example

$$V \leftarrow 4 \quad 3 \quad \bar{1} \quad 2$$

so that

$$I \leftarrow Y + . \times V$$
$$I$$
$$12 \quad 23 \quad \bar{4}2 \quad 7$$

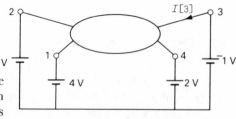

Fig. 1.7
The circuit of Fig. 1.6 with voltage excitation

Thus, we have found the currents flowing into the nodes from the external sources; e.g., $I[3]$, which equals $^-42$ A, is the current flowing into node 3. Note that

$$+/I$$
$$0$$

the sum of all the currents flowing into the circuit is zero, in accordance with Kirchhoff's current law.

1.4   Circuit Response to Current Exitation

In practice, it is the vector of nodal currents (I) that is known. For example, with a filter (Fig. 1.8) having 5 internal (i.e., inaccessible) nodes, we know that

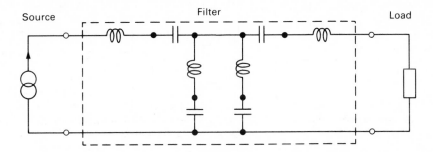

Fig. 1.8
A filter circuit, with source and load terminations. No external current sources can be applied to the inaccessible (shaded) nodes within the box.

the current injected into each of these nodes from an external source is zero; by contrast, we do not know their voltages. Thus Y and I are given, and V is to be found. In other words, a relation which is the *inverse* of Equation (1.6) is sought.

While one could propose the relation

$$V \leftarrow Z +.\times I$$

any attempt to find Z by inverting Y

$$\boxplus\ Y$$
$$DOMAIN\ ERROR$$
$$\boxplus\ Y$$
$$\land$$

will fail, because Y is a singular matrix (remember, its rows and columns sum to zero — see Example 1.1). A physical interpretation of the reason that $\tilde{Y}$ cannot be inverted is that the reference voltage for the circuit is arbitrary, so that for the same nodal currents there is an infinity of sets of possible nodal voltages.

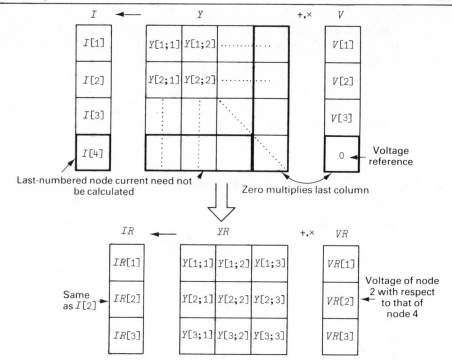

Fig. 1.9
Illustrating the derivation of the reduced nodal condutance matrix from the complete nodal conductance matrix

A restriction to one set of nodal voltages is achieved by selecting *one* of the nodes as the reference point for nodal voltages. This approach will be illustrated in the context of the circuit of Fig. 1.7. We choose node 4 , the highest numbered node, as the voltage reference point, so

$$0 = V[4]$$

Since this voltage multplies column 4 of Y in Equation (1.6), there is no need to retain this column. We have also seen that the elements of I sum to zero, and therefore it is only necessary to determine three elements of I : let us therefore choose not to determine directly the value of the nodal current $I[4]$ associated with the reference node. As illustrated in Fig. 1.9, we then have a new set of nodal equations

$$IR \leftarrow YR + . \times VR \tag{1.8}$$

where YR is the reduced nodal conductance matrix obtained by dropping the last row and column from Y, IR is the reduced vector of nodal currents, and VR is the reduced vector of nodal voltages as follows:

$$YR \leftarrow {}^{-}1 \ {}^{-}1 \downarrow Y$$
$$IR \leftarrow {}^{-}1 \downarrow I$$
$$VR \leftarrow {}^{-}1 \downarrow V - V[4]$$

Example 1.2

For the circuit of Fig. 1.6

$$\square\leftarrow YR\leftarrow \bar{\ }1\quad \bar{\ }1\downarrow Y$$

$$
\begin{array}{rrr}
3 & 0 & \bar{\ }2 \\
0 & 8 & \bar{\ }5 \\
\bar{\ }2 & \bar{\ }5 & 11
\end{array}
$$

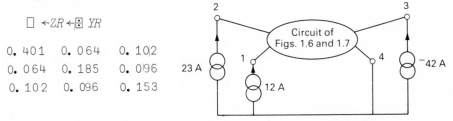

Fig. 1.10 shows the selection of a voltage reference node.

Thus, if the circuit is excited by the voltage sources shown in Fig. 1.10 (an excitation which is equivalent to that of Fig. 1.7)

$$VR\leftarrow 2\quad 1\quad \bar{\ }3$$

$$\square\leftarrow IR\leftarrow YR+.\times VR$$

$$12\quad 23\quad \bar{\ }42$$

This current response is identical with that obtained earlier in Example 1.1: this is expected since, if we take the complete nodal voltage vector

$$VR,0$$

$$2\quad 1\quad \bar{\ }3\quad 0$$

and add the previous voltage of node 4,

$$(VR,0)+2$$

$$4\quad 3\quad \bar{\ }1\quad 2$$

we obtain the earlier nodal voltage vector.

We now return to the task of finding the voltage response to current excitation. The inverse of expression (1.8) is

$$VR\leftarrow ZR+.\times IR \tag{1.9}$$

and, if ZR and IR are known, this form allows the voltage response VR to be calculated. The matrix ZR is the inverse of YR which (in contrast to Y) is nonsingular. For the example of Fig. 1.6 and Fig. 1.7,

$$\square\leftarrow ZR\leftarrow\boxplus YR$$

$$
\begin{array}{ccc}
0.401 & 0.064 & 0.102 \\
0.064 & 0.185 & 0.096 \\
0.102 & 0.096 & 0.153
\end{array}
$$

Fig. 1.11
The circuit of Fig. 1.6 with nodal current excitation

Thus, for the nodal current excitation shown in Fig. 1.11,

$$IR \leftarrow 12 \ 23 \ ^-42$$
$$\square \leftarrow VR \leftarrow ZR + . \times IR$$
$$2 \quad 1 \quad ^-3$$

It is seen that, in this example, the excitation currents of Fig. 1.11 were taken to be the response currents of the circuit of Fig. 1.7, so we would expect the voltage responses and excitations to be identical as well. Equation (1.9) for voltage response can also be written as

$$VR \leftarrow IR \boxplus YR$$

which, for a *circuit*, is a generalized form of the corresponding relation

$$V \leftarrow I \div G$$

for a *component*, where G is the component conductance.

1.5   Nodal Analysis Summarized

The relations discussed above are summarized in Fig. 1.12. For completeness,

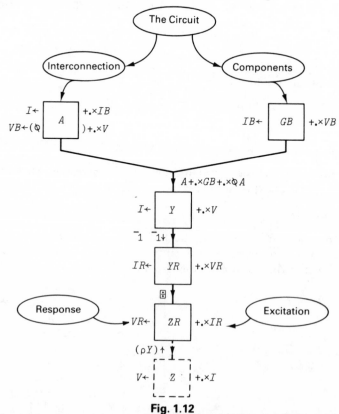

Fig. 1.12
Representation of the relation between the expressions describing the voltage response, to current excitation, of a circuit containing linear components

one may wish to add the further stage (see dotted section) to include a *complete* nodal resistance matrix Z relating the complete nodal voltage and current vectors. This is simply done: for the example of Fig. 1.6

$$Z \leftarrow (\rho Y) \uparrow ZR$$
$$Z$$

0.401	0.064	0.102	0
0.064	0.185	0.096	0
0.102	0.096	0.153	0
0	0	0	0

1.6   Functions Easing the Analysis of a Circuit Containing Two-terminal Linear Resistors

To facilitate the analysis of a circuit containing only linear two-terminal resistors we make use of two simple functions, $INCID$ and $COND$ (Table 1.1) which, respectively, ease the formation of the incidence and branch conductance matrices.

Table 1.1

	$\nabla A \leftarrow INCID\ M$		
[1]	$N \leftarrow \lceil /,M$	N	is the highest node number
[2]	$AP \leftarrow (\iota N)\circ . = M[1;]$	AP	contains the $+1$ entries
[3]	$AN \leftarrow (\iota N)\circ . = M[2;]$	AN	contains the $^-1$ entries
[4]	$A \leftarrow AP - AN$	A	is the incidence matrix
	∇		
	$\nabla GB \leftarrow COND\ \ G$		
[1]	$B \leftarrow \rho G$	B	is the number of branches
[2]	$GB \leftarrow ((B,B)\rho G) \times UNIT\ B$	GB	is the branch conductance
	∇		matrix
	$\nabla R \leftarrow UNIT\ N$		
[1]	$R \leftarrow (\iota N)\circ . = \iota N \nabla$		Generation of the unit matrix

---

The circuit of Fig. 1.13 will now be analyzed with the aid of the functions in Table 1.1. First form a branch connection matrix M, each of whose columns indicates the 'out of' and 'into' nodes of the branches arranged in order:

$$\square \leftarrow M \leftarrow \lozenge 5\ 2 \rho 1\ 2\ 3\ 2\ 1\ 3\ 3\ 4\ 2\ 4$$
$$1\ 3\ 1\ 3\ 2$$
$$2\ 2\ 3\ 4\ 4$$

from which the incidence matrix is generated by using the function $INCID$:

```
     A←INCID M
     A
 1   0   1   0   0
¯1  ¯1   0   0   1
 0   1  ¯1   1   0
 0   0   0  ¯1  ¯1
```

The vector of component conductances

```
G←5 2.5 1 4 2
```

is then used as the argument of $COND$ to generate the branch conductance matrix GB

```
     GB←COND G
     GB
5    0    0    0    0
0    2.5  0    0    0
0    0    1    0    0
0    0    0    4    0
0    0    0    0    2
```

The voltage response to the applied current excitation

```
        IR
14   0   ¯7
```

is then obtained according to the algorithm depicted in Fig. 1.12:

```
     VR←IR⊟ ¯1 ¯1↓A+.×GB+.×⌽A
     VR
4.5   2.5   0.5
```

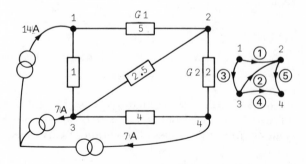

Fig. 1.13
A circuit containing linear two-terminal
conductances, and its directed graph

The branch voltages can be found from Kirchhoff's voltage law (1.3):

$$VB \leftarrow (\Diamond A) + . \times VR, 0$$
$$VB$$
$$2 \quad {}^-2 \quad 4 \quad 0.5 \quad 2.5$$

and then the branch currents from the branch relations:

$$IB \leftarrow GB + . \times VB$$
$$IB$$
$$10 \quad {}^-5 \quad 4 \quad 2 \quad 5$$

Finally, as a check, we compute the expected nodal currents according to Kirchhoff's current law.

$$I \leftarrow A + . \times IB$$
$$I$$
$$14 \quad {}^-1.8208E{}^-14 \quad {}^-7 \quad {}^-7$$

and see that they agree with the actual nodal excitation currents.

1.7   **Alternative Formation of** Y

The generation of the complete nodal conductance matrix Y by means of the expression

$$Y \leftarrow A + . \times GB + . \times \Diamond A$$

involves many products in which one or both factors are zero. For this reason an alternative method described below is usually preferred, since it avoids the redundant products.

Consider the circuit of Fig. 1.13 containing only linear resistive two-terminal components. Let the nodal voltages be designated by the four-element vector V . It is clear from inspection that, due to these voltages, a current

$$G1 \times V[1] - V[2]$$

flows through $G1$ and must, therefore, be supplied by the external excitation currents applied at nodes 1 and 2. Thus, if the structure of the equations describing the circuit is as set out in Fig. 1.14, the connection of $G1$ between nodes 1 and 2 introduces the four elements shown into the complete nodal conductance matrix.

Next, consider the conductance $G2$ to be connected between nodes 2 and 4, thereby drawing a current $G2 \times V[2] - V[4]$ which must be supplied by the external current applied at nodes 4 and 2. Thus, four new elements must be *added* to the complete nodal conductance matrix, as shown in Fig. 1.15. This process continues until the contributions due to all conductances have been

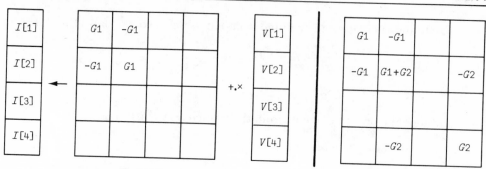

Fig. 1.14
The contribution, to the complete
nodal conductance matrix, of a
conductance $G1$ **connected**
between nodes 1 and 2

Fig. 1.15
The contribution of conductance
$G2$ **to the complete nodal**
conductance matrix is added to the
already existing partially formed
matrix

added into the matrix, whereupon the complete nodal conductance matrix is obtained.

It is not difficult to formulate the general rule for forming the Y matrix of an *N*-node circuit composed of linear two-terminal resistive components. First, a zero matrix of the appropriate shape is generated:

$$Y \leftarrow (N, N) \rho 0$$

Then, for each conductance G , we form a 2×2 matrix

$$\Delta Y \leftarrow 2\ 2\rho\ 1\ ^-1\ ^-1\ 1\ \times G$$

which is that component's contribution to the Y matrix. The elements of this matrix must then be added to the appropriate elements of Y . If the conductance G is connected between nodes P and Q, and we let $I \leftarrow P, Q,$ this addition is executed by the expression

$$Y[I; I] \leftarrow Y[I; I] + \Delta Y$$

To incorporate the above algorithm in a function, we must first describe the circuit in a suitable form. As an alternative to the incidence and branch conductance matrices we employ a single matrix GT having three columns and as many rows as there are components. The first and second columns define the nodes between which the conductances are connected, while the third column contains their values. Thus, the circuit of Fig. 1.13 is described by the matrix GT

```
        GT
    1   2   5
    3   2   2.5
    1   3   1
    3   4   4
    2   4   2
```

From the above discussion the following function can be developed: it takes the matrix GT describing the circuit as its argument.

```
        ∇ Y←FORM  GT
[1]       N←⌈/,GT[;1 2]
[2]       B←1↑ρ GT
[3]       Y←(N,N)ρ0
[4]       K←1
[5]     L:ΔY←2 2 ρ 1 ¯1 ¯1 1×GT[K;3]
[6]       NDS←GT[K;1 2]

[7]       Y[NDS; NDS]←Y[NDS; NDS]+ΔY
[8]       →L×(K←K+1)≤B
```

N is the number of nodes
B is the number of branches (components)
Initial value of Y
Component index
Addition to Y due to component K
Nodes between which component is connected
Updated value of Y
Termination if all components considered

Thus, for the circuit example of Fig. 1.13,

```
        Y←FORM  GT
        Y
  6        ¯5       ¯1        0
 ¯5         9.5     ¯2.5     ¯2
 ¯1        ¯2.5      7.5     ¯4
  0        ¯2       ¯4        6
```

we can confirm that $FORM$ returns the correct complete nodal conductance matrix.

Note that, after the nodal voltages have been computed in the usual way:

```
        IR←14 0 ¯7
        VR←(⊟ ¯1 ¯1↓Y)+.×IR
        VR
4.5  2.5  0.5
```

the branch voltages, previously obtained by executing (1.3)

$$VB←(⍉A)+.×V←VR,0$$

can instead be generated by an indexing operation:

```
        VB←-/V[GT[;1 2]]
        VB
 2  ¯2  4   0.5  2.5
```

Exercises

1.1 Form the branch conductance matrix GB for the collection of conductances shown in Fig. E1.1.

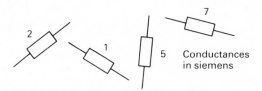

Fig. E1.1

1.2 The conductances of Exercise 1.1 are connected according to the graph of Fig. E1.2. (Note that you have already decided the number of each conductance by ordering them in Exercise E1.1.) Derive an incidence matrix A for this graph.

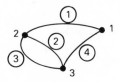

Fig. E1.2

1.3 Obtain the complete nodal conductance matrix Y of the circuit.

1.4 Check that every row and column of Y sums to zero.

1.5 Find the shape of Y .

1.6 The circuit is now excited by voltage sources as shown in Fig. E1.3. Calculate the values of the currents $I[1]$, $I[2]$ and $I[3]$.

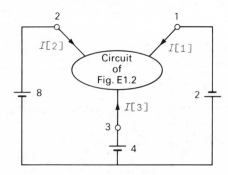

Fig. E1.3

1.7 Check, by a single calculation on the APL terminal, that these currents sum to zero.

1.8 Choose node 3 as the reference node, and obtain the reduced nodal conductance matrix YR of the circuit of Fig. E1.3.

1.9 Check the shape YR.

1.10 Obtain the reduced nodal resistance matrix ZR of the circuit of Fig. E1.3.

1.11 Define the reduced nodal current vector IR describing the nodal current excitation shown in Fig. E1.4.

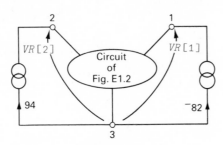

Fig. E1.4

1.12 Calculate the values of $VR[1]$ and $VR[2]$ (see Fig. E1.4).

1.13 From VR, calculate the vector VB of branch voltages.

1.14 From VB, calculate the vector IB of branch currents.

1.15 From IB, compute the expected injected nodal currents. Do they agree with the actual values?

1.16 Index the $Z|R$ matrix to obtain the voltage at node 2 (with respect to the reference node) due to a 1A current source applied at node 1. (This current is returned via the reference node).

1.17 Index the $Z|R$ matrix to obtain all the nodal voltages (except that of the reference node) due to a 1A current source applied at node 2.

1.18 Compute the voltage response (i.e., VR) for the same circuit (Fig. E1.3) but with the new current excitation shown in Fig. E1.5.

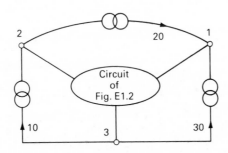

Fig. E1.5

1.19 You have now considered two different current excitations of the same circuit. Combine these two current excitations into a 2×2 matrix and then compute the corresponding 2×2 matrix of nodal voltage responses using the relation
$VR \leftarrow ZR + . \times IR$

1.20 Above, you chose the last-numbered node as the reference node. Now choose node 2 as the reference node and obtain the new *YR* from *Y* by means of logical reduction, as illustrated below:

```
            M
    1   2   3    4
    5   6   7    8
    9  10  11   12
   13  14  15   16
         L V ← 1  1  0  1
          L V / [ 1 ] M
    1   2   3    4
    5   6   7    8
   13  14  15   16
```

1.21 For the *YR* matrix you obtained in Exercise 1.20, obtain the reduced nodal resistance matrix *ZR*. Is it identical with the previous *ZR* ?

1.22 Using the new value of *ZR*, calculate the nodal voltage response to the current excitation shown in Fig. E1.4. Does your answer agree with that obtained for Exercise 1.12 (it should !)?

1.23 In Exercise 1.20 you made use of a logical vector. Use this same vector to perform expansion on the vector *VR* to obtain the complete nodal voltage vector *V*.

1.24 Generate a matrix containing all possible nodal current excitation vectors, each containing only one nonzero source of one ampere (for the circuit of Fig. E1.3 this will be a 2×2 matrix). In obtaining this matrix try to use the outer product operator.

1.25 Using the matrix obtained in Exercise 1.24, generate the voltage response to each current vector. By inspecting your answer and, if necessary, comparing it with *ZR*, say what interpretation can be placed on the separate elements of *ZR* concerning voltage response to current excitation.

Solutions

```
        ∇ A←INCID M;N;AP;AN
[1]     N←⌈/,M
[2]     AP←(⍳N)∘.=M[1;]
[3]     AN←(⍳N)∘.=M[2;]
[4]     A←AP-AN∇
        ∇ GB←COND G;B
[1]     B←ρ G
[2]     GB←((B,B)ρ G)×UNIT B∇
        ∇R←UNIT N
[1]     R←(⍳N)∘.=⍳N∇
```

It may be convenient to use the functions defined in Table 1.1.

If they are not already in your workspace they are easy to enter.

1.1
```
        X←7 1 5 2
        X
7   1   5   2
        GB←COND X
        GB
7   0   0   0
0   1   0   0
0   0   5   0
0   0   0   2
```

By ordering the conductance values in the vector X we have implicitly assigned branch numbers (e.g., the 5S conductance is branch 3).

1.2
```
        M←⍉4 2ρ2 1 2 3 3 2 3 1
        M
2   2   3   3
1   3   2   1
        A←INCID M
        A
¯1   0   0   ¯1
 1   1  ¯1   0
 0  ¯1   1   1
```
Number of branches

Before defining the branch connection matrix M we must (arbitrarily) choose branch reference directions (see Fig. E1.6).

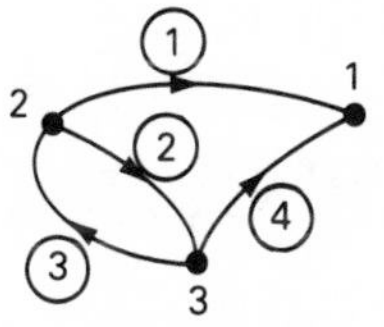

Fig. E1.6

The incidence matrix

```
        T∆INCID←⍳4
        A←INCID M
INCID[1]   3
INCID[2]
0   0   0   0
1   1   0   0
0   0   1   1
INCID[3]
1   0   0   1
0   0   1   0
0   1   0   0
INCID[4]
¯1   0   0  ¯1
 1   1  ¯1   0
 0  ¯1   1   1
        T∆INCID←⍳0
```

We take the opportunity to explore the trace facility. This expression indicates that we want to examine the result of lines 1–4.

Number of nodes

+1 entries in incidence matrix

Location of ¯1 entries in incidence matrix

The incidence matrix

Removal of trace on *INCID*

1.3
```
        Y←A+.×GB+.×⍉A
        Y
 9  ¯7  ¯2
¯7  13  ¯6
¯2  ¯6   8
```

1.4
$$+/[1]Y$$
  0   0   0
All columns sum to zero

$$+/[2]Y$$
  0   0   0
All rows sum to zero

$$+/Y$$
  0   0   0
Reduction over *last* dimension of an array need not be stated explicitly.

1.5
$$\rho Y$$
  3   3
There are 3 nodes in the circuit.

1.6
$$V \leftarrow {}^-2\ 8\ 4$$
$$V$$
 ${}^-2$ 8   4
The nodal voltage excitation

$$I \leftarrow Y+.\times V$$
$$I$$
 ${}^-82$ 94 ${}^-12$
The nodal current response

1.7
$$+/I$$
  0
The currents obey Kirchhoff's Current Law

1.8
$$YR \leftarrow {}^-1\ {}^-1 \downarrow Y$$
$$YR$$
      9    ${}^-7$
     ${}^-7$ 13
The reduced nodal conductance matrix corresponding to the choice of the *last* numbered node as the reference node.

1.9
$$\rho YR$$
  2   2
The circuit has 3 nodes, 2 in addition to the reference node

1.10
$$ZR \leftarrow \boxplus YR$$
$$ZR$$
  0.19118   0.10294
  0.10294   0.13235
Matrix inversion yields ZR

1.11
$$IR \leftarrow {}^-82\ 94$$
Note that IR is chosen to be identical to the previous current *response*

1.12
$$VR \leftarrow ZR+.\times IR$$
$$VR$$
 ${}^-6$ 4
The value of VR corresponds to the excitation shown in Fig. E1.3 (see Note immediately above)

$$VR[1]$$
 ${}^-6$

$$VR[2]$$
  4

1.13
$$V \leftarrow VR,0$$
$$V$$
 ${}^-6$ 4   0
Remember to include the voltage (zero) of the reference node in V.

$$VB \leftarrow (\lozenge A)+.\times V$$
$$VB$$
  10   4 ${}^-4$ 6

1.14
$$IB \leftarrow GB+.\times VB$$
$$IB$$
  70   4 ${}^-20$ 12

1.15
$$I \leftarrow A+.\times IB$$
$$I$$
 ${}^-82$ |94 ${}^-12$
The branch currents agree with the injected nodal currents

1.16
$$ZR[2;1]$$
  0.10294

1.17 |Z|R[;2]   All rows are indexed.
          0.10294 0.13235

1.18   IRN←50 ¯10   Current excitation for Fig. E1.5
          VRN←ZR+.×IRN   Voltage response
          VRN
          8.5294 3.8235

1.19   IRX←⍉2 2ρIR,IRN   The 2×2 array combining the two excitations
          IRX
      ¯82   50
      94   ¯10

          VRX←ZR+.×IRX.   Exactly the same relation is used to compute
          VRX   voltage response
      ¯6   8.5294
      4   3.8235

1.20   LV ←1   0   1   Row 2 and column 2 are to be removed
          YR ←LV/[1]LV/Y
          YR   Remember that in reduction over the last dimension
      9   ¯2   the dimension need not be stated.
     ¯2   8

1.21   ZR ← ⌹ YR   This matrix is not identical with the previous ZR
          ZR
      0.11765   0.029412
      0.029412   0.13235

1.22   IR← ¯82   ¯12
          VR←ZR +.×IR   The voltage response agrees with that obtained in
          VR   the answer to Exercise 1.12.
      ¯10   ¯4

1.23   V ←LV\VR   Complete vector of nodal voltages.
          V
      ¯10 0 ¯4

1.24   IR←1   2∘.=1   2
          IR
     1   0
     0   1   Alternative ways of obtaining a 2×2 unit matrix
          IR←UNIT 2
          IR
     1   0
     0   1

1.25   VR←ZR+.×IR
          VR   The voltage response to the current vectors is identi-
      0.11765   0.029412   cal with the reduced nodal resistance matrix.
      0.029412   0.13235
          ZR   The ZR matrix elements are the voltage response,
      0.11765   0.029412   at the node corresponding to a *row*, to a current of 1
      0.029412   0.13235   A applied at the node corresponding to a *column*.

Quiescent Analysis I

The majority of manufactured electronic circuits contain *nonlinear* components such as diodes and transistors. In general, however, a resistive circuit containing one or more nonlinear components does not admit of a direct solution, and an iterative algorithm must be employed. In such an approach, an initial guess at the solution is successively refined until the actual solution is achieved within a specified tolerance.

It is possible to illustrate many aspects of quiescent (i.e., d.c.) analysis with the simple circuit of Fig. 2.1 containing *one* nonlinear resistive component, *one* constant excitation current source and *one* unknown response voltage. In a later part of the text (Chapter 7) the approach will be extended to a general nonlinear circuit containing more than one component, and more than one excitation and response, and where any component may have more than two terminals.

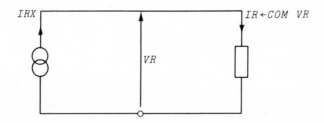

Figure 2.1
A nonlinear resistive component excited by an independent current source

2.1 The Newton–Raphson Algorithm

The most common iterative algorithm for quiescent (i.e., d.c.) analysis is called the Newton–Raphson algorithm: it will be described by reference to Fig. 2.2 which shows the current-voltage characteristics $(IR \leftarrow COM\ VR)$ of the nonlinear component.

The function COM (of which an illustrative example is in Table 2.1) takes a scalar voltage argument and returns the corresponding component current. Shown in this figure are the excitation current IRX and the corresponding value of VR.

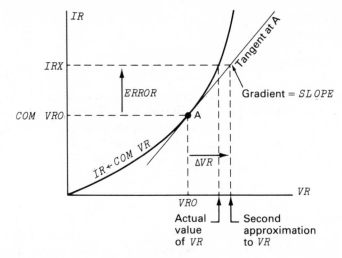

Figure 2.2
Graphical illustration of one iteration of the Newton–Raphson iteration

The function embodying the Newton–Raphson algorithm takes the current IRX and the initial guess VRO at the unknown voltage as its arguments, and returns VR as its result:

$$VR \leftarrow IRX \ ITER \ VRO \qquad\qquad (2.1)$$

We shall consider the few steps that constitute the function $ITER$ separately. First, since the initial guess might be correct, we assign its value to VR

$$VR \leftarrow VRO \qquad\qquad (2.2)$$

A test is then carried out to see if the component current associated with this voltage ($COM \ VR$) differs from the current that is known to flow from the excitation source IRX, by forming as $ERROR$ the difference between the two:

$$ERROR \leftarrow IRX - | COM \ VR$$

This quantity is sometimes known as the *nodal imbalance current*. If the magnitude of the error ($|ERROR$) is within a specific tolerance (say $1E^{-}6$), the current value of VR is accepted as being sufficiently close to the actual value, the function branches to 0 (i.e. it terminates), and the current value of VR is returned as the result. If not, branching to line $L2$ is caused:

$$\rightarrow L2 \times 1E^{-}6 < | ERROR \leftarrow IRX - COM \ VR \qquad\qquad (2.3)$$

If the error is too large, the correction to VR is based on the assumption that the current–voltage function COM is linear around the current operating point. Thus, a new function is used to generate the slope ($SLOPE$)of the function COM for the current value of VR:

$$SLOPE \leftarrow DIF_|COM \ VR \qquad\qquad (2.4)$$

If the function COM *were* linear around the current value of VR, the correction ΔVR to VR would be given by

$$VR \leftarrow VR + \Delta VR \leftarrow ERROR \div SLOPE \qquad (2.5)$$

Despite any approximation involved, this correction is, in fact, accepted. By means of a branching command the new value is then tested according to (2.3), and the iterations continue until the magnitude of $ERROR$ is less than the prescribed tolerance, here chosen to be 1 μA for illustration.

A function $ITER$ embodying the above steps is:

```
      ∇VR←IRX ITER VRO;ERROR;SLOPE;ΔVR
[1]     VR←VRO
[2]     L1:→L2×1E‾6<|ERROR←IRX-COM VR
[3]     L2:SLOPE←DIFCOM VR
[4]     VR←VR+ΔVR←ERROR÷SLOPE
[5]     →L1 ∇
```

$$(2.6)$$

Here, for the first time, we illustrate the *localization* of the variables generated within a function. In this case they are $ERROR$, $SLOPE$ and ΔVR. Such localization has two advantages that we wish to identify here. The first is educational: the localized variable names constitute a reminder of the essential steps in the algorithm. The other will later be seen to be associated with good programming practice, in that if variables are not localized, the workspace can quickly be filled with unwanted variables, and lead to the danger that some variables might have their values changed accidentally.

Example 2.1

A specific example can be used for a brief illustration of the function $ITER$. If the nonlinear resistor is described by a cubic equation (see Table 2.1 for details of the functions COM and $DIFCOM$), and the current

```
         Table 2.1

      ∇  IR←COM VR          Function simulating a
[1]      IR←VR+0.1×VR*3      nonlinear component
      ∇

      ∇  S←DIFCOM VR         Function which generates the
[1]      S←1+0.3×VR*2        component slope conductance
      ∇
```

excitation IRX is equal to 4 A, execution of the function with a guessed value for VR of 11 V

```
    4 ITER 11
2.4781
```

returns a value for VR of 2.4781V, which is correct within the tolerance defined in line 2 of $ITER$. Confirmation of this result is straightforward:

```
    COM 4 ITER 11
4
```

Use of a new guessed value of VR

```
    4 ITER ‾16
2.4781
```

is seen to return the same value of VR as before. Such a brief demonstration of $ITER$ does not in any way constitute a rigorous test of that function, or provide sufficient insight into its operation. Further consideration is suggested as an exercise for the student (see Exercises).

2.2   Taylor-series Derivation of the Newton–Raphson Algorithm

The circuit equation can be expressed in conventional notation as:

$$0 = F(VR) \tag{2.7}$$

For example, in the circuit of Fig. 2.1:

$$0 = (COM\ VR) - IRX$$

Let $VR[J+1]$ denote a Taylor series expansion of (2.7) around the Jth guess at VR yielding[†]

$$F(VR[J+1]) = F(VR[J]) + \left.\frac{dF(VR)}{dVR}\right|_{VR=VR[J]} \times (VR[J+1] - VR[J]) +$$

$$\left.\frac{d^2 F(VR)}{dVR^2}\right|_{VR=VR[J]} \times (VR[J+1] - VR[J])^2 \quad + \dots \left.\right\} \tag{2.8}$$

Now assume that $VR[J]$ is quite a good guess at the solution, so that high-order terms can be neglected and

$$F(VR[J+1]) = F(VR[J]) + \left.\frac{dF(VR)}{dVR}\right|_{VR=VR[J]} \times (VR[J+1] - VR[J]) \left.\right\} \tag{2.9}$$

---

† APL does not include notation related to differentiation, so conventional symbolism is used here

It is now our intention to choose $VR[J+1]$ such that it is a solution of (2.7); in other words,

$$F(VR[J+1]) = 0 \tag{2.10}$$

Substitution in (2.9) yields

$$VR[J+1] - VR[J] = -\left.\frac{F(VR[J])}{\dfrac{\mathrm{d}\,F(VR)}{\mathrm{d}\,VR}}\right|_{VR=VR[J]} \qquad\Biggr\} \tag{2.11}$$

or

$$VR[J+1] = VR[J] - F(VR[J]) \div J(VR[J]) \tag{2.12}$$

where J denotes the gradient of F and is a function of VR. A geometrical interpretation of (2.12) appears in Fig. 2.3. Comparison with the function (2.6) shows that $J(VR) = SLOPE$ and $((-IRX) + COM \;\; VR) = -ERROR = F(VR)$

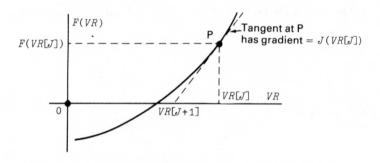

Figure 2.3
A geometric interpretation of equation 2.12

2.3 Circuit Interpretation of the Newton–Raphson Algorithm

Figure 2.4 illustrates a circuit interpretation of the Newton–Raphson algorithm, involving the assumption of linear behaviour around the current value of VR. The 'companion model', shown in Fig. 2.4(c), possesses precisely the same current–voltage relation as the dashed line (Fig. 2.4(b)) which is tangential to the nonlinear relation COM at the current value of voltage. Thus, when excited by an external current source it will develop a voltage VR, which is an approximation to the actual value of VR, the approximation improving as the difference between the estimated and actual values decreases in magnitude.

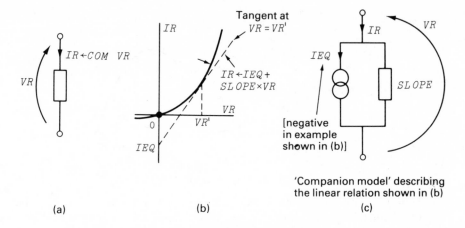

Figure 2.4
A circuit interpretation of the Newton–Raphson algorithm

The circuit interpretation can also be embodied in the following function:

```
     ∇VR←IRX ITER1 VRO; ERROR; IR; SLOPE ; IEQ
[1]   VR←VRO
[2]   L 1:→L 2×1E ‾6<|ERROR←IRX- IR← COM  VR
[3]   L 2:SLOPE←DIFCOM  VR
[4]    IEQ←IR- VR×SLOPE
[5]    VR←(IRX-IEQ)÷SLOPE
[6]    →L 1     ∇
```

2.4   Convergence

From Fig. 2.3 it can be seen that for the $F(VR) = 0$ chosen for illustration, the algorithm will converge if the initial guess is close to the solution. It can be shown (Chua and Lin (1975), p. 217) that, if convergence occurs, the rate of convergence for the Newton–Raphson algorithm is *quadratic*. In other words, if $E[J]$ is the difference between the J th approximation to VR and the exact solution, then,

$$E[J+1] \leq K \times E[J] * 2$$

Thus, if the error is small, each error will decrease to a constant times the square of the previous error. In this way, the number of correct decimal places in successive approximations to the solution is approximately doubled at each iteration.

But convergence cannot always be guaranteed. Consider, for example, Fig. 2.5, which illustrates a case in which there are two possible solutions.

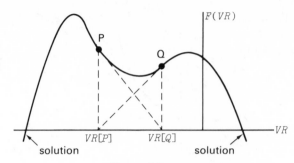

Figure 2.5
Illustrating numerical oscillation

Whereas the *actual* circuit would move towards and remain at one of these solutions, the nature of the Newton–Raphson algorithm is such that neither of these solutions is attained. Rather, a continuous numerical oscillation occurs, as illustrated in the figure.

There are other – commonly encountered – situations for which convergence of the Newton–Raphson algorithm does not only not occur, but where one of the successively computed values of VR lies outside the domain (range) handled by the computer. Within APL this leads to a $DOMAIN\ ERROR$, and execution of the algorithm is halted. It is not difficult to find a practical example. The semiconductor diode is typically modelled by a mathematical model – called the 'exponential diode' – described by the relation

$$IR \leftarrow IS \times {}^-1 + \star 40 \times VR \tag{2.13}$$

The quantity IS is called the reverse saturation current because it is the largest negative value of IR that can be obtained. The current expressed by this relation exhibits a very rapid rise, typically for voltages approaching and exceeding 1V (Fig. 2.6) so that, even if the actual voltage is quite modest (0.5 V, say), a small positive correction to an intermediate value of VR can easily be associated with an extremely large current, and give rise to a $DOMAIN$

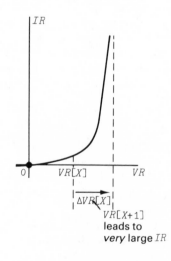

Figure 2.6
A possible cause of
$DOMAIN\ ERROR$

ERROR. Thus, while the *actual* circuit has a well-defined behavior, it is the fault of the *algorithm* that this behavior is not predicted.

Since semiconductor diodes, as well as other devices giving rise to the same computational problem, are found in the great majority of electronic circuits, a reliable means of ensuring convergence of the Newton–Raphson algorithm must be found. Three common approaches are described below, and illustrated extensively in the solutions to the exercises for this chapter.

2.4.1   Linearization beyond a Critical Voltage

For the exponential diode (Fig. 2.7: full curve) the current-voltage characteristic is approximated, for voltages greater than some critical value VT, by a

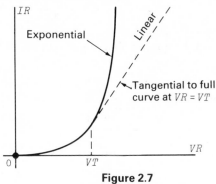

Figure 2.7
Linearization beyond
a critical voltage

straight line (Fig. 2.7: dashed curve) which is tangential to the full curve at $VR = VT$. With this model, an increase (correction) ΔVR in VR will rarely lead to a *DOMAIN ERROR.* Naturally, the value of VT is chosen to exceed any likely solution for VR, though it is advisable to check the eventual solution in this respect. Convergence is quite sensitive to the initial value of VR as well as to the choice of VT.

2.4.2   Very Small Slope Conductance

Very large reverse voltages associated with an exponential diode can lead to a value of $SLOPE$ which is so small that $ERROR \div SLOPE$ lies outside the range of

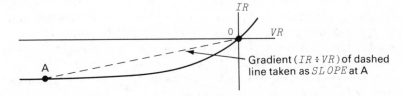

Figure 2.8
Illustrating a method of overcoming a problem associated with a very small
slope conductance

the computer, and a $DOMAIN$ $ERROR$ is returned. A possible solution is to approximate the slope conductance as $IR \div VR$ for VR less than some critical value (Fig. 2.8).

Another simple approach is to allow the value of the slope ($SLOPE$) to be computed, but to take the maximum of $SLOPE$ and some specified minimum value $SMIN$ as the value to be used in the next stage of the algorithm:

$$SLOPE \leftarrow SMIN \lceil SLOPE$$

2.4.3   Control of the Magnitude of Voltage Corrections

An approach which has been found to be effective in practice limits component voltage excursions within the high-current region, with emphasis on the situation in which current increases. Thus, increases in voltage are limited to VM (say 0.026 V) if the *final* voltage lies within the high-current region, defined by a voltage greater than a critical value VT (say, 0.52 V). Decreases are similarly limited in magnitude if the *initial* voltage lies in the high current region ($VR>VT'$).

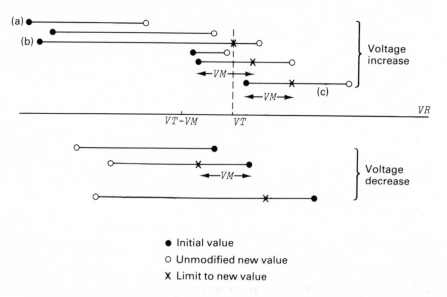

● Initial value
○ Unmodified new value
X Limit to new value

Figure 2.9
A method of controlling the magnitude of voltage corrections

The details of the algorithm are summarized in Fig. 2.9, which addresses all possible categories of voltage change, and indicates the limits actually imposed on these changes. Thus, example (a) of Fig. 2.9 illustrates a case in which no modification is made to the computed voltage correction. In (b), since the computed correction would make the new value of VR exceed VT, the correction is modified so that the new VR is equal to VT. In (c), VR is already greater than VT, so the actual correction to VR is limited to VM, even though the initially computed correction may have been greater than VM.

2.5   Comment

It is useful to recall the purpose of this chapter: it is to introduce the algorithmic techniques employed in the quiescent analysis of a nonlinear *circuit*. To illustrate these techniques in the simplest possible way we have, however, chosen to consider the quiescent analysis of a single *component*, knowing that generalization to handle a circuit is (as we see in Chapter 7) a straightforward matter. Thus, the fact that the voltage VR across an exponential diode due to an excitation current IR can easily be obtained from the inverse of (2.13)

$$VR \leftarrow (\div 40) \times \circledast 1 + IR \div IS$$

rather than by application of the Newton–Raphson algorithm, is irrelevant: for a nonlinear *circuit* such an inverse relation will usually be impossible to obtain.

Exercises

2.1 Explore the application of the Newton–Raphson algorithm to the analysis of a single two-terminal component in the following way:

(a)   Assume that the current of the component is related to the voltage across it by the relation $IR \leftarrow VR + .1 \times VR * 3.$ Define one function (COM) for this relation and another ($DIFCOM$) which generates the slope of this relation. Thoroughly test the Newton–Raphson algorithm, for example by employing a very wide range of initial guesses at the result.

(b)   Incorporate in $ITER$ an iteration counter, and investigate the effect on the number of iterations (and hence the cost of executing the function), of the initial guess VRO and the tolerance which was initially set (on line 2 of $ITER$) at 1 μA.

(c)   Use the function COM to generate that excitation which corresponds to a component voltage of exactly 1 V (i.e. $IRX \leftarrow COM$ 1). Then execute $ITER$, but with a trace to display successive values of VR. By observing the number of zeros following the decimal point observe the quadratic convergence. (It may be necessary to set print precision to a rather high value, for example $\square PP \leftarrow 15$).

(d)   Test the Newton–Raphson iteration for the case of a linear resistor, and check that the number of iterations is what you would expect.

2.2 An exponential diode imposes a relation between current and voltage of the form

$$IR \leftarrow IS \times {}^{-}1 + * 40 \times VR$$

Embody this relation in a defined function $DIODE$, and also define a function which generates the slope of the current–voltage relation: use the name **DIFDIODE** for the latter function. Introduce these functions appropriately into the Newton–Raphson function. Use the Newton–Raphson function to compute the diode voltage from knowledge of the current injected into the diode. Note any problems which arise, and decide *why* they arise.

2.3 To overcome the problems encountered with the exponential diode expression, it is convenient to describe the relation between diode current and voltage as before up to a threshold voltage VT, and by a linear relation thereafter, but with the requirement that the slope of the relation between current and voltage be continuous. Define a suitable function relating diode current and voltage, of the form

$$IR \leftarrow COMLIN \ VR$$

where VT is a global variable which can be chosen at will. Normally, it is set to a value such as 1 V, which is higher than any expected value of diode voltage. Also, define a function $DIFCOMLIN$ which takes the diode voltage as its argument and returns the slope of the current–voltage relation. Before proceeding, test the functions $COMLIN$ and $DIFCOMLIN$. Incorporate these functions within the Newton–Raphson iterative algorithm, and test this algorithm for a variety of positive diode voltages.

2.4 Test the above algorithm with, as the initial guess at the diode voltage, a very large negative value such as $^{-}100.$ Comment on the reason for any problem encountered.

2.5 To alleviate the problem encountered in 2.4 above one may take, as the value of $SLOPE$ when the diode voltage is negative, the ratio of the absolute values of diode current and voltage. In other words, when $V<0,$ $SLOPE \leftarrow I \div V.$ Appropriately redefine $COMLIN$ and $DIFCOMLIN$ and test the Newton–Raphson algorithm for a very wide range of guessed initial voltages. Note the number of iterations involved in achieving a solution.

2.6 Define the function $ITER1$ and test it on the diode example.

2.7 Test the Newton–Raphson algorithm developed in Exercise 5 for the case of a *reverse* biassed diode for which the magnitude of the excitation current is less than the tolerance (e.g. $1E^{-}6$) placed on the nodal imbalance current, and devise a means of overcoming the associated problem.

Solutions

2.1(a)

```
     ∇   VR←IRX ITER VRO
[1]      VR←VRO
[2]    L1:→L2×1E¯6<|ERROR←IRX-COM VR
[3]    L2:SLOPE←DIFCOM VR
[4]      VR←VR+ΔVR←ERROR÷SLOPE
[5]      →L1
     ∇
```
Function to execute the Newton–Raphson algorithm.

```
     ∇   IR←COM VR
[1]      IR←VR+0.1×VR*3
     ∇
```
Function simulating a component.

```
     ∇   S←DIFCOM VR
[1]      S←1+0.3×VR*2
     ∇
```
Function which generates the component's slope conductance.

```
     4 ITER 5
2.4781
     COM 2.4781
3.9999
     COM 4 ITER 5
4
```
Use of the function $ITER$. With a 4 A excitation, component voltage is 2.478 V.

Result checked.

```
     4 ITER 1000
2.4781
     4 ITER ¯2000
2.4781
```
Same result using different initial guesses of VR.

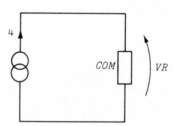

Figure E2.1

(b)

```
     ∇   VR←IRX ITER VRO
[1]      W←0
[2]      VR←VRO
[3]    L1:→L2×TOL<|ERROR←IRX-COM VR
[4]    L2:SLOPE←DIFCOM VR
[5]      VR←VR+ΔVR←ERROR÷SLOPE
[6]      W←W+1
[7]      →L1
     ∇
```
Modification of $ITER$ to count number of iterations.

```
     TΔITER←5
```
A trace is put on line 5 of $ITER$ to display successive values of VR.

```
        4 ITER 5
ITER[5] 3.4118
ITER[5] 2.6586
ITER[5] 2.4863
ITER[5] 2.4782
ITER[5] 2.4781
2.4781
        W
5
```
                                                   An example, for which five iterations were involved.

Experiment to investigate effect of tolerance and initial guess at VRO on the number of iterations required. See Fig. E2.2.

$TOL \leftarrow 1E^-3$	$TOL \leftarrow 1E^-6$	$TOL \leftarrow 1E^-9$
` 4 ITER 5`	` 4 ITER 5`	` 4 ITER 5`
`2.4782`	`2.4781`	`2.4781`
`   W`	`   W`	`   W`
`4`	`5`	`5`
` 4 ITER 50`	` 4 ITER 50`	` 4 ITER 50`
`2.4781`	`2.4781`	`2.4781`
`   W`	`   W`	`   W`
`10`	`10`	`11`
` 4 ITER 500`	` 4 ITER 500`	` 4 ITER 500`
`2.4783`	`2.4781`	`2.4781`
`   W`	`   W`	`   W`
`15`	`16`	`17`
` 4 ITER 5000`	` 4 ITER 5000`	` 4 ITER 5000`
`2.4781`	`2.4781`	`2.4781`
`   W`	`   W`	`   W`
`21`	`22`	`22`

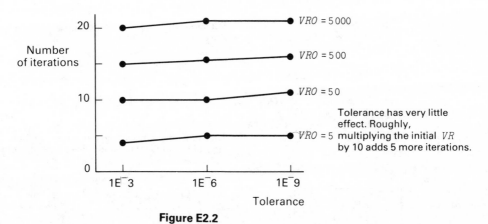

Number of iterations vs **Tolerance**

- $VRO = 5000$
- $VRO = 500$
- $VRO = 50$
- $VRO = 5$

Tolerance has very little effect. Roughly, multiplying the initial VR by 10 adds 5 more iterations.

Figure E2.2

(c)

```
IRX←COM 1
IRX
```
`1.1`
```
TOL←1E^-15

□PP←15
```

The current excitation corresponds to a response voltage of exactly 1V.

```
      IRX ITER 5
ITER[5]   3.07058823529412
ITER[5]   1.79969156313312
ITER[5]   1.14918015931102
ITER[5]   1.00525747739771
ITER[5]   1.00000638553001
ITER[5]   1.00000000000941
ITER[5]   1
1
```

Roughly, the number of zeros after the decimal point doubles at each iteration, suggestive of quadratic convergence.

(d)

```
    ∇   IR←COM VR
[1]     IR←3×VR
    ∇
```

A linear two-terminal conductance.

```
    ∇   S←DIFCOM VR
[1]     S←3
    ∇
```

DIFCOM generates the slope conductance of the linear conductance.

```
T∆ITER←ι0
□PP←5
TOL←1E¯6
```

The trace is removed from *ITER*.
The print precision is reduced.

```
    3 ITER 6
1
    W
1
    ERROR
0
```

For the case of a linear conductance, only one iteration is required.

2.2

```
    ∇   IR←DIODE VR
[1]     IR←1E¯11×¯1+*40×VR
    ∇
```

Function describing the current–voltage characteristic of an exponential diode.

```
    ∇   S←DIFDIODE VR
[1]     S←1E¯11×40×*40×VR
    ∇
```

Function which generates the slope conductance of the exponential diode.

```
    ∇   VR←IRX ITERDIODE VRO
[1]     VR←VRC
[2]  L1:→L2×1E¯6<|ERROR←IRX-DIODE VR
[3]  L2:SLOPE←DIFDIODE VR
[4]     VR←VR+∆VR←ERROR÷SLOPE
[5]     →L1
    ∇
```

The function embodying the Newton–Raphson algorithm modified to handle an exponential diode.

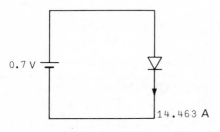

Figure E2.3

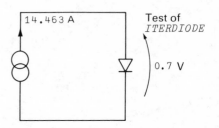

Figure E2.4

```
        DIODE .7
14.463
        14.463 ITERDIODE .7
0.7
        14.463 ITERDIODE .8
0.7

        14.463 ITERDIODE .5      The initial guess of 0.5 V for VRO leads to a
                                 DOMAIN ERROR.
DOMAIN ERROR
DIODE[1]IR←1E⁻11×⁻1+*40×VR       The second value of VR is so large(75.001 V) that
              ∧                  an attempt to raise e to this power results in a
        VR                       DOMAIN ERROR.
75.001
        ERROR
14.458
```

2.3

```
    ∇    IR←COMLIN VR
[1]      IR←(DIODE VRⵊVT)+(DIFDIODE VT)×0⌈(VR-VT)
    ∇
```
The function $COMLIN$ describes a new current–voltage relation which is identical to $DIODE$ for $VR \leq V$ but linear thereafter. There is no change in slope at the threshold voltage VT.

```
        VT←.7
        COMLIN .1 .2 .3 .4 .5 .6 .7 .8 .9 1.0
5.3598E⁻10 2.98E⁻8 1.6275E⁻6 8.8861E⁻5 0.0048517 0.26489
        14.463 72.313 130.16 188.01

        A←COMLIN .1×ι10
        B←DIODE .1×ι10
        A-B
0 0 0 0 0 0 ⁻2.7311E⁻14 ⁻717.32 ⁻42982 ⁻2.3537E6
        A=B
1 1 1 1 1 1 1 0 0 0
```
$COMLIN$ agrees with $DIODE$ for $VR \leq VT$.

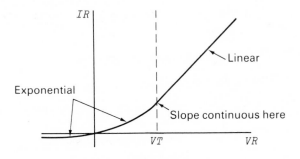

Figure E2.5

```
    ∇    R←DIFF X
[1]      R←(⁻1↓X)-1↓X
    ∇

        DIFF A
⁻2.9264E⁻8 ⁻1.5977E⁻6 ⁻8.7234E⁻5 ⁻0.0047628 ⁻0.26004 ⁻14.198
        ⁻57.85 ⁻57.85 ⁻57.85
```
Constant slope for $VR \geq VT$.

```
     ∇   S←DIFCOMLIN VR
[1]      S←DIFDIODE VR⌊VT
     ∇
```
Function which generates slope conductance of
COMLIN.

```
     DIFCOMLIN .1×ι10
2.1839E‾8 1.1924E‾6 6.5102E‾5 0.0035544 0.19407 10.596
     578.5 578.5 578.5 578.5
```
Test of *DIFCOMLIN.*

```
     ∇   VR←IRX ITERSPLINE VRO
[1]      W←0
[2]      VR←VRO
[3]   L1:→L2×1E‾6<|ERROR←IRX-COMLIN VR
[4]   L2:SLOPE←DIFCOMLIN VR
[5]      VR←VR+ΔVR←ERROR÷SLOPE
[6]      W←W+1
[7]      →L1
     ∇
```
The Newton–Raphson function modified to handle
the component described by *COMLIN*.

```
     VT
0.7
```
The threshold voltage is set at 0.7 V.

```
     COMLIN .7
14.463
     14.463 ITERSPLINE 1
0.7
     W
1
     14.463 ITERSPLINE 10
0.7
     W
1
```
At the threshold voltage, the component current is
14.463 A.
  With these guesses at *VR*, (which are greater than
VT) the result is achieved in one iteration, since
only a linear function is encountered.

```
     TΔITERSPLINE ←5
```
A trace is put on line 5 of *ITERSPLINE*.

```
     IRX←DIODE .6
     IRX
0.26489
```
A current is generated which, when applied as excitation to a diode, will cause a response voltage of
0.6 V, which is below the threshold of *VT*.

```
     IRX ITERSPLINE .8
ITERSPLINE[5]   0.67546
ITERSPLINE[5]   0.65168
ITERSPLINE[5]   0.62984
ITERSPLINE[5]   0.61242
ITERSPLINE[5]   0.60263
ITERSPLINE[5]   0.60013
ITERSPLINE[5]   0.6
ITERSPLINE[5]   0.6
0.6
```
Satisfactory convergence.

```
     IRX ITERSPLINE 10
ITERSPLINE[5]   0.67546
ITERSPLINE[5]   0.65168
ITERSPLINE[5]   0.62984
ITERSPLINE[5]   0.61242
ITERSPLINE[5]   0.60263
ITERSPLINE[5]   0.60013
ITERSPLINE[5]   0.6
ITERSPLINE[5]   0.6
0.6
```
A much larger initial guess at *VR* does not increase
the number of interations.

2.4

```
      IRX ITERSPLINE ‾33
DOMAIN ERROR
ITERSPLINE[5]  VR←VR+ΔVR←ERROR÷SLOPE
                  ∧
```

```
      ERROR
0.26489
      SLOPE
0
```
A large negative value of VR results in a *DOMAIN ERROR* because the slope conductance is zero and *ERROR÷SLOPE* is infinite.

2.5

```
    ∇   S←NEWDIFCOMLIN VR
[1]     →(VR>0)/L1
[2]     S←(COMLIN VR)÷VR
[3]     →0
[4]   L1:S←DIFCOMLIN VR
    ∇
```
The function *DIFCOMLIN* which generates slope conductance is replace by a function *NEWDIFCOMLIN* which ensures that, for negative values of the voltage VR, the value of *SLOPE* becomes $IR÷VR$.

```
    ∇   VR←IRX ITERSPLINE VRO
[1]     W←0
[2]     VR←VRO
[3]   L1:→L2×1E‾6<|ERROR←IRX-COMLIN VR
[4]   L2:SLOPE←NEWDIFCOMLIN VR
[5]     VR←VR+ΔVR←ERROR÷SLOPE
[6]     W←W+1
[7]     →L1
    ∇
```
The Newton–Raphson function is appropriately modified.

```
      IRX ITERSPLINE ‾33
ITERSPLINE[5]  8.7414E11
ITERSPLINE[5]  0.67567
ITERSPLINE[5]  0.65189
ITERSPLINE[5]  0.63002
ITERSPLINE[5]  0.61255
ITERSPLINE[5]  0.60268
ITERSPLINE[5]  0.60014
ITERSPLINE[5]  0.6
ITERSPLINE[5]  0.6
0.6
      IRX ITERSPLINE 33
ITERSPLINE[5]  0.67546
ITERSPLINE[5]  0.65168
ITERSPLINE[5]  0.62984
ITERSPLINE[5]  0.61242
ITERSPLINE[5]  0.60263
ITERSPLINE[5]  0.60013
ITERSPLINE[5]  0.6
ITERSPLINE[5]  0.6
0.6
```
The new Newton–Raphson function is tested, and now appears to be quite robust. Over a wide range of initial voltages, the number of iterations required is about 8.

```
      TΔITERSPLINE←10
```

```
      IRX ITERSPLINE ‾200000
0.6
      W
10
```
Not many more iterations are involved even though an extremely large negative initial value of VR is employed.

2.6

```
        ∇VR←IRX ITER1 VRO;IR;ERROR;SLOPE;IEQ
[1]     VR←VRO
[2]     L1:→L2×1E¯6<|ERROR←IRX-IR←COM VR
[3]     L2:SLOPE←DIFCOM VR
[4]     IEQ←IR-VR×SLOPE
[5]     VR←(IRX-IEQ)÷SLOPE
[6]     →L1∇
        ∇IR←COM VR
[1]     IR←VR-.1×VR*3∇


      ∇ S←DIFCOM VR
[1]      S←1+0.3×VR*2
      ∇
        4 ITER1 5
ITER1[5] 5.4118
ITER1[5] 2.6586
ITER1[5] 2.4863
ITER1[5] 2.4782
ITER1[5] 2.4781
2.4781
        4 ITER1 50
ITER1[5] 33.294
ITER1[5] 22.142
ITER1[5] 14.688
ITER1[5] 9.7041
ITER1[5] 6.385
ITER1[5] 4.2373
ITER1[5] 3.0089
ITER1[5] 2.5425
ITER1[5] 2.4792
ITER1[5] 2.4781
2.4781
```

Agreement with solution 1(b)

Numbers of iterations agree with solution 1(b)

3

Multiterminal Components

Chapter 1 has provided a theoretical and algorithmic basis for the analysis of circuits containing only linear two-terminal components. We must now decide how to handle the analysis of circuits containing, in addition, devices such as transistors and operational amplifiers having more than two terminals. A valuable outcome of our investigation will be the realization that circuits on the one hand, and multiterminal components on the other, possess an identical form of parametric description. In other words, the equations describing their properties are identical in form. This result is of considerable practical and conceptual importance.

As with two-terminal components, we shall seek a node–branch description for a multiterminal component. For convenience, the basis of the node–branch description of an N-terminal component will be illustrated for the example of a three-terminal component.

3.1 Node–Branch Representation

Three alternative node–branch representations of a three-terminal component are shown in Fig. 3.1. Each ensures that Kirchhoff's current law is obeyed by the choice of appropriate currents, and nodes are available at which to observe voltages. Whichever representation is chosen (and the most convenient will shortly be discussed), the two branches are each treated in exactly the same manner as in Chapter 1 when defining the incidence matrix. The only difference between the new situation and that treated in Chapter 1 is that a branch current may now be controlled, not only by the voltage of the same branch, but additionally by the voltage of another branch. This essential difference is now considered.

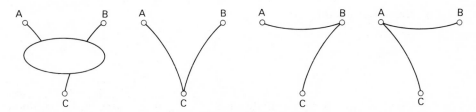

Fig. 3.1
Alternative node-branch representations of a three-terminal component

3.2   Branch Relations

The three-terminal component of Fig. 3.2(a) will be described by the relations between the two-vector IB of branch currents and the two-vector VB of branch voltages. (For a more detailed discussion of the material that follows,

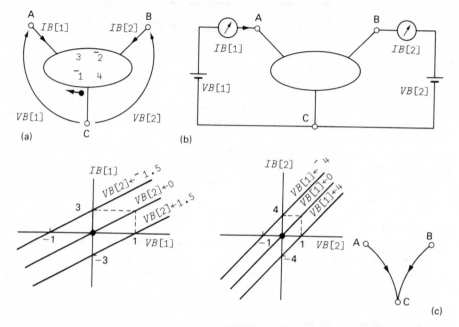

Fig. 3.2
Measurements on a three-terminal component

see Spence (1979).) If the measurements shown in Fig. 3.2(b) were made on the component, then it is described by the linear homogeneous equations

$$IB[1] \leftarrow (3 \times VB[1]) + {}^-2 \times VB[2]$$
$$IB[2] \leftarrow ({}^-1 \times VB[1]) + 4 \times VB[2]$$

or, in array form,

$$IB \leftarrow GB + . \times VB \qquad\qquad (3.1)$$

where, in this example,

$$GB \;=\; \begin{matrix} 3 & {}^-2 \\ {}^-1 & 4 \end{matrix}$$

The matrix GB is called the branch conductance matrix of the component, and is often inscribed within the component symbol, as in Fig. 3.2(a). This value of GB is, as we shall see later, particular to the description in which node

C is the reference point for nodal currents and voltages, and where the elements of IB and VB are ordered clockwise; these facts are indicated by the dot and arrow convention shown in Fig. 3.2(a). Since the elements of IB and VB can be regarded as branch currents and voltages (they obey Kirchhoff's laws), the node–branch representation of the component, for the value of GB given above, is as shown in Fig. 3.2(c).

At this point it is useful to note that we have, here, a description of a three-terminal *component* which is *identical* in form with the description of a three-terminal *circuit*: see, for example, the circuit of Fig. E1.2, reproduced here as Fig. 3.3. In each case the current–voltage characteristics of a 'black box' having three terminals is described by a 2 × 2 conductance matrix. In the former case the description is referred to as the component's branch conductance matrix,

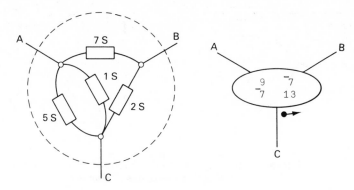

Fig. 3.3
The circuit of Fig. E1.2 (see exercises for Chapter 1) and its desciption by a 2 × 2 reduced nodal conductance matrix

and in the latter as the circuit's reduced nodal conductance matrix. To compute the voltage response of the measured component (a) of Fig. 3.2 to current excitation (Fig. 3.4) we would, therefore, merely refer to Chapter 1: first the (reduced nodal) resistance matrix (⊟ GB) is computed, and then post-multiplied by the reduced vector of nodal currents:

```
    IB ←10  5
    VB ←(⊟ GB )+.×IB
    VB
  5    25
```

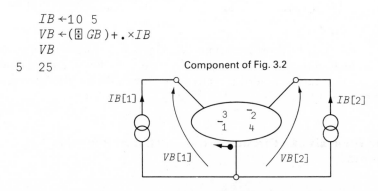

Fig. 3.4
Current excitation applied to the circuit of Fig. 3.2

If the three-terminal component is one of a number of components in a circuit, then its branch conductance matrix will appear as part of the branch conductance matrix describing the entire collection of components. Thus, with reference to the circuit of Fig. 3.5(a), whose directed graph is shown in Fig. 3.5(b), the branch conductance matrix relating branch currents to branch voltages is

$$G = \begin{matrix} 2 & 0 & 0 & 0 \\ 0 & 10 & \bar{1} & 0 \\ 0 & 7 & 5 & 0 \\ 0 & 0 & 0 & 4 \end{matrix}$$

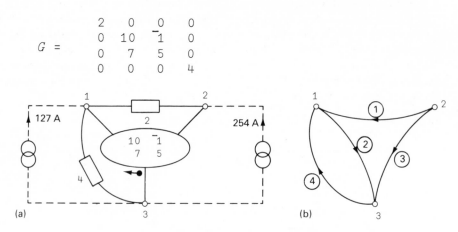

Fig. 3.5
A three-terminal component embedded within a circuit
(all conductance values in siemens)

and contains the matrix GB describing the three-terminal component placed on the main diagonal. However, the matrix GB can be placed directly in this position only if

(a) the branches representing the three-terminal component meet at the reference terminal of the component (indicated by a dot);
(b) the branch reference currents of the three-terminal component are directed *towards* the reference terminal;
(c) the branches of the three-terminal component are numbered *consecutively*, and in the *order* indicated by the arrow associated with the reference dot.

Other node–branch representations (see Fig. 3.1), branch reference directions and branch numbering are possible, but do not allow the component's conductance matrix to be placed directly on the main diagonal of the branch conductance matrix.

3.3 Complete Analysis of a Circuit Containing Linear Resistive Components

In the following analysis of the circuit of Fig. 3.5, whose directed graph also appears in that figure, some defined functions are used, for convenience, to

---

Table 3.1
Functions allowing the straightforward formation of the incidence and branch conductance matrices

$$\nabla A \leftarrow INCID \ M; N$$
[1]　　$A \leftarrow ((\iota N)\circ .= M[1;]) - (\iota N \leftarrow \lceil / , M)\circ , = M[2;] \nabla$

$$\nabla \ YB \leftarrow X \ AA \ Y; XX; YY; I$$
[1]　　$XX \leftarrow (\bar{2} \uparrow 1 \ 1, \rho X)\rho X$
[2]　　$YY \leftarrow (\bar{2} \uparrow 1 \ 1, \rho Y)\rho Y$
[3]　　$I \leftarrow (\rho XX) + \rho YY$
[4]　　$YB \leftarrow (I \uparrow XX) + (-I) \uparrow YY \nabla$

---

generate matrices which would otherwise be tedious to enter directly. These functions are listed in Table 3.1.

The directed graph is easily described by a branch connection matrix M

$$\square \leftarrow M \leftarrow \lozenge \ 4 \ 2 \ \rho \ 2 \ 1 \ 1 \ 3 \ 2 \ 3 \ 3 \ 1$$
2　1　2　3
1　3　3　1

from which the incidence matrix can be generated:

$$\square \leftarrow A \leftarrow INCID \ M$$
$\bar{1}$　1　0　$\bar{1}$
　1　0　1　0
　0　$\bar{1}$　$\bar{1}$　1

The branch conductance matrix YB is generated by means of the function AA:

$$\square \leftarrow YB \leftarrow 2 \ AA \ (2 \ 2 \ \rho \ 10 \ \bar{1} \ 7 \ 5) \ AA \ 4$$
2　　0　　0　　0
0　10　$\bar{1}$　　0
0　　7　　5　　0
0　　0　　0　　4

The complete nodal conductance matrix can now be found:

$$Y \leftarrow A + . \times YB + . \times \lozenge \ A$$

and from it the reduced nodal conductance matrix. Since the highest-numbered node is chosen as the circuit reference node,

$$YR \leftarrow \bar{1} \ \bar{1} \downarrow Y$$
$$YR$$
16　　　$\bar{3}$
　5　　　　7

The nodal current excitation is described by

$$IR \leftarrow 127\ 254$$

so that the nodal voltage response vector VR is

$$VR \leftarrow IR \boxplus YR$$
$$VR$$
$$13\quad 27$$

By reference to Kirchhoff's voltage law the branch voltages can be obtained:

$$VB \leftarrow (\lozenge A) + . \times VR, 0$$

(note the zero value of the reference node voltage)

$$VB$$
$$14\quad 13\quad 27\quad {}^{-}13$$

and hence the branch currents

$$\square \leftarrow IB \leftarrow YB + . \times VB$$
$$28\quad 103\quad 226\quad {}^{-}52$$

.Finally, as a check, we derive the injected nodal curents from Kirchhoff's current law

$$A + . \times IB$$
$$127\quad 254\quad {}^{-}381$$

and observe that they are identical to the actual values: we therefore have a check on our calculations.

3.4 *N*-terminal Components

The result we have derived for a three-terminal component can be generalized to an *N*-terminal component or circuit. For example, its node–branch representation will contain *N* nodes and *N*-1 branches, all branches being incident upon the node designated as the reference point for the component description (by an (*N*-1) × (*N*-1) branch conductance matrix). This generalization is illustrated by an example in Exercise 3.5 and the associated solution.

3.5 The Voltage-controlled Current Source

We have described a linear homogeneous three-terminal component by a 2 × 2 matrix relating its terminal voltages and currents. Nevertheless a different —

though equivalent — form of this description is in frequent use. It is called the equivalent-circuit description.

Consider the equations

$$IB[1] \leftarrow (3 \times VB[1]) + {}^-2 \times VB[2]$$
$$IB[2] \leftarrow ({}^-1 \times VB[1]) + 4 + VB[2]$$

describing the three-terminal component of Fig. 3.6(a). These equations also describe the circuit shown in Fig. 3.6(b), a circuit containing two two-terminal conductances and two current sources. The latter differ from the independent current sources encountered previously in that their value, instead of being constant, is proportional to a voltage occurring elsewhere in the circuit.

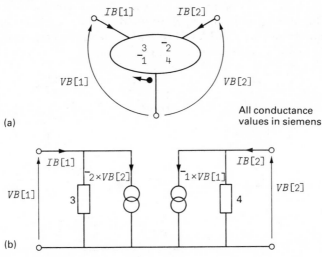

(a)

(b)

Fig. 3.6
A three-terminal component and a model containing voltage-controlled current sources

An isolated voltage-controlled current source (VCCS) is shown in Fig. 3.7(a). A special case of a VCCS, in which the controlling voltage appears across the current source, is indistinguishable from a two-terminal resistor.

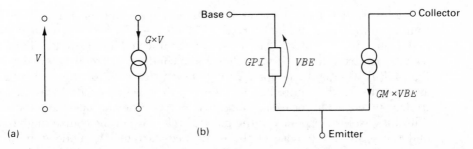

(a)   (b)

Fig 3.7
(a) A voltage-controlled current source; (b) a transistor model containing a voltage-controlled current source and a resistor

The equivalent circuit (Fig. 3.6(b)) representation of a three-terminal component is sometimes favored because it can permit an alternative understanding of circuit behavior, especially if the resistors and VCCSs are each related to some physical effect. It is of the same form as a simple equivalent circuit commonly used to represent a transistor (Fig. 3.7(b)).

3.6   Alternative Formation of the Complete Nodal Conductance Matrix

In Chapter 1 it was shown how the complete nodal conductance matrix can be formed in a step-by-step procedure in which each conductance is examined in turn. The same approach can be adopted for a circuit containing both linear two-terminal resistors and VCCSs, with little modification, as can be demonstrated using the example shown in Fig. 3.8(a).

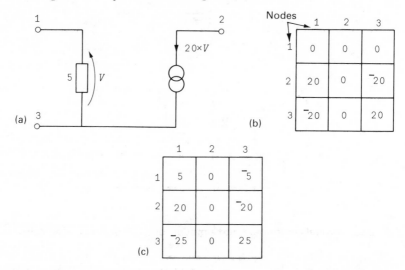

Fig. 3.8
Formation of the complete nodal conductance matrix of a circuit containing a voltage-controlled current source

For given values of the nodal voltages, the voltage-controlled current source requires an additional current of $20 \times V[1] - V[3]$ to injected into node 2 and a current of $-20 \times V[1] - V[3]$ to be injected into node 3. Thus, the VCCS's contribution to the complete nodal conductance matrix is as shown in Fig. 3.8(b). A useful rule for locating the entries is to identify the columns associated with the controlling voltage and the rows associated with the controlled currents. The location of a *positive* entry of the mutual conductance (20 S in this case) corresponds to the positive reference of the controlling voltage and the positive reference of the controlled current (the node from which current flows away). If the two-terminal components are now taken into account in the manner described in Chapter 1, the resulting complete nodal conductance matrix is as shown in Fig. 3.8(c).

3.7 The Port Description of a Circuit

Occasionally it is necessary to obtain a circuit description which is valid under special circumstances. One such circumstance is when only the input and output currents and voltages of a large circuit are of interest, and when the form of any externally connected circuit is rigidly prescribed. An example will serve as an illustration.

Consider the six-terminal circuit shown in Fig. 3.9(a); the relation between its nodal currents and voltages will, as we know, be described by a 6-row, 6-column nodal conductance matrix Y (or — see Figure 1.12 — by the reduced nodal conductance or resistance matrices YR and ZR respectively, or by the complete nodal resistance matrix Z).

But suppose the circuit is designed to be connected between a source and a load, and with a two-terminal component between a particular pair of nodes, as shown in Fig. 3.9(b). Then, it follows that

$$
\begin{aligned}
-I[1] \quad &= \quad I[2] \quad &= \quad IP[A] \\
-I[6] \quad &= \quad I[5] \quad &= \quad IP[B] \\
-I[3] \quad &= \quad I[4] \quad &= \quad IP[C]
\end{aligned}
$$

and

$$
\begin{aligned}
VP[A] \quad &= \quad -V[1] \quad &+ \quad V[2] \\
VP[B] \quad &= \quad -V[6] \quad &+ \quad V[5] \\
VP[C] \quad &= \quad -V[3] \quad &+ \quad V[4]
\end{aligned}
$$

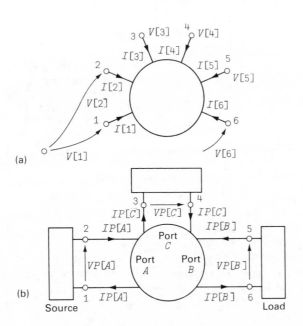

Fig. 3.9
(a) A general 6-terminal circuit; (b) constraints imposed by the nature of the allowed terminations reduce the number of independent currents

Notwithstanding the fact that the original nodal description (e.g. Y) still applies, the most appropriate, and simpler, characterization may now be the 3×3 *port* resistance matrix ZP relating the three-element vector VP of port voltages to the three-element vector IP of port currents according to

$$VP = ZP + . \times IP \qquad (3.2)$$

The port resistance matrix ZP is, in fact, simply related to the complete nodal resistance matrix Z.

Useful insight into the relation can be obtained if it is first quoted and then interpreted. The relation between ZP and Z is

$$ZP = (\lozenge A) + . \times Z + . \times A \qquad (3.3)$$

where A is a special incidence matrix related to the circuit. The matrix A is, in fact, a node–port incidence matrix, and can be illustrated by reference to the situation of Fig. 3.9. For the port connections shown in that figure, Fig. 3.10 shows the necessary directed graph. It contains a node for every circuit node, but only the port terminations are represented by branches. Thus, for this example, A is given by

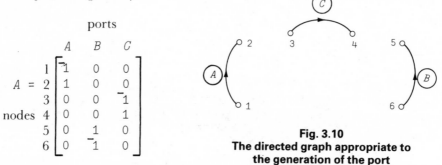

$$
\begin{array}{c}
\quad\quad\quad\quad \text{ports} \\
\quad\quad\quad\quad A \quad B \quad C \\
A = \begin{array}{c} 1 \\ 2 \\ 3 \\ \text{nodes } 4 \\ 5 \\ 6 \end{array}
\left[
\begin{array}{ccc}
1 & 0 & 0 \\
1 & 0 & 0 \\
0 & 0 & 1 \\
0 & 0 & 1 \\
0 & 1 & 0 \\
0 & \bar{1} & 0
\end{array}
\right]
\end{array}
$$

Fig. 3.10
The directed graph appropriate to the generation of the port resistance matrix of the circuit of Fig. 3.9

To interpret the relation (3.3) we substitute it into (3.2) to obtain

$$VP = (\lozenge A) + . \times Z + . \times A + . \times IP \qquad (3.4)$$

Beginning on the right, we observe that $A + . \times IP$ transforms a single port current into the corresponding *pair* of nodal currents. In the general case, $A + . \times IP$ is therefore the vector of nodal currents equivalent to the given set of port currents. The inner product $Z + . \times A + . \times IP$ therefore yields the *n*-vector of nodal voltages. Finally, the transformation from nodal voltages to port voltages (the latter being simple differences between nodal voltage pairs) is achieved by premultiplication by $\lozenge A$, since each column of A contains the necessary +1 and $\bar{1}$ elements.

A useful alternative to (3.3) is the relation

$$ZP = (\lozenge AR) + . \times ZR + . \times AR \qquad (3.5)$$

where AR is the reduced incidence matrix obtained from A by the deletion of the row of A associated with the reference terminal (see Exercises).

Situations often arise in which not every circuit node is also one of the two nodes comprising a port. As an extreme example we may wish to characterize a circuit containing many nodes as a two-port so that the circuit description involves, at most, only four of those nodes. The approach is straightforward: for the two-port just referred to, the node–port incidence matrix would have as many rows as the circuit has nodes, and two columns; each column contains just two elements, a $+1$ and a $^-1$ indicating the node pair constituting that port.

3.8 Voltage Excitation

The scheme illustrated in Fig. 1.12 for the analysis of a linear resistive circuit is relevant to the situation in which the circuit's *current* excitation is known. In other words, the current injected into *every* circuit node is known, even though many of these currents are zero. Suppose, however, the circuit is connected to one or more voltage sources (Fig. 3.11): how should this case be handled?

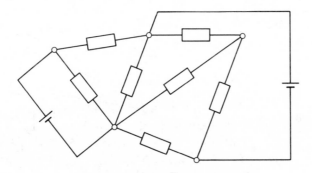

Fig. 3.11
A circuit excited by voltage sources

The first point to make is that with a situation such as is shown in Fig. 3.11 *we do not know* the voltage at all nodes (i.e. the nodal voltage vector V). If we did, the calculation

$$I \leftarrow Y + . \times V$$

would lead to the values of the currents into each node and

$$GB + . \times (\lozenge A) + . \times V$$

would yield the component currents, the whole calculation being quite straightforward. But it is an extremely rare circuit that is designed to have a voltage applied at every node!

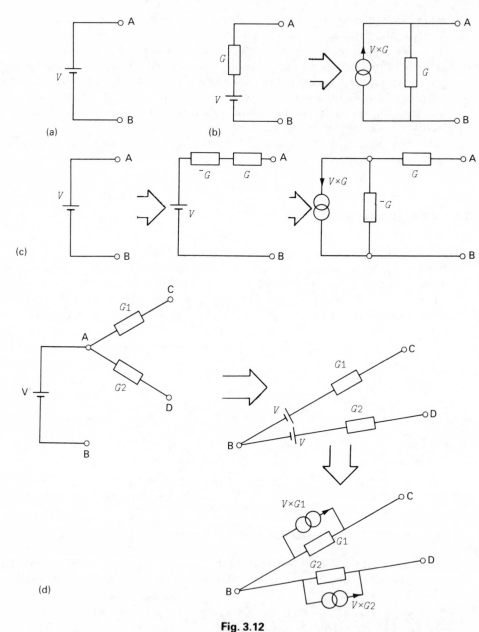

Fig. 3.12
Approximate and exact models to allow a voltage excitation to be replaced with current excitation

There are three simple ways of handling a voltage excitation (Fig. 3.12(a)) applied between two terminals A and B, and their common objective is to transform the voltage excitation to the current excitation we already know how to handle. The first is shown in Fig. 3.12(b): the voltage excitation is approximated by a current source and low-value resistor. The product of the current and resistance is chosen to be equal to the voltage of the source, and the resistance is chosen to be low enough to minimize the error involved in the approximation but sufficiently high not to cause numerical problems.

The second approach is shown in Fig. 3.12(c): it is exact, but has the disadvantage of introducing an additional node into the circuit. The third approach, called the Blakesley transformation, is shown in Fig. 3.12(d), and again is exact. Briefly, the voltage source is 'moved' to be connected in series with every two-terminal element connected directly to node A. Then, a Norton transformation on each such source and series conductance leads to the creation of an excitation current.

Exercises

Ensure that the functions *INCID* and *AA* (Table 3.1) are in your workspace.

3.1 At the APL terminal, follow through the analysis of the circuit of Fig. 3.5(a).

3.2 Repeat the analysis with the branch- and node- numbering, and branch reference directions shown in Fig. E3.1. Branches 1 and 2 refer to the three-terminal component, and branches 4 and 3 to the 4 S and 2 S conductances. Take terminal 3 as the reference node. Check that your answers agree with those obtained in the notes. As a check, the complete nodal conductance matrix should be

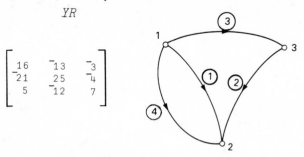

$$YR$$

$$\begin{bmatrix} 16 & \overline{1}3 & \overline{3} \\ \overline{2}1 & 25 & \overline{4} \\ 5 & \overline{1}2 & 7 \end{bmatrix}$$

Fig. E3.1

3.3 Using the value of YR obtained in Exercise 3.2, obtain the reduced nodal resistance matrix $ZR.$ Index $ZR,$ and perform any other calculation necessary, to find the voltage V in each of the conditions shown in Fig. E3.2.

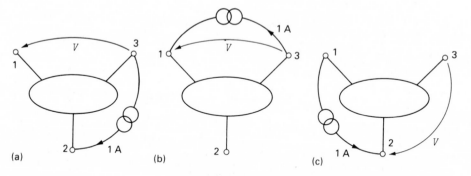

(a)   (b)   (c)

Fig. E3.2

3.4 Following the procedure shown diagrammatically in Fig. 1.12, find the voltage response to the current excitation for the circuit shown in Fig. E3.3.

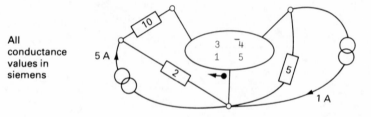

All
conductance
values in
siemens

Fig. E3.3

3.5　Let the circuit shown in Fig. E3.3 be called N, as shown in Fig. E3.4(a). Using the reduced nodal conductance matrix YR calculated for N in Exercise 3.4, calculate the value of V in the circuit shown in Fig. E3.4(b).

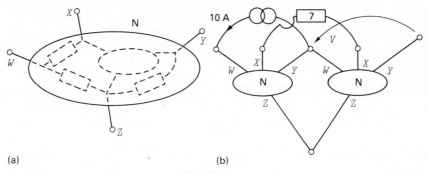

(a)　　　　　　　　　　　　　　　　(b)

Fig. E3.4

3.6　Define a function, similar in algorithmic detail to $FORM$ (see Chapter 1) which adds, to an existing complete nodal conductance matrix, those terms which are appropriate to the connection, within the existing circuit, of a number of voltage-controlled current sources. Test your function with an example involving at least two controlled sources.

3.7　By using, in turn, each of the transformations illustrated in Fig. 3.12(b)—(d) determine the response voltage V in the circuit shown in Fig. E3.5.

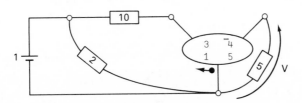

Fig. E3.5

3.8　The circuit of Exercise 3.4 is shown in Fig. E3.6(a) with three ports identified. Obtain the 3×3 port impedance matrix. Use this matrix to determine the voltage response V to the current excitation shown in Fig. E3.6(b). Compare the results with those of Exercise 3.4.

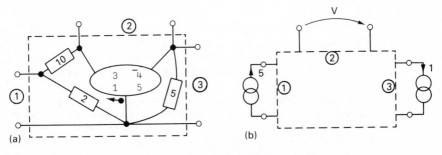

(a)　　　　　　　　　　　　　　　　(b)

Fig. E3.6

3.9 What is the implication, for the calculated values of the following, of an interchange of the entries 1 and $\bar{1}$ in the incidence matrix of a circuit:

   (a)   the branch conductance matrix
   (b)   the complete nodal conductance matrix
   (c)   the reduced nodal resistance matrix
   (d)   the predicted nodal response voltages
   (e)   the predicted branch voltages and currents?

Test your answers by means of a circuit example.

3.10 Attempt to compute the nodal voltage response, and then the branch voltages and currents, of the current excited circuits shown in Fig. E3.7. Comment on any difficulties or special characteristics of the analyses you perform.

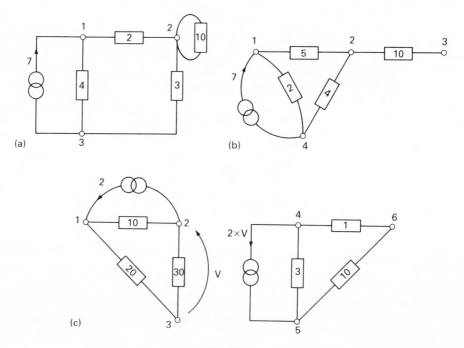

Fig. E3.7

3.11 The circuit shown in Fig. E3.8 contains linear homogeneous two-terminal components whose conductance values are shown. Obtain a branch conductance matrix and an incidence matrix for the circuit, and hence the complete nodal conductance matrix (use the functions $COND$ and $INCID$ if you wish). Determine the currents flowing into terminals 1, 2 and 3 via the voltage sources. Then determine, by the systematic application of Kirchhoff's laws and Ohm's law, the branch currents. What change will occur in the nodal and branch currents if (a) each voltage source is increased in value by 100 V? (b) each voltage source has its value doubled? (c) each conductance has its value multiplied by ten? Test your answers.

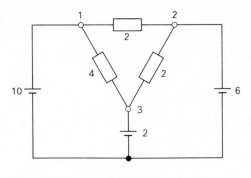

Fig. E3.8

3.12 The circuit of Exercise 3.11 is subjected to the nodal current excitation shown in Fig. E3.9. Compute the values of $V[1]$ and $V[2]$, and hence the component voltages and currents. Check that Kirchhoff's laws are obeyed. What power is dissipated in each component? What is the total power dissipated by the components? Should this be equal to the power supplied by the current sources? Test your answer. What change will occur in the response voltages if all conductance values are multiplied by ten? Check your answer by means of a calculation.

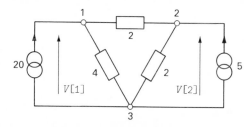

Fig. E3.9

3.13 You are handed a piece of paper on which is printed the following array:

$$
\begin{array}{cccccccc}
1 & 1 & 0 & 0 & 0 & 1 & \bar{1} & 0 \\
0 & 0 & 1 & 1 & 0 & 0 & 1 & 0 \\
0 & \bar{1} & 0 & \bar{1} & 0 & 0 & 0 & \bar{1} \\
\bar{1} & 0 & \bar{1} & 0 & 1 & 0 & 0 & 0
\end{array}
$$

You are told that it was intended to print the whole of the incidence matrix of a 5-node, 8-branch directed graph, but that the last row was omitted by mistake. If possible, find the missing last line and hence draw the directed graph.

3.14 It was found in Exercise 3.13 that an incidence matrix from which a row has been omitted, which is called a *reduced incidence matrix*, AR, is a sufficient description of a directed graph. Under the assumption that it is the last row of the incidence matrix that has been omitted to obtain AR, express Kirchhoff's laws in terms of AR for a circuit with nodal current excitation. For a circuit with N nodes and B branches, what are the dimensions of the vectors of branch and nodal voltage and current involved in your new statements of Kirchhoff's laws? Test these statements by means of a numerical example.

3.15 Can *any* row of an incidence matrix be removed and still allow the remainder to carry all the information about the directed graph?

3.16 In Exercise 3.14 it was discovered that Kirchhoff's laws could be expressed in terms of the reduced incidence matrix AR as

$$IR \leftarrow AR + . \times IB$$
$$VB \leftarrow VR + . \times AR$$

where AR is normally obtained by deleting the last row of the incidence matrix, and IR and VR are called the reduced vectors of nodal current and voltage. Combine the two Kirchhoff laws so expressed with the branch relations $IB \leftarrow YB + . \times VB$ to obtain relations, similar to equations (1.6) and (1.7), but expressing the *reduced* nodal conductance matrix YR as a function of AR and YB. Test the relation you propose by using the example of Fig. 3.5.

Solutions

```
      ∇A←INCID M;N
[1]   A←((ιN)∘.=M[1;])-(ιN←⌈/,M)∘.=M[2;]∇
```
                                                                    The functions required

```
      ∇ GB ←X AA Y;XX;YY;I
[1]   XX←(¯2↑1 1,ρX)ρX
[2]   YY←(¯2↑1 1,ρY)ρY
[3]   I←(ρXX)+ρYY
[4]   GB ←(I↑XX)+(-I)↑YY∇
```

3.1   Analysis of the circuit of Fig. 3.5(a) Chapter 3.

```
          M←⍉4 2ρ2 1 1 3 2 3 3 1
          M
2 1 2 3
1 3 3 1
          A←INCID M
          A
¯1  1  0 ¯1
 1  0  1  0
 0 ¯1 ¯1  1
          GB ←2 AA(2 2ρ10 ¯1 7 5)AA 4
          GB
2  0  0  0
0 10 ¯1  0
0  7  5  0
0  0  0  4
          Y←A+.×GB+.×⍉A
          Y
 16  ¯3 ¯13
  5   7 ¯12
¯21  ¯4  25
          YR←¯1 ¯1↓Y
          YR
16 ¯3
 5  7

          IR←127 254
          VR←IR⌹ YR
          VR
13 27
          VB ←(⍉A)+.×VR,0
          VB
14 13 27 ¯13
          IB ←GB+.×VB
          IB
28 103 226 ¯52
          A+.×IB
127 254 ¯381
```

3.2   Repeat of the analysis of the circuit of Fig. 3.5(a) (repeated as Fig. E3.10) using different branch and node numbering.

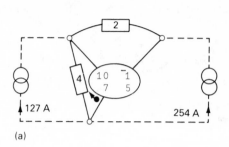

(a)

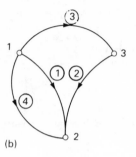

(b)

Fig. E3.10

```
      M←Q4 2ρ1 2 3 2 1 3 1 2
      M
1 3 1 1
2 2 3 2
      A←INCID M
      A
 1  0  1  1
‾1 ‾1  0 ‾1
 0  1 ‾1  0
```
The incidence matrix of the directed graph is generated

```
      GB←(2 2ρ10 ‾1 7 5)AA 2 AA 4
      GB
10 ‾1  0  0
 7  5  0  0
 0  0  2  0
 0  0  0  4
```
Branch conductance matrix

```
      Y←A+.×GB+.×QA
      Y
 16 ‾13  ‾3
‾21  25  ‾4
  5 ‾12   7
```
Complete nodal conductance matrix

```
      YR←‾1 ‾1↓Y
      YR
 16 ‾13
‾21  25
```
Reduced nodal conductance matrix

```
      ZR←⌹ YR
      ZR
0.1968503937   0.1023622047
0.1653543307   0.125984252
```
Reduced nodal resistance matrix

```
      IR←127 ‾381
      VR←ZR+.×IR
      VR
‾14 ‾27
```
Current excitation

Reduced vector of nodal voltages

```
      VB←(QA)+.×VR,0
      VB
13 27 ‾14 13
      IB←GB+.×VB
      IB
103 226 ‾28 52
      A+.×IB
127 ‾381 254
```
A check on the result

1 ○ ◄———VR[1]——— ○ 3

VR[2]

○
2   **Fig. E3.11**

3.3 $ZR \leftarrow \boxdot \ YR$ The reduced nodal resistance matrix
              ZR
0.1968503937   0.1023622047
0.1653543307   0.125984252

        $ZR[1;2]$ (a)
0.1023622047

        $ZR[1;1]$ (b)   The value of V in the different
0.1968503937   excitation conditions

        $-/ZR[2;2 \ 1]$ (c)
‾0.03937007874

3.4 $M \leftarrow \mathbb{Q} \ 5 \ 2\rho 1 \ 2 \ 1 \ 4 \ 2 \ 4 \ 3 \ 4 \ 3 \ 4$
              M
1 1 2 3 3
2 4 4 4 4
        $A \leftarrow INCID \ M$
        A
 1   1   0   0   0
‾1   0   1   0   0
 0   0   0   1   1
 0 ‾1 ‾1 ‾1 ‾1
        $GB \leftarrow 10 \ AA \ 2 \ AA \ (2 \ 2\rho 3 \ \ ‾4 \ 1 \ 5) \ AA \ 5$
        GB
10   0   0   0   0
 0   2   0   0   0
 0   0   3 ‾4   0
 0   0   1   5   0
 0   0   0   0   5
        $Y \leftarrow A + . \times GB + . \times \mathbb{Q} \ A$
        Y
 12 ‾10   0   ‾2
‾10   13   ‾4   1
  0   1   10 ‾11
 ‾2   ‾4   ‾6 ‾12
        $YR \leftarrow \ ‾1 \ \ ‾1 \downarrow Y$
        YR
 12 ‾10   0
‾10   13 ‾4
  0   1 10

        $ZR \leftarrow \boxdot YR$
        ZR
 0.2203947368   0.1644736842   0.06578947368
 0.1644736842   0.1973684211   0.07894736842
‾0.01644736842 ‾0.01973684211   0.09210526316

        $IR \leftarrow 5 \ 0 \ ‾1$ Nodal current excitation
        $VR \leftarrow ZR + . \times IR$
        VR Nodal voltage response
1.036184211 0.7434210526 ‾0.1743421053

        $VB \leftarrow (\mathbb{Q} A) + . \times VR, 0$
        $IB \leftarrow GB + . \times VB$ Check on calculation
        $A + . \times IB$
5 ‾6.661338148E‾16 ‾1 ‾4

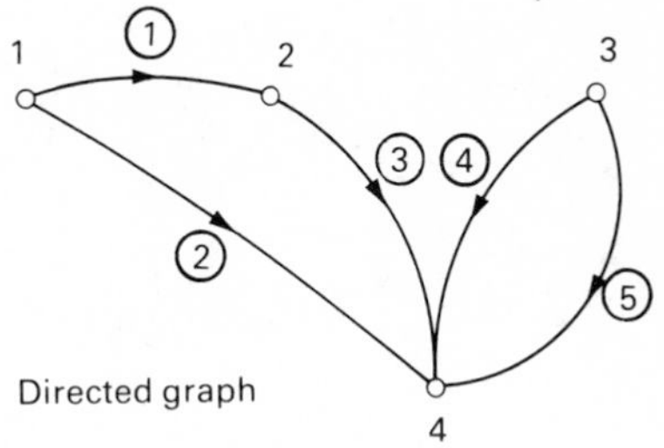

Directed graph

Fig. E3.12

3.5 The reduced nodal conductance matrix YR (see answer to 3.4) describes the *component* shown below, a component which can be represented by the directed graph shown.

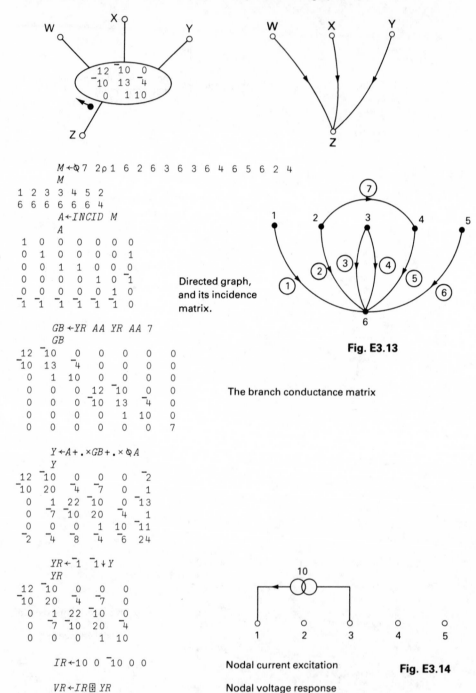

$$M \leftarrow Q7 \ 2\rho1 \ 6 \ 2 \ 6 \ 3 \ 6 \ 3 \ 6 \ 4 \ 6 \ 5 \ 6 \ 2 \ 4$$
$$M$$

1	2	3	3	4	5	2
6	6	6	6	6	6	4

$$A \leftarrow INCID \ M$$
$$A$$

1	0	0	0	0	0	0
0	1	0	0	0	0	1
0	0	1	1	0	0	0
0	0	0	0	1	0	‾1
0	0	0	0	0	1	0
‾1	‾1	‾1	‾1	‾1	‾1	0

Directed graph, and its incidence matrix.

$$GB \leftarrow YR \ AA \ YR \ AA \ 7$$
$$GB$$

12	‾10	0	0	0	0	0
‾10	13	‾4	0	0	0	0
0	1	10	0	0	0	0
0	0	0	12	‾10	0	0
0	0	0	‾10	13	‾4	0
0	0	0	0	1	10	0
0	0	0	0	0	0	7

Fig. E3.13

The branch conductance matrix

$$Y \leftarrow A + . \times GB + . \times \lozenge A$$
$$Y$$

12	‾10	0	0	0	‾2
‾10	20	‾4	‾7	0	1
0	1	22	‾10	0	‾13
0	‾7	‾10	20	‾4	1
0	0	0	1	10	‾11
‾2	‾4	‾8	‾4	‾6	24

$$YR \leftarrow \ ‾1 \ \ ‾1 \downarrow Y$$
$$YR$$

12	‾10	0	0	0
‾10	20	‾4	‾7	0
0	1	22	‾10	0
0	‾7	‾10	20	‾4
0	0	0	1	10

$$IR \leftarrow 10 \ 0 \ \ ‾10 \ 0 \ 0$$

Nodal current excitation

$$VR \leftarrow IR \boxplus YR$$
$$VR$$

1.238124069   0.4857488823   ‾0.515788003 · ‾0.08615871833   0.008615871833

Nodal voltage response

Fig. E3.14

```
        -/VR[3  5]
⁻0.5244038748
```
The voltage of interest $VR[3] - VR[5]$

```
        VB ←(⍉A)+.×VR,0
        IB ←GB+.×VB
        A+.×IB
10 ⁻4.218847494E⁻15  ⁻10  ⁻1.776356839E⁻15  ⁻2.359223927E⁻16  1.110223025E⁻15
```
Check on the calculation

3.6 The function $ADDGM$, shown below, takes the existing complete nodal conductance matrix as its left argument, and a matrix GM describing the newly connected voltage-controlled current sources as its right argument. The nature of the matrix GM is explained below. The function is applied to the circuit shown.

```
     ∇    R ←Y ADDGM  GM; C; VNDS; INDS; ΔY
[1]       R ←Y
[2]       C ←1
[3]       L : VNDS←GM [C ;  1  2]
[4]       INDS←GM [C ;  3  4]
[5]       Δ Y← 2  2 ρ  1  ⁻1  ⁻1  1  ×GM [C ; 5]
[6]       R[INDS; VNDS]←R[INDS; VNDS]+Δ Y
[7]       →L ×(1↑ρ GM )≥C← C +1
     ∇
```
$VNDS$ is nodes associated with controlling voltage
$INDS$ is nodes associated with controlled current
addition to Y due to controlled source

```
         GMT
    1    2    2    4    10
    4    3    3    2    20
```
Description of controlled sources added to existing two-terminal component circuit

```
         Y
    6          ⁻5         ⁻1         0
   ⁻5          9.5        ⁻2.5       ⁻2
   ⁻1         ⁻2.5        7.5       ⁻4
    0         ⁻2         ⁻4          6
```
Complete nodal conductance matrix of existing two-terminal component circuit (see Fig. 1.12)

```
       Y ADDGM  GMT
     6          ⁻5         ⁻1         0
     5         ⁻0.5       17.5       ⁻22
    ⁻1         ⁻2.5       12.5       16
   ⁻10          8         ⁻4          6
```
The complete nodal conductance matrix of the entire circuit

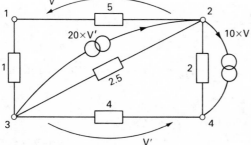

Conductances in siemens

Fig. E3.15

3.7 **(b)**

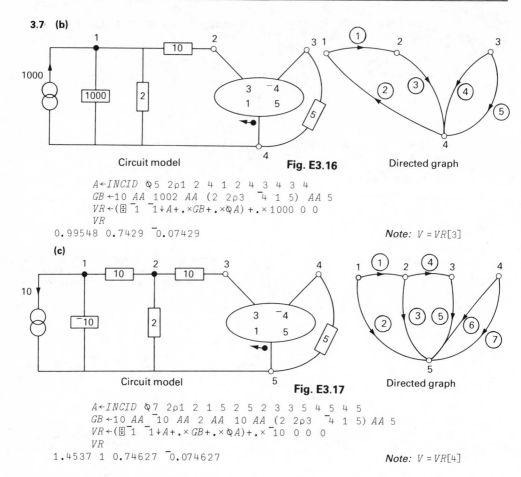

Circuit model   **Fig. E3.16**   Directed graph

```
A←INCID Q5 2ρ1 2 4 1 2 4 3 4 3 4
GB←10 AA 1002 AA (2 2ρ3 ¯4 1 5) AA 5
VR←(⊟¯1 ¯1↓A+.×GB+.×QA)+.×1000 0 0
VR
0.99548 0.7429 ¯0.07429
```
                                                                   Note: $V = VR[3]$

(c)

Circuit model   **Fig. E3.17**   Directed graph

```
A←INCID Q7 2ρ1 2 1 5 2 5 2 3 3 5 4 5 4 5
GB←10 AA ¯10 AA 2 AA 10 AA (2 2ρ3 ¯4 1 5) AA 5
VR←(⊟¯1 ¯1↓A+.× GB+.× QA)+.× ¯10 0 0 0
VR
1.4537 1 0.74627 ¯0.074627
```
                                                                   Note: $V = VR[4]$

(d) First, note that the 2 siemen conductance is redundant, since it is connected directly across an ideal voltage source.

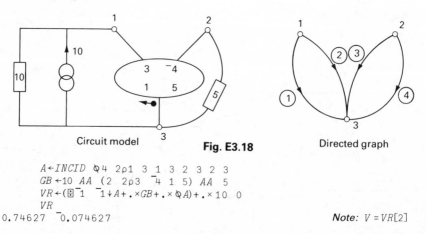

Circuit model   **Fig. E3.18**   Directed graph

```
A←INCID Q4 2ρ1 3 1 3 2 3 2 3
GB←10 AA (2 2ρ3 ¯4 1 5) AA 5
VR←(⊟¯1 ¯1↓A+.×GB+.× QA)+.×10 0
VR
0.74627 ¯0.074627
```
                                                                   Note: $V = VR[2]$

Comment: note that the calculated values of V in (c) and (d) are identical, and that in (b) differs by about 0.5%.

3.8

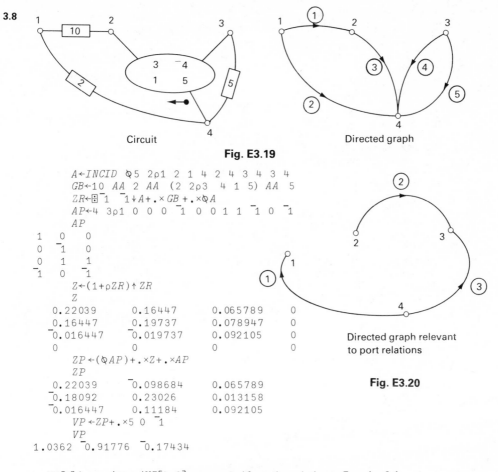

Circuit　　　　　　　　　　　　　Directed graph

Fig. E3.19

```
A←INCID ⍴5 2ρ1 2 1 4 2 4 3 4 3 4
GB←10 AA 2 AA (2 2ρ3 ‾4 1 5) AA 5
ZR←⊟‾1 ‾1↓A+.×GB +.×⍉A
AP←4 3ρ1 0 0 0 ‾1 0 0 1 1 ‾1 0 ‾1
AP
```

```
 1   0   0
 0  ‾1   0
 0   1   1
‾1   0  ‾1
```

```
    Z←(1+ρZR)↑ZR
    Z
 0.22039      0.16447      0.065789    0
 0.16447      0.19737      0.078947    0
‾0.016447    ‾0.019737     0.092105    0
 0            0            0           0
    ZP←(⍉AP)+.×Z+.×AP
    ZP
 0.22039     ‾0.098684     0.065789
‾0.18092      0.23026      0.013158
‾0.016447     0.11184      0.092105
    VP←ZP+.×5 0 ‾1
    VP
 1.0362  ‾0.91776  ‾0.17434
```

Directed graph relevant
to port relations

Fig. E3.20

$VP[2]$ is equal to $-/VR[3\ 2]$ computed from the solution to Exercise 3.4.

3.9　Let us choose the same circuit as in Exercise 3.4, with a single excitation current source.

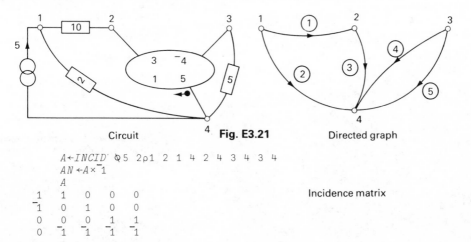

Circuit　　　**Fig. E3.21**　　　Directed graph

```
A←INCID ⍴5 2ρ1 2 1 4 2 4 3 4 3 4
AN←A×‾1
A
```
　　　　　　　　　　　　　　　　　　　Incidence matrix
```
‾1   1   0   0   0
‾1   0   1   0   0
 0   0   0   1  ‾1
 0  ‾1  ‾1  ‾1  ‾1
```

```
     AN
 ¯1  ¯1   0   0   0
  1   0  ¯1   0   0
  0   0   0  ¯1  ¯1
  0   1   1   1   1
     GB←10 AA 2 AA (2 2ρ3 ¯4 1 5) AA 5
     Y←A+.×GB+.×⍉A
     Y
 12  ¯10   0   ¯2
¯10   13  ¯4    1
  0    1  10  ¯11
 ¯2   ¯4  ¯6   12
     YN←AN+.×GB+.×⍉AN
     YN
 12  ¯10   0   ¯2
¯10   13  ¯4    1
  0    1  10  ¯11
 ¯2   ¯4  ¯6   12
     ZR←⊞¯1 ¯1↓Y
     ZR
 0.22039     0.16447     0.065789
 0.16447     0.19737     0.078947
¯0.016447   ¯0.019737    0.092105
     ZR+.×5 0 0
1.102 0.82237 ¯0.082237
     VR←ZR+.×5 0 0
     VR
1.102 0.82237 ¯0.082237
     VB←(⍉A)+.×VR,0
     VB
0.27961 1.102 0.82237 ¯0.082237 ¯0.082237
     VBN←(⍉AN)+.×VR,0
     VBN
¯0.27961 ¯1.102 ¯0.82237 0.082237 0.082237
```

Incidence matrix with
signs reversed

Branch conductance matrix
cannot be affected

The complete nodal
conductance matrix is not
affected . . .

and therefore neither is the
reduced nodal resistance
matrix

The nodal voltages cannot
therefore be affected

The calculated branch voltages
differ in their sign, and therefore
so also will the branch currents

3.10 (a)

```
     A←INCID ⍉4 2ρ1 2 1 3 2 3 2 2
     A
 1   1   0   0
¯1   0   1   0
 0  ¯1  ¯1   0
     GB←2 AA 4 AA 3 AA 10
     Y←A+.×GB+.×⍉A
     Y
 6  ¯2  ¯4
¯2   5  ¯3
¯4  ¯3   7
     VR←(⊞¯1 ¯1↓Y)+.×7 0
     VR
1.3462 0.53846
     VB←(⍉A)+.×VR,0
     VB
0.80769 1.3462 0.53846 0
     IB←GB+.×VB
     IB
1.6154 5.3846 1.6154 0
     A+.×IB
7 2.2204E¯16 ¯7
```

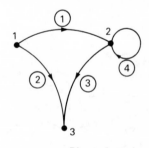

Directed graph

Fig. E3.22

No problems, even thought the directed
graph has a 'self-loop'

(b)
```
    A←INCID ⋄4 2ρ1 2 2 4 2 3 1 4
    GB←5 AA 4 AA 10 AA 2
    Y←A+.×GB+.×⋄A
    Y
  7   ‾5    0   ‾2
 ‾5   19  ‾10   ‾4
  0  ‾10   10    0
 ‾2   ‾4    0    6
    YR←‾1 ‾1↓Y
    YR
  7   ‾5    0
 ‾5   19  ‾10
  0  ‾10   10
    ZR←⊞ YR
    ZR
   0.23684      0.13158      0.13158
   0.13158      0.18421      0.18421
   0.13158      0.18421      0.28421
    VR←ZR+.×7 0 0
    VR
1.6579 0.92105  0.92105
    VB←(⋄A)+.×VR,0
    VB
0.73684 0.92105 ‾1.9429E‾16 1.6579
    IB←GB+.×VB
    IB
3.6842 3.6842 ‾1.9429E‾15 3.3158
    A+.×IB
7 ‾4.2188E‾15 1.9429E‾15 ‾7
```

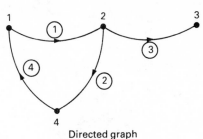

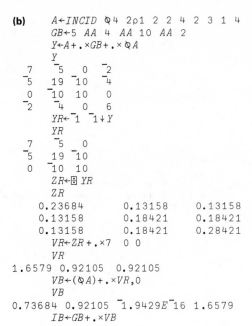

Directed graph

Fig. E3.23

The 'hanging node' (node 3) does not introduce any problems. All component voltages and currents are correctly computed

(c) For simplicity, we first disconnect the voltage-controlled current source

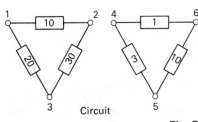

Circuit

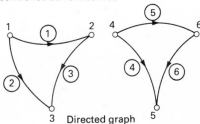

Directed graph

Fig. E3.24

```
    A←INCID ⋄ 6 2ρ1 2 1 3 2 3 4 5 4 6 6 5
    GB←10 AA 20 AA 30 AA 3 AA 1 AA 10
    Y←A+.×GB+.×⋄A
    Y
 30 ‾10 ‾20   0   0   0
‾10  40 ‾30   0   0   0
‾20 ‾30  50   0   0   0
  0   0   0   4  ‾3  ‾1
  0   0   0  ‾3  13 ‾10
  0   0   0  ‾1 ‾10  11
    YR←‾1 ‾1↓Y
    YR
 30 ‾10 ‾20   0   0
‾10  40 ‾30   0   0
‾20 ‾30  50   0   0
  0   0   0   4  ‾3
  0   0   0  ‾3  13
```

We choose node 6 as the reference node

```
      ⊞ YR
DOMAIN ERROR
      ⊞ YR
      ∧
      YRN←1 1↓YR
      YRN
 40  ‾30   0    0
‾30   50   0    0
  0    0   4   ‾3
  0    0  ‾3   13
```

YR cannot be inverted: we must additionally provide a reference voltage for the left-hand part of the circuit: we set $V[1]=0$

```
      ZRN←⊞ YRN
      IRN←‾2  0  0  0
      VRN←ZRN+.×IRN
      VRN
‾0.090909 ‾0.054545  0  0
      V←0,VRN,0
      V
0 ‾0.090909 ‾0.054545  0  0  0
      VB←(⍉A)+.×V
      VB
0.090909 0.054545 ‾0.036364  0  0  0
      IB←GB+.×VB
      IB
0.90909 1.0909 ‾1.0909  0  0  0
      A+.×IB
2 ‾2 6.6613E‾16  0  0  0
```

The resulting (4 × 4) reduced nodal conductance matrix *can* now be inverted, and the voltage response computed normally (note that *IRN* is a 4-element vector, whereas the circuit has 6 nodes). The correct branch voltages and currents are computed

```
      GMT←1 5ρ2 3 4 5 2
      YA←Y ADDGM GMT

      YA
 30 ‾10 ‾20   0   0   0
‾10  40 ‾30   0   0   0
‾20 ‾30  50   0   0   0
  0   2  ‾2   4  ‾3  ‾1
  0  ‾2   2  ‾3  13 ‾10
  0   0   0  ‾1 ‾10  11
```

Now let us add the voltage-controlled current source to the circuit. The new complete nodal conductance matrix is found by use of the function *ADDGM* defined in Exercise 3.6

```
      ⊞‾1 ‾1↓YA
DOMAIN ERROR
      ⊞ ‾1 ‾1 ↓YA
      ∧
      ZRA←⊞1 1↓‾1 ‾1↓YA
```

Again it is not sufficient to select a single reference node: one must be selected for each separate part of the circuit: nodes 1 and 6 (first and last) are chosen

```
      IRA←‾2  0  0  0
      VRA←ZRA+.×IRA
      V←0,VRA,0
      V
0 ‾0.090909 ‾0.054545 0.016913 ‾0.0016913 0
      VB←(⍉A)+.×V
      VB
0.090909 0.054545 ‾0.036364 0.018605 0.016913 0.0016913
      IB←GB+.×VB
      IB
0.90909 1.0909 ‾1.0909 0.055814 0.016913 0.016913
      A+.×IB
2 ‾2 1.3323E‾15 0.072727 ‾0.072727 ‾3.2092E‾17
```

The calculation of nodal and branch voltages, and branch currents, proceeds normally

The calculation $A+.\times IB$ reveals excitation currents at nodes 4 and 5 due to the controlled source, because a corresponding branch is not defined in *A*.

```
      2×-/V[2 3]
‾0.072727
```

These excitation currents are shown to be correctly related to their controlling voltage

4

Frequency-domain Analysis

In the discussion of linear resistive circuits it was shown how, starting from a description of the components (the branch conductance matrix G,) and their interconnection (the incidence matrix A), one can compute, in turn, the complete nodal conductance matrix (Y), the reduced nodal conductance matrix (YR)relevant to a specific reference node, and finally the reduced nodal resistance matrix (ZR). From the ZR matrix the nodal voltage response VR to current excitation IR can be computed (Fig. 4.1(a)).

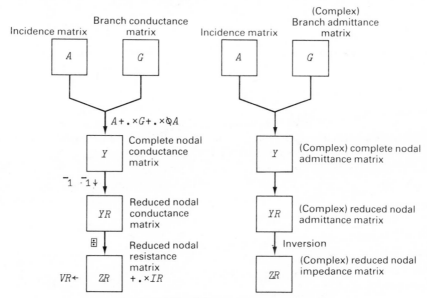

Fig. 4.1

.**The analysis of (a) resistive and (b) reactive circuits**

A parallel development can be associated with the frequency-domain behavior of a circuit containing *reactive* as well as resistive components, if the VR and IR are the complex[†] amplitudes of the voltages and currents at a particular frequency. In place of the branch conductance matrix one has a branch *admittance* matrix for each frequency of interest (and similarly the

---

† It is assumed that the reader is familiar with the use of complex numbers to describe and analyze the frequency-domain behavior of a circuit.

complete and reduced nodal *admittance* matrices), and in place of the reduced nodal resistance matrix one has the reduced nodal *impedance* matrix; thus, G, Y, YR and ZR now have complex elements (Fig. 4.1(b)).

The analysis of a circuit containing both reactance and resistance *could* proceed along the steps outlined in Figs. 1.12 and Fig. 4.1, but taking account of the complex nature of most of the variables involved.[†] In this introduction, however, we shall form the complete nodal admittance matrix Y by the scheme developed at the end of Chapter 1 (and illustrated in Fig. 4.2), wherein the contribution of each component to Y is calculated from the 3-column conductance table GT and added to Y in turn. For convenience we shall first consider circuits containing only two-terminal resistance, capacitance and inductance.

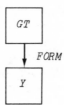

Fig. 4.2
Direct generation of Y **from** GT

4.1   Circuit Description

The circuit will be described by three matrices G, C and IL, corresponding respectively to the conductances, capacitances and inductances: in the latter case, for ease of exposition, we record the inverse (reciprocal) of the inductance values (hence IL). Each matrix has three columns, the first two containing the numbers of the nodes to which the component is connected, the third containing its value. There are as many rows as there are components of the relevant type: thus, if a circuit contains two conductances, one capacitor and no inductors (Fig. 4.3), the matrices G, C and IL are as shown in that figure.

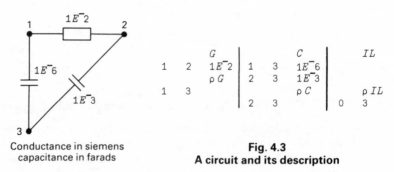

Conductance in siemens
capacitance in farads

Fig. 4.3
A circuit and its description

[†] This is complicated, at present, by the fact that although the APL *notation* has been extended to complex numbers, an implementation is available only on a limited number of systems. Therefore, for most of this chapter, it will be assumed that no complex APL implementation is available.

It is a simple matter to initialize any of the three matrices. Thus, in forming G for Fig. 4.3 one can enter the expressions

```
      G← 0 3ρ0
      G

      G← G,[1] 1  2  1E¯2
      G
\ 1    2   1E¯2
        ρG
  1    3
```

4.2 Formation of the Real and Imaginary Parts of the Nodal Admittance Matrix

The real part (RY) of Y is determined by the circuit conductances, and can be constructed by use of the function $FORM$ (which is a slightly modified version of a function of the same name developed in Chapter 1):

```
       ∇     R←Y FORM G; K; I
[1]          R←Y
[2]          K←0
[3]          S: → L× (1↑ρG) ≥ K←K+1
[4]          L: I←G[K;  1  2]
[5]          R[I; I] ←R[I; I] +  2  2 ρ  1  ¯1  ¯1  1  ×G[K; 3]
[6]          → S
       ∇
```

The left-hand argument of $FORM$ is the initialized value of RY containing only zero elements; for the four-node example of Fig. 4.4, now taken as a working illustrative example,

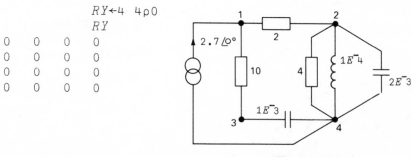

```
           RY←4  4ρ0
           RY
  0    0    0    0
  0    0    0    0
  0    0    0    0
  0    0    0    0
```

Fig. 4.4
An example circuit

The right-hand argument is the matrix G. For the same example (Fig. 4.4),

```
            G
   1    2    2
   2    4    4
   1    3    10
```

By executing $FORM$

```
        RY←RY FORM G
        RY
    12   ‾2  ‾10    0
    ‾2    6    0   ‾4
   ‾10    0   10    0
     0   ‾4    0    4
```

we obtain the nodal conductance matrix we would expect from inspection of the circuit.

The imaginary part (IY) of Y is a function of the capacitors and inductors, described by the arrays C and IL:

```
            C                           IL
   3    4   0.001            2    4    10000
   2    4   0.002
```

It is also frequency dependent, and in frequency-domain analysis is constructed separately for each frequency. The contribution of each capacitor to IY is the same as that of a similar-valued conductance to RY, except that the capacitance is first multiplied by the angular frequency W. Thus, after initializing the matrix IY to contain zero elements

```
    IY ← 4   4ρ 0
```

and defining the frequency of interest (say)

```
    W←1000
```

we can add the capacitive contributions to IY by executing

```
        IY←IY FORM C[;1 2],C[;3]×W
        IY
   0    0    0    0
   0    2    0   ‾2
   0    0    1   ‾1
   0   ‾2   ‾1    3
```

Similarly, the contribution due to the inductances may be added to obtain the entire imaginary part of the complete nodal admittance matrix:

```
        IY←IY FORM IL[;1 2],IL[;3]÷-W
        IY
   0    0    0    0
   0   ‾8    0    8
   0    0    1   ‾1
   0    8   ‾1   ‾7
```

It is convenient to represent the complete nodal admittance matrix of the circuit of Fig. 4.4 by a three-dimensional array Y (see Fig. 4.5): here

```
        Y←RY,[.5]IY
        Y
  12   ‾2  ‾10    0
  ‾2    6    0   ‾4
 ‾10    0   10    0
   0   ‾4    0    4

   0    0    0    0
   0   ‾8    0    8
   0    0    1   ‾1
   0    8   ‾1   ‾7
```

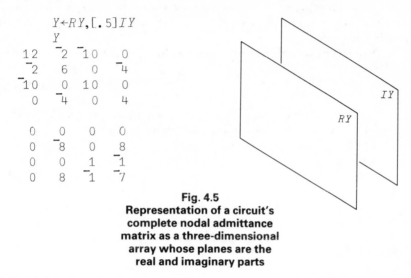

Fig. 4.5
Representation of a circuit's
complete nodal admittance
matrix as a three-dimensional
array whose planes are the
real and imaginary parts

If the highest numbered node is chosen as the reference node, the corresponding reduced nodal admittance matrix is obtained by dropping the last row and column from each plane of Y:

```
        YR←0 ‾1 ‾1↓Y
        YR
  12   ‾2  ‾10
  ‾2    6    0
 ‾10    0   10

   0    0    0
   0   ‾8    0
   0    0    1
```

4.3   Equation Solution

We have just obtained the representation (YR) of a complex matrix relating vectors of complex voltages and currents in a manner parallel to the relation

$$IR←YR+.×VR$$

which we have already encountered with resistive circuits. We could therefore proceed to invert the complex matrix and, from knowledge of the current excitation, find the voltage response by the same procedure used for resistive circuit analysis:

$$VR \leftarrow (\boxminus YR) + . \times IR$$

There would be no difficulty in doing so. However, we choose instead to illustrate a different approach which more closely resembles the methods commonly used in practice, and which are described in detail in Chapters 5 and 6. To do so we *temporarily*, and only for convenience, confine our attention to resistive circuits.

A purely resistive circuit is, as we know, described by the relation

$$IR \leftarrow YR + . \times VR \tag{4.1}$$

If, as is commonly the case, YR and IR are known, then VR has up to now been found by execution of the following expressions

$$ZR \leftarrow \boxminus YR$$
$$VR \leftarrow ZR + . \times IR \tag{4.2}$$

However, if only VR is required, and ZR is of no interest (in other words if we merely require the solution of the set of equations (4.1)), then VR may be obtained *without* the intermediate generation of ZR by executing the expression

$$VR \leftarrow IR \boxminus YR \tag{4.3}$$

Thus, $\boxminus$ used *dyadically* denotes a vector result that is the solution of the set of equations (4.1). The essential difference between the execution of (4.2) and (4.3) is illustrated in Fig. 4.6, and an example of the application of (4.3) is

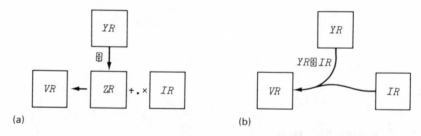

Fig. 4.6
Representation of the execution of 4.2(a) and 4.3(b)

shown in Fig. 4.7. One advantage of the approach to solution embodied in (4.3) is that it is computationally less expensive than (4.2), as we shall see in Chapters 5 and 6. Another is that there is no need to store ZR. The syntax of (4.3) is easy

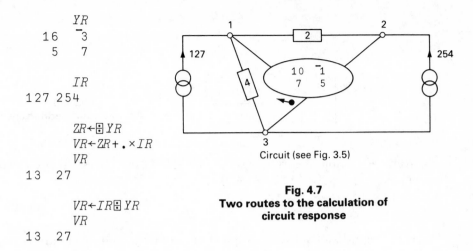

$$YR$$
$$16 \quad ^-3$$
$$5 \quad 7$$

$$IR$$
$$127 \; 254$$

$$ZR \leftarrow \boxdot YR$$
$$VR \leftarrow ZR + . \times IR$$
$$VR$$
$$13 \quad 27$$

$$VR \leftarrow IR \boxdot YR$$
$$VR$$
$$13 \quad 27$$

Circuit (see Fig. 3.5)

Fig. 4.7
Two routes to the calculation of
circuit response

to remember: it is a generalization of the relation

$$VR \leftarrow IR \div YR$$

characterizing the voltage–current relation of a two-terminal conductance of value YR (Fig. 4.8).

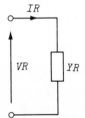

Fig. 4.8
A two-terminal conductance

If the circuit contains *reactive* components, the pertinent relations are of the same *form* as (4.1), but the elements of the matrix and vectors involved are complex numbers. If a complex number implementation of APL is available, the solution can again be obtained by execution of (4.3), and is discussed later in Section 4.6. If it is not, an alternative approach must be devised, as described in the next section.

4.4   The Function $REALSOLV$

One method of solving a set of *complex* equations involves combining the real and imaginary coefficients in such a way that we have a new set of equations containing only *real* coefficients. For illustration, consider a simple two-terminal admittance (Fig. 4.9) with real and imaginary parts RY and IY respec-

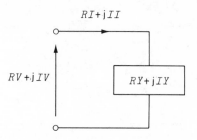

Fig. 4.9
A two-terminal admittance

tively. A complex[†] current excitation $(RI+jII)$ gives rise to a complex voltage response $(RV+jIV)$ so that

$$(RI+jII) \leftarrow (RY+jIY) \times (RV+jIV)$$ (4.4)

If the right-hand side is multiplied out, and real and imaginary coefficients equated on both sides, this *single complex* equation becomes a pair of *real* equations which, expressed in array form, become

$$\begin{bmatrix} RI \\ II \end{bmatrix} \leftarrow \begin{bmatrix} RY & -IY \\ IY & RY \end{bmatrix} +.\times \begin{bmatrix} RV \\ IV \end{bmatrix}$$ (4.5)

This equation can, of course, be solved using either (4.2) or (4.3).

The discussion above generalizes, with no difficulty, from the situation of a single component to that of a circuit. For a circuit, the equations have the form for which a suitable representation might be

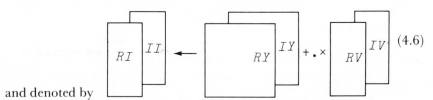

(4.6)

and denoted by

$$IR \leftarrow YR +.\times VR$$

If the same procedure adopted to obtain (4.5) from (4.4) is now applied to (4.6) we obtain

$$\begin{bmatrix} RI \\ II \end{bmatrix} \leftarrow \begin{bmatrix} RY & -IY \\ IY & RY \end{bmatrix} +.|\times \begin{bmatrix} RV \\ IV \end{bmatrix}$$ (4.7)

which will be denoted[††] by $I \leftarrow Y +.\times V$

---

†   The prefix j is used here according to conventional notation. The use of j in complex APL notation is discussed in Section 4.6.

†† Temporarily we use I, V and Y to denote the arrays involved in equation (4.7).

Note that if the circuit has N nodes in addition to the reference node, so that both RY and IY have the shape N, N, then the set of equations (4.7) we must now solve is characterized by a matrix whose shape is $2 \times N, N$.

To solve the set of equations (4.6) we define $REALSOLV$ (Fig. 4.10). Its left argument YR is the 3-dimensional array discussed earlier whose first (RY) and second (IY) planes are the real and imaginary parts of the circuit's reduced nodal admittance matrix. Its right argument is the 3-dimensional array IR (see (4.6) . With reference to (4.7) and (4.6) we see that

```
        ∇    VR←YR REALSOLV IR;YT;YB;Y;V
    [1]      YT←YR[1;;],¯1×YR[2;;]
    [2]      YB←YR[2;;],YR[1;;]
    [3]      Y←YT,[1]YB
    [4]      V←(,IR)⊟Y
    [5]      VR←(ρIR)ρV
        ∇
```

Fig. 4.10
The function $REALSOLV$

$$RY←YR[1;;]$$
$$IY←YR[2;;]$$

Thus, lines 1 and 2 of $REALSOLV$ generate the top (YT) and bottom (YB) halves of the matrix in (4.7), and the matrix is then formed on line 3. On line 4 we generate the solution (V) of (4.7) by the use of dyadic ⊟. Its right argument is Y. Its left (vector) argument I (see 4.7) is generated from the three-dimensional array IR according to

$$I←,IR$$

Having generated the vector V of (4.7), it is then reformed into the required shape by

$$VR←(ρIR)ρV$$

on line 5. The function $REALSOLV$ is usefully illustrated by an example.

Example Consider again the circuit of Fig. 4.4, repeated here for convenience as Fig. 4.11. We have already generated the array YR:

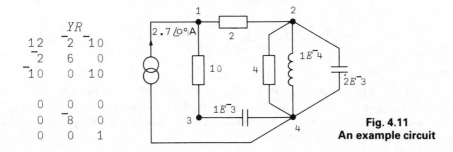

	YR	
12	¯2	¯10
¯2	6	0
¯10	0	10
0	0	0
0	¯8	0
0	0	1

Fig. 4.11
An example circuit

The current excitation applied at node 1 (and removed via node 4, see Fig. 4.11) has an amplitude of 2.7 A and a phase angle of zero, described by real and imaginary parts of 2.7 and 0, so that

```
        IR←2 1 3ρ2.7 0 0 0 0 0
        IR
  2.7 0   0
  0   0   0
```

The complete voltage response of the circuit of Fig. 4.11 is then obtained by

```
        VR←YR REALSOLV IR
        VR
  ‾1.2854      0.22838      ‾1.2268
  ‾0.46333     0.15006      ‾0.586
```

Thus, with respect to the reference terminal, the voltage of node 3 has real and imaginary parts of 1.2268 and ‾0.586 V respectively.

The complete vector of nodal voltages can easily be formed

```
        VREF←2 1 1ρ0 0
        V←VR,VREF
        V
  ‾1.2854      0.22838      ‾1.2268   0
  ‾0.46333     0.15006      ‾0.586    0
```

From this vector, by appropriate indexing, we can find the voltages across the conductances by making use of the information in the first two columns of G:

```
          G
  1   2   2
  2   4   4
  1   3   10
          V[;;G[;1 2]]
  ‾1.2854      0.22838
  0.22838      0
  ‾1.2854      1.2268

  ‾0.46333     0.15006
  0.15006      0
  ‾0.46333     ‾0.586
          ρV[;;G[;1 2]]
  2 1 3 2
```

This array contains the real and imaginary parts of the nodal voltages associated with the node pairs corresponding to each branch. Minus reduction over the last dimension then yields the real and imaginary parts of the conductance branch voltages:

```
    VG←- /[4]V[;;G[;1 2]]
    VG
 1.057      0.22838      0.0586
‾0.61339   0.15006      0.12268
```

In the same way we find the complex capacitor voltages:

```
        VC←-/V[;;C[;1 2]]
        VC
 1.2268     0.22838
‾0.586      0.15006
```

4.5 Frequency-domain Analysis

Normally, a designer wishes to compute the response of a circuit at a *number* of sample frequencies, and to examine a plot of some function of a nodal voltage (its magnitude, for example) versus frequency. The foregoing development leading to the function *REALSOLV* may easily be applied to this problem. We illustrate the approach that can be adopted, taking as our example the circuit of Fig. 4.11 which contains only two-terminal components.

    The circuit of Fig. 4.11 (also Fig. 4.4) is described by three three-column arrays G, C and IL

	G			C			IL	
1	2	2	3	4	0.001	2	4	10000
2	4	4	2	4	0.002			
1	3	10						

and its current excitation by the array IR

```
            IR
 2.7  0   0
 0    0   0
```

    We assume that a vector F is defined whose elements are the sample frequencies in hertz:

```
 F←10 20 40 80 160 320 640 1280
```

Thus the radian frequencies of interest are

```
 W←2×○F
 W
62.832  125.66  251.33  502.65  1005.3  2010.6  4021.2
        8042.5
```

In computing the circuit's complete nodal admittance matrix we recall that both its real (RY) and imaginary (IY) parts are of the same shape, and must each be initialized to contain zero elements. First, however, we initialize only the real part RY:

```
N←⌈/(,G[;1 2]),(,C[;1 2]),,IL[;1 2]
N
```
4
```
RY←(N,N)ρ0
```

The real part (RY) of the admittance matrix is common to all frequencies and need only be generated once:

```
     RY←RY FORM G
     RY
 12   ¯2 ¯10    0
 ¯2    6    0   ¯4
¯10    0   10    0
  0   ¯4    0    4
```

The imaginary part is particular to the radian frequency, for which we define an index K: the ensuing generation of IY will be re-executed every time K is incremented by means of a looping activity:

```
K←1
LOOP: IY←(N,N)ρ0
    IY←IY FORM C[; 1 2],C[;3]×W[K]
    IY←IY FORM IL[; 1 2],IL[;3]÷-W[K]
```

Note (second line) that IY must be re-initialized for each new frequency. The complete and reduced nodal admittance matrices are then formed:

```
YR←0 ¯1 ¯1↓RY,[.5]IY
```

Use of $REALSOLV$ then leads to the three-dimensional array of complex nodal voltages

```
     VR←YR REALSOLV IR
     VR
 1.3499      0.00095935   1.3497
¯0.025461    0.01694     ¯0.033941
```

where

```
        ρVR
 2   1   3
```

However, since an array VR will be generated for each frequency of interest it is convenient to initialize the array $VREIM$:

$$NR \leftarrow N-1$$
$$VREIM \leftarrow (2,NR,(\rho F))\rho 0$$
$$VREIM$$

```
0    0    0    0    0    0    0    0
0    0    0    0    0    0    0    0
0    0    0    0    0    0    0    0

0    0    0    0    0    0    0    0
0    0    0    0    0    0    0    0
0    0    0    0    0    0    0    0
```

and then ensure that the computed response for a particular frequency is placed in the correct location within $VREIM$:

$$VREIM[;;K] \leftarrow YR \quad REALSOLV \quad IR$$

If, after being incremented, the frequency index K is found to be greater than ρW execution is terminated: if not, the process continues by returning to the line labelled $LOOP$:

$$\rightarrow LOOP \times (\rho W) \geq K \leftarrow K+1$$

A function $ACSOLVE$ embodying the steps we have just examined is shown in Fig. 4.12. It generates the global variable $VREIM$. Thus, for the circuit of Fig. 4.11, and the frequencies defined by the vector F, execution of the expression

$$ACSOLVE$$

generates an array $VREIM$ of the expected size:

$$\rho VREIM$$
```
2    3    8
```

```
       ∇    ACSOLVE;W;N;NR;IY;RY;K;YR
 [1]        W←,2×○F
 [2]        N←⌈/(,G[; 1 2]),(,C[; 1 2]),,IL[; 1 2]
 [3]        NR←N-1
 [4]        RY←(N,N)ρ0
 [5]        VREIM←(2,NR,(ρW))ρ0
 [6]        RY←RY FORM G
 [7]        K←1
 [8]        LOOP:IY←(N,N)ρ0
 [9]        IY←IY FORM C[; 1 2],C[;3]×W[K]
[10]        IY←IY FORM IL[; 1 2],IL[;3]÷-W[K]
[11]        YR← 0 ¯1 ¯1 ↓RY,[0.5]IY
[12]        VREIM[;;K]←YR REALSOLV IR
[13]        →LOOP×(ρW)≥K←K+1
       ∇
```

Fig. 4.12
The function $ACSOLVE$ **which carries out the frequency-domain analysis of
a two-terminal element R, C, L circuit**

Rather than examine the 3-dimensional array *VREIM*, the designer usually wishes to generate a plot showing the frequency dependence of some property (magnitude, for example) of a particular nodal voltage. To this end, we can define 'utility' functions such as *VOLTAGE* which selects the voltage response at a given node (*NN* = node number):

```
       ∇    Z←VOLTAGE  NN
  [1]       Z←VREIM[;NN;]
       ∇
```

and functions such as *MAG* to compute specific properties of the selected nodal voltage:

```
       ∇    Z←MAG  X
  [1]       Z←(+/[1]  X*2)*0.5
       ∇
```

Then, if we have conveniently named the vector of sample frequencies

FREQUENCY←F

and ensured the availability of a simple plot function such as that shown in Fig. 4.13

```
       ∇    X PLOT Y;SS;SX;SY
  [1]       SS← 40 14 ,(⌊/X),(⌈/X),(⌊/Y),(⌈/Y)
  [2]       SX←⌊0.5+(X-SS[3])×SS[1]÷SS[4]-SS[3]
  [3]       SY←⌊0.5+(Y-SS[5])×SS[2]÷SS[6]-SS[5]
  [4]   LP:'|',' *'[1+(0,ιSS[1])∈(SY∈SS[2])/SX]
  [5]       →(0≤SS[2]←SS[2]-1)/LP
  [6]       'o',(SS[1]ρ'-'),'→'
       ∇
```

Fig. 4.13
A simple plot function

a simple expression such as

FREQUENCY PLOT MAG VOLTAGE 2

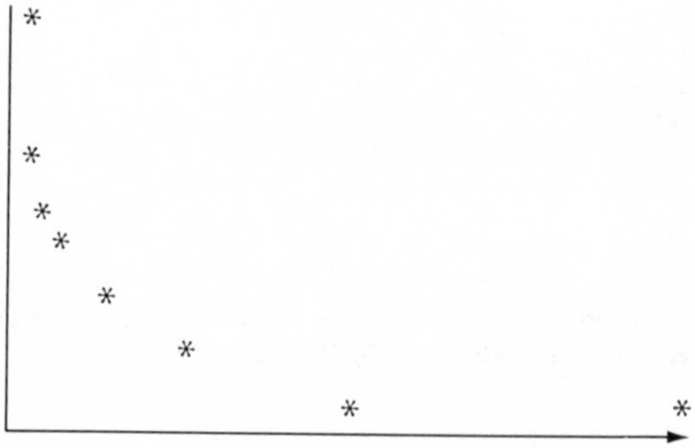

Fig. 4.14

results in a plot, versus frequency, of the magnitude of the voltage at node 2, as shown in Fig. 4.14. Naturally, more sophisticated — and readily available — plotting functions will provide a plot with scaled axes.

4.6 Complex Numbers

Within the APL notation, the character J is used to associate the real and imaginary parts of a complex number, and an extension of the circular ($\circ$) function is employed in complex number representation. In what follows, we shall largely present this recent extension of APL by means of an illustrative example.

If two real scalar numeric constants are connected by the letter J, as in

        3J4

this is interpreted as the rectangular representation of a complex number, with the real part first and the imaginary part second. Most of the primitive functions we have already encountered extend[†] to complex arguments:

            3J4+¯1J2.5+0
        2J6.5
                |3J4
        5

and also to arrays of complex numbers

            CV←3J4 4.1 0J1E2 ¯5J¯1.2 0
            ρCV
        5
            +/CV
        2.1J102.8

To transform a complex scalar into the vector whose components are its real and imaginary parts the dyadic circular function is used with 9 11 as its left vector argument.

            9 11○3J¯14
        3 ¯14
            ρ9 11○3J¯14
        2

The inverse transformation is effected with ¯9 ¯11 as the left argument:

            ¯9 ¯11+.○3 ¯14
        3J¯14

---

† Primitive functions which either cannot be, or have not yet been, extended to complex numbers include the dyadic functions <, ≤, >,≥,⌈, ⌊, and ⊤, and monadic functions such as ↑,↓,⌊,and⌈.

In preparation for our study of a complex matrix we examine the same transformation applied to a complex vector:

```
      R←9  11∘.○CV
      R
3       4.1      0       ‾5      0
4         0    100     ‾1.2      0
      ρR
2   5
```

Frequency-domain analysis using complex APL can proceed in precisely the same manner as before as far as the formation of YR, representing the reduced nodal admittance matrix. For the circuit of Fig. 4.4, we recall the value and shape of YR:

```
            YR
   12    ‾2  ‾10
    ‾2     6    0
   ‾10     0   10

    0    0    0
    0   ‾8    0
    0    0    1
          ρYR
  2    3    3
```

Use of the dyadic circular function (but in an inner product) with ‾9 ‾11 as its left argument generates a two-dimensional array of complex numbers which is the complex nodal admittance matrix:

```
      YR←‾9 ‾11+.○YR
      YR
   12      ‾2      ‾10
    ‾2   6J‾8        0
   ‾10      0     10J1
          ρYR
  3    3
```

Using the new complex number representation, the current excitation of Fig. 4.4 is

```
      IR←2.7J0 0 0
      IR
2.7   0   0
```

Just as dyadic ⊟ was used in (4.3), with *real* arguments, to obtain the voltage response of a resistive circuit, it can now be used with complex arguments

```
      VR←IR⊟YR
      VR
1.2854J⁻0.46333 0.22838J0.15006 1.2268J⁻0.586
      ρVR
3
```

to obtain the complex voltage response of the circuit of Fig. 4.11. The expression

```
      9 11∘.○VR
 1.2854    0.22838    1.2268
⁻0.46333  0.15006   ⁻0.586
```

generates the real and imaginary parts of the nodal voltages: if only the imaginary parts are needed, then

```
      11○VR
⁻0.46333 0.15006 ⁻0.586
```

will suffice. Since the magnitude and arc of currents and voltages are often of interest, we note that

```
      10 12∘.○VR
 1.3663    0.27327  1.3596
⁻0.34596  0.58133  ⁻0.44563
```

generates these quantities.

From the above discussion it follows that the function ACSOLVE may be modified as follows if complex APL is available:

```
[5]      VR←(NR,(ρW))ρ0
[11.1]   YR← ⁻9 ⁻11 +.○YR
[12]     VR[;K]←IR⊟YR
```

Exercises

4.1 For the circuit shown in Fig. E4.1 and with the node numbering shown, create matrices describing the conductances and the capacitances and their connection within the circuit.

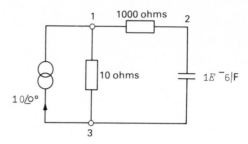

Fig. E4.1

4.2 (Continued from Exercise 4.1) Use the function *FORM* defined below to create both the real and imaginary parts of the complete nodal admittance matrix Y for an angular frequency of 1000 radians per second.

```
        ∇R←Y FORM G; K; I
[1]     R←Y
[2]     K←0
[3]     S: →L×(1↑ρ G) ≥K←K+1
[4]     L: I←G[K; 1 2]
[5]     R[I; I]←R[I; I]+2 2ρ1 ¯1 ¯1 1×G[K; 3]
[6]     →SV
```

4.3 (Continued from Exercise 4.2) Form the three-dimensional array *Y* whose planes are the real and imaginary parts of the complete nodal admittance matrix, and hence obtain the reduced nodal admittance matrix *YR*.

4.4 (Continued from Exercise 4.3) For the current excitation shown in Fig. E4.1 determine the nodal voltage response using the function *REALSOLV* given in the text and repeated here:

```
        ∇V←Y REALSOLV I; YT; YB
[1]     YT←Y[1;;],(¯1×Y[2;;])
[2]     YB←Y[2;;],Y[1;;]
[3]     V←(ρI)ρ(,I)⊞YT,[1] YBV
```

4.5 (Continued from Exercise 4.4) For the circuit of Fig. E4.1 calculate the magnitude and phase of the voltage at node 2 relative to the current excitation at node 1.

4.6 Compute, for the circuit of Fig. E4.1, the magnitude of the voltage at node 2 for a number of frequencies in a range sufficiently wide to show the effect of the capacitor. Provide both a table of voltages and a plot versus frequency. For the plot, use logarithmic scaling in order to place in evidence the straight-line asymptotes of the dependence of the magnitude of the node 2 voltage on frequency. Select one of the sample frequencies, and see if the real and imaginary parts of *YR* are each, separately, capable of being inverted.

4.7 Compute the input impedance ZIN of the circuit of Fig. E4.2 for about five sample frequencies over quite a wide frequency range. Be sure to choose the values of the resistance (R), capacitance (C) and inductance (L) to satisfy the relation $R = (L \div C) * 0.5$. Also determine the voltage V at each of the sample frequencies. If V turns out to be zero, whatever the frequency, can a short circuit be connected between nodes 2 and 3 without affecting ZIN? Test your answer.

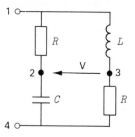

Fig. E4.2

4.8 Compute the input impedance ZIN of the circuit of Fig. E4.3 for an arbitrarily selected value of G, and decide if it corresponds to a pure inductance. The three-terminal device shown is called a gyrator, and it can be realized in microelectronic form. Since inductors cannot easily be realized in such a form, the answer to this problem illustrates one of the principal applications of the gyrator.

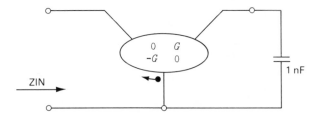

Fig. E4.3

4.9 Some circuits, especially filters, have a 'ladder' structure as shown in Fig. E4.4, in which each of the two-terminal circuits can contain G, L and C in series or parallel connection. With such a structure it is possible to compute (say) the voltage gain $(VOUT \div VIN)$ by assuming an output voltage of $1 + j0$ V and then computing, successively and very simply, the current through $Z9$, the current through $Z8$, the voltage across $Z8$, the voltage across (and hence the current through) $Z7$, etc., until the value of VIN is found. Write a function or functions to undertake such an analysis (for a circuit containing G, L and C) and test it. Compare its behavior with $ACSOLVE$.

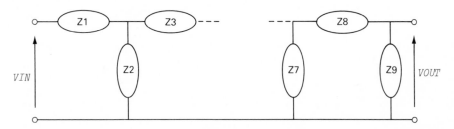

Fig. E4.4

4.10   Using complex APL (if available), define a set of functions that will access the description of a circuit containing resistance, capacitance, inductance and voltage-controlled current sources and, given the current excitation and a vector F of sample frequencies, compute the magnitude and phase of the voltage at a selected node. The circuit should be described by three three-column tables (for R, L, C) and one five-column table (for $VCCS$) defining the node connections and parameter values of the components.

Test your functions thoroughly: in particular, use the test circuit of Fig. E4.5 for the case where only *one* frequency is of interest.

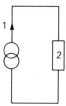

Fig. E4.5

4.11   Repeat Exercise 4.10 under the assumption that complex APL is not available.

Solutions

4.1
```
      G←2 3ρ1 3  .1  1  2  .001
      G
  1              3              0.1
  1              2              0.001
      C 1←3ρ2 3 1E¯6
      C
  2.0000E0     3.0000E0     1.0000E¯6
```

4.2
```
         N←⌈/,G[; 1  2] ,[1] C[; 1  2]
         N
  3
```
N, the number of nodes, is the largest number to be found in columns 1 and 2 of the matrices describing the components

```
         RY←(N,N)ρ0
         RY
  0 0 0
  0 0 0
  0 0 0
```
Initialization of the real part of the complete nodal admittance matrix

```
         RY←RY FORM G
         RY
   0.101        ¯0.001          ¯0.1
  ¯0.001         0.001           0
  ¯0.1           0               0.1
```
The real part of the complete nodal admittance matrix is a function only of the conductances

```
      W←1000
      IY←(N,N)ρ0
      IY
  0 0 0
  0 0 0
  0 0 0
```
The imaginary part *IY* of the complete nodal admittance matrix is a function of frequency.

The angular frequency *W* is defined

IY is initialized

```
      IY←IY FORM C[;1  2] ,C[;3]×W
      IY
  0           0            0
  0           0.001       ¯0.001
  0          ¯0.001        0.001
```
The contribution due to the capacitor is added.

4.3
```
      Y←RY,[.5]IY
      Y
   0.101      ¯0.001        ¯0.1
  ¯0.001       0.001         0
  ¯0.1         0             0.1
```
The complete nodal admittance matrix

real part

```
  0           0            0
  0           0.001       ¯0.001
  0          ¯0.001        0.001
```
imaginary part

```
      YR←0 ¯1 ¯1↓Y
      YR
   0.101      ¯0.001
  ¯0.001       0.001
```
The reduced nodal admittance matrix

```
  0           0
  0           0.001
```

4.4
```
      IR←2 1 2ρ10 0 0 0
      IR
  10  0

   0  0
```
The complex nodal current excitation

         VR←YR REALSOLV IR   The nodal voltage response. The voltage at terminal
         VR   1, with respect to terminal 3, is 99.5-j0.495 V
     99.5   49.502

      ¯0.49502   ¯49.998
         ρVR   The shape of VR
     2 1 2

4.5   MAGV2←(+/(,VR[;;2])∘.*2)*.5   The magnitude of the voltage of node 2
         MAGV2
     70.358

         (360÷○2)×¯30÷/,VR[;;2]   The phase, in degrees, of the voltage of node 2, tak-
     ¯44.715   ing the phase of the current source as reference

                                                     The phase is close to 45°, as would be expected from
                                                     inspection of the circuit

4.6   The values of *G*, *C* and *IR* are unchanged. We define a frequency vector

             F
     10 100 1000 10000 100000 1000000

Execution of *ACSOLVE*

     ACSOLVE

yields a reduced nodal voltage array of the expected shape

         ρVREIM
     2 2 6

We select the real and imaginary parts (i.e., both planes) of the voltage at node 2 (row 2) at all frequencies:

     X←VREIM[;2;]

The magnitude of the node 2 voltage at each frequency is easily generated

     M←(+/[1]X*2)*.5
     M
     99.799 81.998 14.055 1.4182 0.14182 0.014182

The required plot is generated by

     (⊕ *FREQUENCY*) *PLOT* ⊕M
     and shown in Fig. E4.6.

                                                            Fig. E4.6

At the frequency of 1000 radians per second, the real part of the reduced nodal admittance matrix
(see *YR* in answer to Exercise 4.3) can be inverted, but the imaginary part is singular and cannot
be inverted.

4.7 (This solution uses complex APL.) The circuit is shown in Fig. E4.7 and the frequencies chosen are given by the vector F:

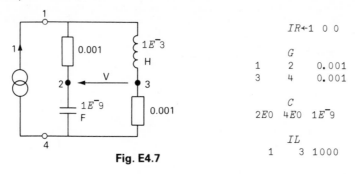

$IR \leftarrow 1 \ 0 \ 0$

G

1	2	0.001
3	4	0.001

C

2E0	4E0	$1E^-9$

IL

1	3	1000

Fig. E4.7

F
10 100 1000 10000 100000

Execution of the $ACSOLVE$ function modified for complex APL (see box below)

     $ACSOLVE$

then computes the two-dimensional array VR

```
VR
1000J1.581E‾9  1000J9.641E‾11  1000J3.051E‾12  1000J‾2.962E‾12  1000J4.075E‾13
1000J‾0.06283  1000J‾0.6283    1000J‾6.283     996.1J‾62.58     717J‾450.5
1000J‾0.06283  1000J‾0.6283    1000J‾6.283     996.1J‾62.58     717J‾450.5
```

from which it appears that the input impedance is resistive, and equal to 1000 ohms. (You should check theoretically that this is expected). It is also seen that $VR[2]$ and $VR[3]$ are identical, so that $V = 0$. Nodes 2 and 3 can therefore be connected by a short circuit without affecting ZIN (check this).

```
       ∇    ACSOLVE;W;N;NR;IY;RY;K;YR
  [1]      W←,2×○F
  [2]      N←⌈/(,G[; 1 2]),(,C[; 1 2]),,IL[; 1 2]
  [3]      NR←N-1
  [4]      RY←(N,N)ρ0
  [5]      VR←(NR,(ρW))ρ0
  [6]      RY←RY FORM G
  [7]      K←1
  [8]  LOOP: IY←(N,N)ρ0
  [9]      IY←IY FORM C[; 1 2],C[;3]×W[K]
  [10]     IY←IY FORM IL[; 1 2],IL[;3]÷-W[K]
  [11]     YR← 0 ‾1 ‾1 ↓RY,[0.5] IY
  [12]     YR← ‾9 ‾11 +.○YR
  [13]     VR[;K]←IR⊞YR
  [14]     →LOOP×(ρW) ≥K←K+1
       ∇


       ∇    R←Y FORM G;K;I
  [1]      R←Y
  [2]      K←0
  [3]  S: →L×(1↑ρ G) ≥K←K+1
  [4]  L: I←G[K; 1 2]
  [5]      R[I;I]←R[I;I]+ 2 2 ρ 1 ‾1 ‾1 1 ×G[K;3]
  [6]      →S
       ∇
```

4.8 (The function *ACSOLVE* of Fig. 4.12 is employed, though modified as shown below.)

The circuit to be analyzed is shown in Fig. E4.8, where the gyrator has been modelled by two voltage-controlled current sources. Since VCCS are now involved, the function *ACSOLVE* (Fig. 4.12) must be modified as shown below to call the function *ADDGM* defined in the solution to Exercise 6 of Chapter 3.

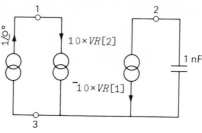

Fig. E4.8

```
        ∇ACSOLVE[6.1]
[6.1]    RY←RY ADDGM GMT∇
        ∇ACSOLVE[□] ∇
   ∇     ACSOLVE; W; N; NR; IY; RY; K; YR
[1]      W←,2×○F
[2]      N←⌈/(,G[; 1 2]),(,C[; 1 2]),,IL[; 1 2]
[3]      NR←N-1
[4]      RY←(N,N)ρ0
[5]      VREIM←(2,NR,(ρW))ρ0
[6]      RY←RY FORM G
[7]      RY←RY ADDGM GMT
[8]      K←1
[9]      LOOP: IY←(N,N)ρ0
[10]     IY←IY FORM C[; 1 2],C[; 3]×W[K]
[11]     IY←IY FORM IL[; 1 2],IL[; 3]÷-W[K]
[12]     YR← 0 ¯1 ¯1 ↓RY,[0.5]IY
[13]     VREIM[;;K]←YR REALSOLV. IR
[14]     →LOOP×(ρW) ≥K←K+1
   ∇
```

The circuit and its excitation are defined below (note that even though no conductances or inductors are present, their (empty) arrays must exist).

```
    GMT                              IR←2 1 2ρ1 0 0 0
2   3   1   3   10                   IR
1   3   2   3  ¯10                1   0

    C
2.0000E0   3.0000E0   1.0000E¯9 |    0   0

    IL←0 3ρ0
    G←0 3ρ0                          F←10*⍳5
```

Execution of *ACSOLVE* and examination of the real and imaginary parts of the voltage of node 1 reveals that the input impedance is reactive:

```
    ACSOLVE
    VREIM[1;1;]
0 0 0 0 0 0
    VREIM[2;1;]
6.2832E¯10 6.2832E¯9 6.2832E¯8 6.2832E¯7 6.2832E¯6 6.2832E¯5
```

Division of the imaginary part of *ZIN* by the radian frequency reveals that *ZIN* corresponds to a pure inductance of value $1E^-11$.

```
    W←2×○F
    VREIM[2;1;]÷W
1E¯11 1E¯11 1E¯11 1E¯11 1E¯11 1E¯11
```

Solution of Linear Equations I

The behavior of an $(N+1)$-node linear resistive circuit is described by a set of N equations relating nodal voltages (the vector VR) and nodal currents (the vector IR) by the reduced nodal conductance matrix YR:

$$IR \leftarrow YR + . \times VR \tag{5.1}$$

If the circuit contains both resistive and reactive components then the arrays in (5.1) have complex elements. The expression (5.1) is directly relevant to the calculation of IR if VR is given. Normally, however, the current excitation of a circuit is specified[†] and the voltage response must be found. Thus, in the example shown in Fig. 5.1 the values of $IR[1]$, $IR[2]$ and $IR[3]$ are given and the value of the vector VR of nodal voltages must be found.

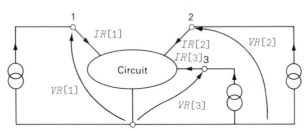

Fig. 5.1
A circuit with current excitation

5.1 Forward Reduction

The solution VR can be obtained systematically by the successive elimination of unknown voltages (i.e., the elements of VR) until only one remains. Then,

---

[†] By this, we do not mean that practical sources approximate closely to current sources. Rather, as explained in Chapter 3, we refer to the fact that most of the nodes of a circuit have no external connection, so that their current excitation (zero) is precisely known: we cannot consider their voltage excitation. If, therefore, the remaining (i.e., accessible) nodes can be assumed to be current excited, then the vector IR is known. Even if an accessible node is *not* connected to a pure current source, then by using a Norton equivalent circuit or one of the transformations illustrated in Fig. 3.12, its current excitation can be assured.

back substitution can be used to obtain each element of VR in reverse order. Take the example of the four-node circuit of Fig. 5.1 described by

$$(YR[1;1] \times VR[1]) + (YR[1;2] \times VR[2]) + (YR[1;3] \times VR[3]) = IR[1] \quad (5.2)$$

$$(YR[2;1] \times VR[1]) + (YR[2;2] \times VR[2]) + (YR[2;3] \times VR[3]) = IR[2] \quad (5.3)$$

$$(YR[3;1] \times VR[1]) + (YR[3;2] \times VR[2]) + (YR[3;3] \times VR[3]) = IR[3] \quad (5.4)$$

First, we divide through equation (5.2) by $YR[1;1]$ to obtain

$$VR[1] + ((YR[1;2] \div YR[1;1]) \times VR[2]) +$$

$$((YR[1;3] \div YR[1;1]) \times VR[3]) = IR[1] \div YR[1;1] \quad (5.5)$$

Next, we multiply this equation by $YR[2;1]$ and subtract the result from (5.3) to obtain

$$(0 \times VR[1]) + ((YR[2;2] - YR[2;1] \times YR[1;2] \div YR[1;1]) \times VR[2]$$

$$+ ((YR[2;3] - YR[2;1] \times YR[1;3] \div YR[1;1]) \times VR[3] \quad (5.6)$$

$$= IR[2] - (YR[2;1] \div YR[1;1]) \times IR[1]$$

By a similar operation, the coefficient of $VR[1]$ in the third equation (5.4) can also be set to zero. In this way we obtain a new set of equations having the form

$$YRN + . \times VR = IRN \quad (5.7)$$

where the first column of YRN should explicitly be noted:

$$\begin{bmatrix} 1 & \\ 0 & \\ 0 & \end{bmatrix} \begin{bmatrix} VR \end{bmatrix} = \begin{bmatrix} IRN \end{bmatrix} \quad (5.8)$$

In general, the elements in the shaded area differ from the elements to be found in the same position in YR. Note that the voltage vector VR is unchanged, but that the current vector (or 'right-hand-side' as it is often loosely called) is in general different from IR. A circuit example is shown in Fig. 5.2.

The importance of (5.7) lies in the fact that we now have to solve a smaller set of simultaneous equations in the sense that row 1 of YRN can now be temporarily ignored, so that we have a 2×2 matrix ($YRN[2\ 3; 2\ 3]$) relating a two-element vector of voltages ($VR[2\ 3]$) to a two-element vector of currents ($IRN[2\ 3]$). Clearly, if this process is repeated until there is only one equation in one unknown, then back substitution can proceed to find the remaining unknowns. Again, for the circuit example of Fig. 5.2, the final form of the equations prior to back substitution is shown in that figure. The process of obtaining the final matrix with ones on the main diagonal and zeros below it is called *forward reduction*.

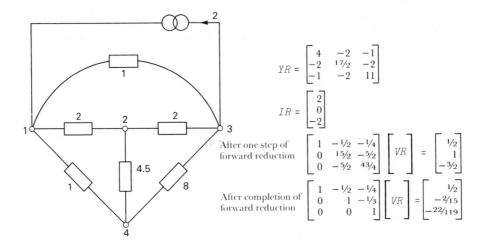

$$YR = \begin{bmatrix} 4 & -2 & -1 \\ -2 & 17/2 & -2 \\ -1 & -2 & 11 \end{bmatrix}$$

$$IR = \begin{bmatrix} 2 \\ 0 \\ -2 \end{bmatrix}$$

After one step of forward reduction

$$\begin{bmatrix} 1 & -1/2 & -1/4 \\ 0 & 15/2 & -5/2 \\ 0 & -5/2 & 43/4 \end{bmatrix} \begin{bmatrix} VR \end{bmatrix} = \begin{bmatrix} 1/2 \\ 1 \\ -3/2 \end{bmatrix}$$

After completion of forward reduction

$$\begin{bmatrix} 1 & -1/2 & -1/4 \\ 0 & 1 & -1/3 \\ 0 & 0 & 1 \end{bmatrix} \begin{bmatrix} VR \end{bmatrix} = \begin{bmatrix} 1/2 \\ -2/15 \\ -22/119 \end{bmatrix}$$

Fig. 5.2
Stages of forward reduction for a circuit example

Since we shall apply the above *row* operations repeatedly to smaller sets of equations, we summarize the steps involved. First, we append the vector IR on to the right-hand side of YR:

$YRI \leftarrow YR, IR$

Then, the first row of this new (non-square) matrix is divided by the element at row 1 column 1:

$YRI[1;] \leftarrow YRI[1;] \div YRI[1;1]$

Finally, operations are carried out on the remaining rows (i.e. 2 and 3) to reduce the elements in the remainder of the first column to zero:

$YRI[2\ 3;] \leftarrow YRI[2\ 3;] - YRI[2\ 3;1] \circ . \times YRI[1;]$

$$\begin{matrix} 1 \\ 0 \\ 0 \end{matrix}$$

Since the above row operations are to be repeated, we now express them more generally prior to defining a suitable function. In general, at the Kth application of the operation, division will occur by the element $YRI[K;K]$, normally referred to as the *pivot element* or simply *pivot*:

$YRI[K;] \leftarrow YRI[K;] \div YRI[K;K]$

Then we must identify the rows which must be operated upon to obtain zeros in the remainder of column K:

$$RR \leftarrow K \downarrow \iota NR$$

where NR is the number of circuit nodes minus one. Finally, the rows identified by the vector RR are modified according to

$$YRI[RR;] \leftarrow YRI[RR;] - YRI[RR;K] \circ . \times YRI[K;]$$

A function for achieving forward reduction can now be defined

```
        ∇YRI←YR FREDUC IR
[1]     YRI←YR,IR
[2]     K←0
[3]     NR←1↑ρYR
[4]  L1:RR←(K←K+1)↓ιNR
[5]     YRI[K;]←YRI[K;]÷YRI[K;K]
[6]     →L2×NR>K
[7]  L2:YRI[RR;]←YRI[RR;]-YRI[RR;K]∘.×YRI[K;]
[8]     →L1
        ∇
```

The content of lines 1, 4, 5 and 7 has been discussed above. The counter K is incremented on line 4. Note that, after the last pivot has been divided by itself to yield unity on the last row of YRI the process should be halted: the appropriate test is carried out on line 6. The operation of $FREDUC$ is easily explored by means of an example based on the resistive circuit encountered in Exercise 3.4.

Example

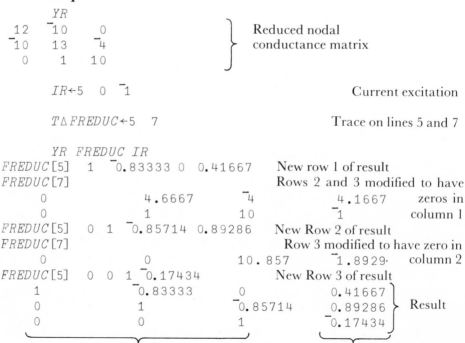

```
      YR
 12  ‾10    0                    Reduced nodal
‾10   13   ‾4          }         conductance matrix
  0    1   10

    IR←5  0  ‾1                  Current excitation

  T∆FREDUC←5  7                  Trace on lines 5 and 7

      YR FREDUC IR
FREDUC[5]  1  ‾0.83333 0 0.41667   New row 1 of result
FREDUC[7]                          Rows 2 and 3 modified to have
     0            4.6667      ‾4           4.1667    zeros in
     0            1           10          ‾1         column 1
FREDUC[5]  0  1  ‾0.85714 0.89286  New Row 2 of result
FREDUC[7]                          Row 3 modified to have zero in
     0            0          10.857   ‾1.8929·  column 2
FREDUC[5]  0  0  1 ‾0.17434         New Row 3 of result
     1           ‾0.83333     0           0.41667  ⎫
     0            1          ‾0.85714      0.89286  ⎬ Result
     0            0           1           ‾0.17434  ⎭
```

                  YRN   IRN   See (5.7)

In the algorithm described for the forward reduction procedure no account has been taken of the fact that we know, *a priori*, that certain elements of the final matrix will be 1 or 0. In practice a modified algorithm may be used which exploits this knowledge if, in so doing, faster operation results.

It is often necessary to compare two algorithms which achieve the same end result, on the basis of computational cost. Although such a comparison can be difficult, it is conventionally assumed that an equal cost (of unity) is assigned to multiplication and division and a cost of zero to addition and subtraction. On this basis it can be shown that, if the *a priori* knowledge referred to above is exploited, then the computational cost (C_{FR}) of Forward Reduction is

$$C_{FR} = ((\div 3) \times N * 3) + (0.5 \times N * 2) + N \div 6$$

5.2 Back Substitution

When the circuit equations have been transformed by forward reduction to the form

it is possible to calculate the elements of VR by back substitution, beginning with the last element of VR and working back. Clearly, for this example,

    VR[6] = IRN[6]

so that the next-to-last equation can now be solved to find $VR[5]$. A suitable function is easy to define:

```
         ∇VR←YR  BKSUB  IR
    [1]     VR←(K←ρ IR)ρ0
    [2] L1: VR[K] ←IR[K] - YR[K;] +.×VR
    [3]     → (0<K←K- 1) /L1
         ∇
```

On line 1, the vector VR is initially set up to contain zero elements: each is replaced by the correct value as it is calculated on line 2. On line 3 the count K is decremented so that the next (lower) element of VR can be corrected. To use $BKSUB$, the result YRI of $FREDUC$ must be partitioned to yield the required (square) matrix YR and vector IR. Below, we return to the example above for which forward reduction has just been carried out.

Example

```
    YRI
1            ⁻0.83333      0           0.41667    Result of
0              1          ⁻0.85714     0.89286    forward
0              0           1          ⁻0.17434    reduction


    X←3 3↑YRI                                              Left argument of
    X                                                      BKSUB is formed
1            ⁻0.83333      0
0              1          ⁻0.85714
0              0           1


    I←VRI[;4]                                              Right argument of
    ρI                                                     BKSUB is formed
3
    X BKSUB I
1.0362  0.74342  ⁻0.17434


    TΔBKSUB←2                                        Trace on line 2
                                                     to show successive
                                                     values of nodal voltages
    X BKSUB I
BKSUB[2]  ⁻0.17434                      ◄────── VR[3]
BKSUB[2]   0.74342                      ◄────── VR[2]
BKSUB[2]   1.0362                       ◄────── VR[1]
1.0362 0.74342 ⁻0.17434
```

On the same basis as for forward reduction, the computational cost of back substitution can be shown to be

$$C_{BS} = N \times (N-1) \div 2$$

5.3 Gaussian Elimination

The application of forward reduction followed by back substitution, to obtain the solution of a set of simultaneous equations is known as Gaussian elimination. The definition of a suitable function for Gaussian elimination emphasizes its two components:

```
    ∇VR←YR GAUSS IR
[1] YRI←YR FREDUC IR
[2] VR←(0 ⁻1↓YRI) BKSUB,YRI[;1+⁻1↑ρYR]
    ∇
```

Below, we use the function we have just defined to obtain the same result as obtained in the examples above.

Example 5.1

(The previous traces on *FREDUC* and *BKSUB* are still active)

```
        YR  GAUSS IR
FREDUC[5]  1  ̄0.83333  0  0.41667
FREDUC[7]
     0                4.6667            ̄4              4.1667
     0                   1              10             ̄1
FREDUC[5]  0  1  ̄0.85714    0.89286
FREDUC[7]
     0                   0              10.857         ̄1.8929
FREDUC[5]  0  0  1  ̄0.17434
BKSUB[2]    ̄0.17434
BKSUB[2]    0.74342
BKSUB[2]    1.0362
1.0362  0.74342   ̄0.17434
```

The computational cost of Gaussian elimination is the sum of C_{FR} and C_{BS}:

$$C_{GE} = ((\div 3) \times N * 3) + (N * 2) - N \div 3 \tag{5.9}$$

Mindful of the somewhat questionable assumptions underlying the calculation of computational cost, the major conclusion to be drawn from the dependence of C_{GE} on N is that the computational cost depends primarily on the cube of the number of circuit nodes.

5.4 Pivoting

As illustrated above, the forward reduction component of Gaussian elimination involves a pivoting action for each row of a matrix, in which other elements of the row are divided by the pivot element. This calculation is of concern for two reasons:

(1) If the value of the pivot element is zero, a *DOMAIN ERROR* will be encountered when an attempt is made to divide by this value.

(2) If the value of the pivot element is very small the solution may proceed, but the numerical accuracy may suffer due to cancellations or the propagation of round-off errors. In this connection it is relevant that the elements of *YR* will in general be complex admittances, with values that are extremely frequency dependent.

The occurrence of a zero-valued pivot does not imply a fundamental difficulty in the solution of $IR \leftarrow YR + . \times VR$; it simply means that, in our systematic elimination of unknowns, we must use a more appropriate ordering of the unknowns.

The problem posed by a zero-valued pivot can easily be overcome by a technique known as pivoting. Row pivoting, in particular, involves the interchange of two rows of $IR \leftarrow YR + . \times VR$ to ensure that the next pivot has a nonzero value. To illustrate row pivoting, let us assume that the equations

$$YR + . \times VR = IR$$

$$
\begin{bmatrix}
1 & -\frac{1}{2} & -\frac{1}{4} \\
0 & 0 & -\frac{5}{2} \\
0 & -\frac{5}{2} & 11
\end{bmatrix}
\begin{bmatrix}
VR[1] \\
VR[2] \\
VR[3]
\end{bmatrix}
=
\begin{bmatrix}
\frac{1}{2} \\
1 \\
-\frac{3}{2}
\end{bmatrix}
$$

are produced after the first application of the forward reduction process. The next regular pivot element (row 2, column 2) is zero, so the solution cannot proceed. We now carry out row pivoting by interchanging rows 2 and 3 to obtain

$$
\begin{bmatrix}
1 & -\frac{1}{2} & -\frac{1}{4} \\
0 & -\frac{5}{2} & 11 \\
0 & 0 & -\frac{5}{2}
\end{bmatrix}
\begin{bmatrix}
VR[1] \\
VR[2] \\
VR[3]
\end{bmatrix}
=
\begin{bmatrix}
\frac{1}{2} \\
-\frac{3}{2} \\
1
\end{bmatrix}
$$

whereupon forward reduction can now proceed. The basic step can be regarded as the choice of a permutation vector P which in the above example is

$$P \leftarrow 1 \quad 3 \quad 2$$

and the appropriate modification of the matrix (YR) and vector (IR) to which foward reduction is applied:

$$YR \leftarrow YR[P;] \quad IR \leftarrow IR[P]$$

Note that no record need be made of the row interchanges, because the order of the unknowns (the elements of VR) is unchanged. This would nót be the case if column interchanges were used.

The modification to $FREDUC$ to allow row pivoting to occur is straightforward. After line [4] and before the row division in line [5] the rows of YRI. must be interchanged if necessary:

$$[4.1] \quad YRI \leftarrow YRI \; PIVCH \; K$$

Within the function that achieves any required interchange of rows, it is first necessary to identify the rows (ROS) of the matrix (now called YI) that can be examined for possible interchange, by dropping from the row numbers (ιNR) those rows $(K-1)$ already processed:

$$ROS \leftarrow (K-1) \downarrow \iota NR$$

Then, within the column (K) associated with the pivot, the magnitudes of the elements in these rows $|YI[ROS;K]$ are examined and grade-down (Ψ) employed to generate the row indices in order of decreasing magnitude. The selection ($1 \uparrow$) of the row number associated with the element of largest magnitude, added to $(K-1)$, will identify the row (M) which must be interchanged with row K:

$$YI[R \; M;] \leftarrow YI[M \; R;]$$

The function $PIVCH$ which incorporates the algorithm just described is:

```
         ∇ YI←YR PIVCH R
[1]        YI←YR
[2]        ROS←(R-1)↓ιNR←1↑ρYR
[3]        M←(R-1)+1↑Ψ|YI[ROS;R]
[4]        YI[R,M;]←YI[M,R;]
         ∇
```

and the modified form of the forward reduction function is:

```
         ∇ YRI←YR FREDUCPIV IR
[1]        YRI←YR,IR
[2]        K←0
[3]        NR←1↑ρYR
[4]      L1:RR←(K←K+1)↓ιNR
[5]        YRI←YRI PIVCH K
[6]        YRI[K;]←YRI[K;]÷YRI[K;K]
[7]        →L2×NR>K
[8]      L2:YRI[RR;]←YRI[RR;]-YRI[RR;K]∘.×YRI[K;]
[9]        →L1
         ∇
```

The use of row pivoting is illustrated in the example below.

Example

First, we modify $PIVCH$ so that the pivot position (R) is printed out

```
         ∇ YI←YR PIVCH R
[1]        YI←YR
[2]        ROS←(R-1)↓ιNR←1↑ρYR
[3]        M←(R-1)+1↑Ψ|YI[ROS;R]
[4]        YI[R,M;]←YI[M,R;]
         ∇
```

The set of equations selected for illustration is

$$
\begin{bmatrix}
1 & 2 & 3 & 4 \\
5 & 6 & 7 & 1 \\
9 & 10 & -5 & 0 \\
13 & 14 & 1 & 3
\end{bmatrix}
\begin{bmatrix} VR \end{bmatrix}
=
\begin{bmatrix}
36 \\
19 \\
-64 \\
-29
\end{bmatrix}
$$

```
         S←Y FREDUCPIV I
PIVCH[2] 1 2 3 4          Rows 1, 2, 3, 4 of Column 1 are to be examined
1                         Pivot position is [1;1]
PIVCH[4] 4               Rows 4 and 1 are interchanged
PIVCH[2] 2 3 4            Rows 2, 3, 4 of Column 2 are to be examined
2                         Pivot position is [2;2]
```

$PIVCH[4]$ 4   Rows 4 and 2 are interchanged
$PIVCH[2]$ 3 4   Rows 3, 4 of Column 3 are to be examined
3   Pivot position is [3 ; 3]
$PIVCH[4]$ 3   No interchange is needed
$PIVCH[2]$ 4
4   }   Again, no interchange of rows is needed
$PIVCH[4]$ 4

S				
1	1.0769	0.076923	0.23077	¯2.2308
0	1	3.1667	4.0833	41.417
0	0	1	0.5	8.5
0	0	0	1	7

The reason for the selection, as the new pivot element, of not only a non-zero element but also the largest magnitude element available is based on considerations of numerical accuracy. It cannot, to our knowledge, be proved that this approach will lead to the best numerical accuracy, but the selection of the largest magnitude pivot is a common technique. We shall content ourselves here merely with a simple illustration, taken directly from Chua and Lin (1975), p. 177. Consider the set of equations

$$\begin{bmatrix} 0.000125 & 1.25 \\ 12.5 & 12.5 \end{bmatrix} \begin{bmatrix} x \end{bmatrix} = \begin{bmatrix} 6.25 \\ 75 \end{bmatrix}.$$

If 3-digit arithmetic is used, forward reduction yields

$$\begin{bmatrix} 1 & 10^4 \\ 0 & -12.5 \times 10^4 \end{bmatrix} \begin{bmatrix} x \end{bmatrix} = \begin{bmatrix} 5 \times 10^4 \\ -6.25 \times 10^5 \end{bmatrix}$$

giving, as the solution,

$$x[1] = 0, \qquad x[2] = 5.$$

If the rows are interchanged, however,

$$\begin{bmatrix} 12.5 & 12.5 \\ 0.000125 & 1.25 \end{bmatrix} \begin{bmatrix} x \end{bmatrix} = \begin{bmatrix} 75 \\ 6.25 \end{bmatrix}$$

forward reduction now yields

$$\begin{bmatrix} 1 & 1 \\ 0 & 1.25 \end{bmatrix} \begin{bmatrix} x \end{bmatrix} = \begin{bmatrix} 6 \\ 6.25 \end{bmatrix}$$

giving, as the solution

$$x[1] = 1, \qquad x[2] = 5,$$

which approximates very closely the correct solution which, to 5 significant figures, is

$$x[1] = 1.0001, \qquad x[2] = 4.9999$$

5.5 Complex Matrices

As discussed in the previous chapter, most circuits of interest contain reactive as well as resistive components, so that the set of equations involved in the analysis of a linear circuit is characterized by a complex matrix, the reduced nodal admittance matrix.

The basic components of the Gaussian elimination solution technique apply equally to a complex set of equations, though two factors are pertinent. First, in searching for a suitable pivot, it is the *magnitude* of an element, and not its real or imaginary part, that is relevant. Thus, if complex APL notation is used, line (3) of $PIVCH$ needs no modification. Second, the computational effort involved when dealing with complex numbers exceeds that associated with the same operation on real numbers. For example, complex multiplication of two scalars is about four times as expensive as for real scalars.

5.6 The Gauss–Jordan Algorithm

The set of equations (5.1) describing a circuit can also be solved by a slight modification of the Gaussian elimination method, called the Gauss–Jordan algorithm. In this algorithm operations are carried out on the rows of YR to convert YR to a unit matrix: simultaneously, the same row operations are applied to the vector of excitation currents:

$$\left[\; YR \; \right] \left[\; VR \; \right] = \left[\; IR \; \right] \xrightarrow[\text{operations}]{\text{Row}} \left[\begin{array}{ccccc} 1 & 0 & 0 & & \\ 0 & 1 & 0 & & \\ 0 & 0 & 1 & & \\ & & & \ddots & \\ & & & 1 & 0 \\ & & & 0 & 1 \end{array} \right] \left[\; VR \; \right] = \left[\; IR' \; \right]$$

It can be shown — but is left as an exercise for the reader — that the resulting modified current vector IR' is, numerically, the required voltage response. The computational cost of the Gauss–Jordan algorithm is

$$0.5 \times (N*3) + (2 \times N*2) - N.$$

5.7   Matrix Inversion

In Fig. 1.12 it was shown that the voltage response to current excitation could be calculated by first *inverting* the reduced nodal conductance matrix to obtain the reduced nodal resistance matrix. One method of finding the inverse of a matrix can be derived from the Gauss–Jordan algorithm described above.

The Kth column of ZR is the voltage response due to a current of 1 A applied at node K (Fig. 5.3), all other nodal currents being zero. Thus, the Kth

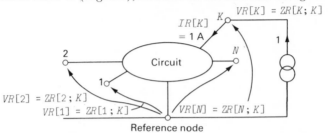

$$VR[K] = ZR[K; K]$$
$$IR[K] = 1\,A$$
$$VR[2] = ZR[2; K]$$
$$VR[1] = ZR[1; K]$$
$$VR[N] = ZR[N; K]$$

Circuit

Reference node

Fig. 5.3

column of ZR can be found by the Gauss–Jordan algorithm if the vector IR contains only zero entries with the sole exception of unity at element K; that is, by solving the set of equations

$$
YR \quad +\cdot\times \quad \begin{bmatrix} \text{Column} \\ K \\ \text{of } ZR \end{bmatrix} = \begin{bmatrix} 0 \\ 0 \\ \cdot \\ \cdot \\ \cdot \\ 0 \\ 1 \\ 0 \\ \cdot \\ \cdot \\ 0 \end{bmatrix} \; K
$$

To obtain *all* columns of ZR one simply applies the Gauss–Jordan algorithm *simultaneously* to the N excitation conditions described by the columns of an $N \times N$ unit matrix:

$$
YR \quad +\cdot\times \quad \begin{bmatrix} ZR \\ (\text{i.e., } all \text{ columns} \\ \text{of } ZR\ ) \end{bmatrix} = \begin{bmatrix} 1 & 0 \ldots 0 & 0 \\ 0 & 1 \ldots 0 & 0 \\ 0 & 0 \ldots 0 & 0 \\ \cdot & \cdots & \cdot \\ \cdot & \cdots & \cdot \\ \cdot & \cdots & \cdot \\ 0 & 0 \ldots 0 & 0 \\ 0 & 0 \ldots 1 & 0 \\ 0 & 0 \ldots 0 & 1 \end{bmatrix}
$$

all current excitations comprising a single 1 A source

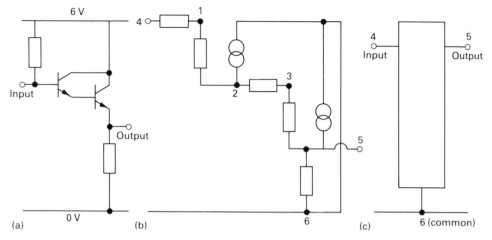

Fig. 5.4
(a) an amplifier (b) its small-signal equivalent circuit (c) with only the accessible terminals shown

Application of the Gauss–Jordan algorithm to the above set of equations is achieved by forming a matrix of size N, $2×N$ which is YR , U (where U is the unit matrix) and then carrying out row operations (as in forward reduction) to reduce YR to a unit matrix. The remainder of the matrix is then the inverse of YR :

$$\begin{bmatrix} YR & \vdots & \begin{matrix} 1 & & & \\ & 1 & & \bigcirc \\ & & 1 & \\ & \bigcirc & & \ddots & \\ & & & & 1 \end{matrix} \end{bmatrix} \xrightarrow{\begin{array}{c} \text{Row} \\ \text{operations} \end{array}} \begin{bmatrix} \begin{matrix} 1 & & & \\ & 1 & & \\ & & 1 & \bigcirc \\ & & & \ddots & \\ \bigcirc & & & & 1 \end{matrix} & \vdots & \begin{matrix} \div YR \\ = ZR \end{matrix} \end{bmatrix}$$

The computational cost of this algorithm is $N*3$.

5.8 Pivotal Condensation

Even if a circuit has many nodes, usually only three or four are externally accessible. Consider for example (Fig. 5.4(a)), an amplifier whose small-signal equivalent circuit (Fig. 5.4(b)) contains six nodes, and is therefore described by the familiar set of equations

$$IR \leftarrow YR + . \times VR \tag{5.10}$$

where YR is a 5×5 matrix. In practice, the amplifier circuit has one node designated as the 'input', another as the 'output', and a third as the 'common' or 'earth' node (Fig. 5.4(c)). All other nodes lie within the boundary of the amplifier: as a consequence, no excitation currents are applied at these 'internal' nodes. Indeed, as in Fig. 5.4(b), certain nodes (1,3) are inaccessible anyway because they are internal to a device model. Similarly, at least to the user of the amplifier, the voltages at these internal nodes are of no concern. Thus, if there are only three nodes of interest to the user, he basically requires a *condensed* description of the circuit of the form

$$IC \leftarrow YC + . \times VC \qquad\qquad (5.11)$$

where YC is a 2×2 matrix. For this and other reasons it is necessary to know how to transform YR of equation (5.10) to YC of equation (5.11). This transformation is known as *pivotal condensation*.

Consider equation (5.10), and note that rows/columns 4 and 5 correspond to the (external) input and output nodes of the amplifier of Fig. 5.4(c). Node 1 is an *internal* node: for this reason, $IR[1] = 0$. We can therefore use the equation represented by the first row of YR to eliminate $VR[1]$ from the remaining equations, thereby obtaining a new set of (four) equations relating the voltages and currents at nodes 2, 3, 4 and 5.

$$IN \leftarrow YN + . \times VN \qquad\qquad (5.12)$$

Thus, YN contains one less row and column than Y. It can easily be shown, but is left to the reader (Exercise 5.11), that the general element of YN is given by

$$YN[R;S] \leftarrow Y[R;S] - Y[R;X] \times Y[X;S] \div Y[X;X]$$

where X is the number associated with the 'suppressed' node. For the case when the node to be suppressed is the first-numbered node, the function $PCOND$:|

```
     ∇  YN ←PCOND Y
[1]  YN← 1   1  ↓ Y- Y[;1] ∘.×Y[1;] ÷ Y[1;1]
     ∇
```

will generate the 'condensed' nodal conductance matrix describing the circuit after the first-numbered node has been suppressed (i.e. made inaccessible). Thus, for the circuit of Fig. 5.4(b), described by the 5×5 nodal conductance matrix YR, the expression

$$YA \leftarrow PCOND \ PCOND \ PCOND \ YR$$

will generate the 2×2 matrix YA describing the three-terminal amplifier. For pivotal condensation leading to a 2×2 matrix the computational cost is:

$$(\div 3) \times (2 \times N * 3) - (3 \times N * 2) + N - 6.$$

Example   The circuit referred to in exercise 3.5 and shown in Fig. 5.5(a) was shown (see Solution 3.5) to be described by the reduced nodal conductance matrix YR:

$$YR$$

12	−10	0	0	0
−10	20	−4	−7	0
0	1	22	−10	0
0	−7	−10	20	−4
0	0	0	1	10

for the case where node 6 is the reference node.

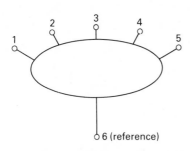

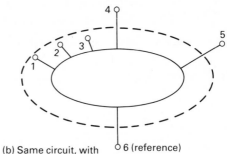

(a) Circuit of Problem E3.5   **Fig. 5.5**   (b) Same circuit, with
                                                     nodes 1, 2 and 3 rendered inaccessible

If, for the convenience, we assume that nodes 4 and 5 (as well as the reference node 6) are to be accessible, and the rest (1, 2 and 3) inaccessible (Fig. 5.5(b)) then the 2×2 reduced nodal conductance matrix obtained by pivotal condensation is

        $YN \leftarrow PCOND\ PCOND\ PCOND\ YR$
        YN
    10.583   ‾4
       1   10

Thus, the voltage at node 5 due to a one-ampere excitation at node 4

        $VN \leftarrow 1\ 0 \boxplus YN$
        $VN[2]$
    ‾0.0091049

should be identical with that obtained using the original matrix YR and the corresponding excitation vector:

        $VR \leftarrow 0\ 0\ 0\ 1\ 0 \boxplus YR$
        $VR[5]$
    ‾0.0091049

and so it is.

Once the 2×2 reduced nodal conductance matrix describing a three-terminal two-port has been obtained, it is a straightforward task to compute various properties of that two-port. Fig. 5.6 shows some examples.

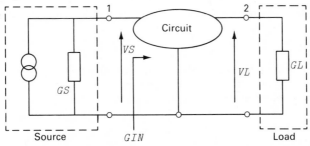

Voltage gain = $VL \div VS = -Y[2;1] \div Y[2;2] + GL$

Input conductance = $GIN = Y[1;1] - Y[1;2] \times Y[2;1] \div Y[2;2] + GL$

Fig. 5.6
A two-port connected between source and load terminations, and expressions for two properties of the two-port in terms of the circuit's reduced nodal conductance matrix Y.

Exercises

5.1   A linear resistive circuit described by the reduced nodal conductance matrix
YR:

$$
\begin{array}{cccc}
& YR & & \\
1 & 2 & 3 & 4 \\
5 & 6 & \bar{7} & 1 \\
9 & 10 & \bar{5} & 0 \\
13 & 14 & 1 & 3
\end{array}
$$

is subjected to a current excitation described by the reduced nodal current
vector IR:

$$IR \leftarrow 36 \ 19 \ \bar{6}4 \ \bar{2}9$$

Enter the expression

$$VR \leftarrow IR \boxdot YR$$

to establish the reduced vector VR of nodal voltage response as a basis for comparison with calculations carried out in response to later questions.

5.2   The function PIV shown below takes as its left argument the array which is to
be subject to one stage of the forward reduction component of Gaussian elimination and, as its right argument, a single integer denoting the pivot position

```
        ∇  R←Y PIV N
[1]        D←1↑ρY
[2]        S←N↓ιD
[3]        Y[N;]←Y[N;]÷Y[N;N]
[4]        Y[S;]←Y[S;]-Y[S;N]∘.×Y[N;]
[5]        R←Y
        ∇
```

First, form the array

$$A \leftarrow YR, IR$$

and then carry out the first step in the forward reduction by entering

$$A1 \leftarrow A \ PIV \ 1$$

and examining the result. Show that even after this single step in the forward
reduction process the system of equations represented by the matrix $A1$ has
the same solution (VR) by generating the new conductance matrix and current
vector:

$$
\begin{array}{l}
YRNEW \leftarrow 0 \ \bar{1} \downarrow A1 \\
IRNEW \leftarrow A1[;5]
\end{array}
$$

and examining the corresponding vector of nodal voltage response:

$$VRNEW \leftarrow IRNEW \boxdot YRNEW$$

5.3 Using the function PIV , proceed with the forward reduction process until it is complete.

5.4 Carry out the back-substitution component of Gaussian elimination on the result of Exercise 5.3, by using the function $BKSUB$ given in the text and repeated below. It is a dyadic function whose left argument is the 'modified YR' and whose right argument is the 'modified IR '.

```
        ∇  V←Y  BKSUB  I;S
  [1]       V←(S←ρI)ρ0
  [2]       L: V[S] ←I[S] - Y[S;] +.×V
  [3]         → (0<S←S- 1) /L
          ∇
```

5.5 Define a function $FREDUC$ to carry out forward reduction for a circuit of any size.

5.6 Define and test a function $GAUSS$ to carry out Gaussian elimination, and incorporating the functions $FREDUC$ and $BKSUB$.

5.7 Extend the function $GAUSS$ to handle more than one current excitation vector simultaneously. It will be found that $FREDUC$ requires no modification, but that $BKSUB$ must be redefined. When testing the new version of $GAUSS$ use the unit matrix to describe all possible excitations comprising a single one-ampere current source, and thereby generate the inverse of the reduced nodal conductance matrix describing the circuit.

5.8 Explore the Gauss–Jordan algorithm for matrix inversion, in which the matrix YR to be inverted is first laminated with a unit matrix:

$$A←YR,UNIT \ 1↑ρYR$$

The left half of A is then transformed to a unit matrix by operations upon its rows, whereupon the resulting right half is then the required inverse. Thereby define a function to perform matrix inversion.

5.9 Define a set of functions to carry out Gaussian elimination for a set of equations describing a circuit containing both reactive and resistive components. Test your functions on the circuit of Fig. 4.4 for a frequency of 1000 radians per second, and check your results by reference to the example in Section 4.4.

5.10 By making use of the function $PCOND$ to carry out pivotal condensation, calculate the voltage V in the circuit of Exercise 5, Chapter 3, and check your answer.

5.11 Derive the expression for the general element of a reduced nodal conductance matrix following the suppression of node X in the pivotal condensation procedure.

Solutions

5.1   *YR*
        1   2   3   4
        5   6   7   1
        9   10   ¯5   0
       13   14   1   3
        IR←36 19 ¯64 ¯29

        VR←*IR*⊟*YR*
        VR   Reduced vector of nodal voltages
     ¯1 ¯3 5 7

5.2   *A*←*YR,IR*
        A
        1   2   3   4   36
        5   6   7   1   19
        9   10   ¯5   0 ¯64
       13   14   1   3 ¯29

        A1←*A PIV* 1   The first stage of forward reduction
        A1
        1   2   3   4   36
        0   ¯4   ¯8   ¯19   ¯161
        0   ¯8   ¯32   ¯36   ¯388
        0   ¯12   ¯38   ¯49   ¯497

        YRNEW←0 ¯1↓*A1*
        YRNEW
    1   2   3   4
    0   ¯4   ¯8   ¯19
    0   ¯8   ¯32   ¯36
    0   ¯12   ¯38   ¯49

        IRNEW←*A1*[; 5]
        IRNEW
   ¯36 ¯161 ¯388 ¯497

        VRNEW←*IRNEW*⊟*YRNEW*
        VRNEW   The same vector of nodal voltages is obtained as in
     ¯1 ¯3 5 7   Exercise 5.1

5.3   *A2*←*A1 PIV* 2   Completion of the forward reduction process
        A3←*A2 PIV* 3
        A4←*A3 PIV* 4

        A4
    1   2   3   4   36
    0   1   2   4.75   40.25
    0   0   1   ¯0.125   4.125
    0   0   0   1   7

        YRNEW←0 ¯1↓*A4*
        YRNEW
    1   2   3   4
    0   1   2   4.75
    0   0   1   ¯0.125
    0   0   0   1
        IRNEW←*A4*[; 5]

```
      IRNEW
36 40.25 4.125 7
```

5.4 *Warning:* This algorithm is modified later, but the same function name is used

```
    ∇ V←Y  BKSUB I;S
[1]    V←(S←ρI)ρ0
[2]  L:V[S]←I[S] - Y[S;] +.×V
[3]    →(0< S←S- 1)/L
    ∇
```

```
      VR←YNEW BKSUB IRNEW
      VR
¯1 ¯3 5 7
```
Back substitution to obtain same voltages as before

5.5
```
    ∇ A←Y  FREDUC I; C; N; S
[1]    A←Y,I
[2]    C←0
[3]  L1: S←(C←C+1) ↓ιN←1↑ρA
[4]    A[C;] ←A[C;] ÷ A[C; C]
[5]    →L2×N> C
[6]  L2:A[S;] ← A[S; ]-A[S; C] ∘.×A[C;]
[7]    →L1
    ∇
```

```
      YR
    1    2    3    4
    5    6    7    1
    9   10   ¯5    0
   13   14    1    3
      IR
36  19  ¯64 ¯29
      YR FREDUC IR
    1            2            3      4          36
    0            1            2      4.75       40.25
    0            0            1      ¯0.125     4.125
    0            0            0      1          7
```
Test of the function *FREDUC*

```
    ∇ A←Y  FREDUC1 I
[1]    A←Y,I
[2]    C←1
[3]  L: A←A PIV C
[4]    →L×(C←C+1) ≤1↑ρA
    ∇
```
An alternative function for forward reduction, incorporating the *PIV* function

```
      YR FREDUC1 IR
    1            2            3      4          36
    0            1            2      4.75       40.25
    0            0            1      ¯0.125     4.125
    0            0            0      1          7
```

5.6
```
    ∇ V←Y  GAUSS I; A
[1]    A←Y FREDUC I
[2]    V←(0 ¯1 ↓A) BKSUB,A[; ¯1↑ρA]
    ∇
```

```
      YR GAUSS IR
¯1  ¯3  5  7
```
The same voltage response as before is obtained

5.7
```
    )ERASE BKSUB
```
The previous *BKSUB* function is erased

```
       ∇ V←Y BKSUB I
[1]       V←(ρI)ρ0
[2]       C←0
[3]    L: S←(1↑ρI) - C
[4]       V[S;]←(÷Y[S;S])×I[S;] - Y[S;] +.×V
[5]       →((1↑ρI) >C←C+1)/L
       ∇
```

The new function will handle an argument I which
is a matrix

```
       ∇ V←Y GAUSS I;A
[1]       A←Y FREDUC I
[2]       V←((ρY)↑A) BKSUB(0,(1↑ρY))↓A
       ∇
```

The function *GAUSS*

```
       YR
    1    2    3    4
    5    6    7    1
    9   10   ⁻5    0
   13   14    1    3
       IR
 1 0 0 0
 0 1 0 0
 0 0 1 0
 0 0 0 1
```

```
       YR GAUSS IR
 ⁻0.55        ⁻0.5         ⁻0.85         0.9
  0.475        0.5          0.825       ⁻0.8
 ⁻0.04         0.1         ⁻0.08         0.02
  0.18        ⁻0.2         ⁻0.14         0.16
```

```
       ⊞YR
 ⁻0.55        ⁻0.5         ⁻0.85         0.9
  0.475        0.5          0.825       ⁻0.8
 ⁻0.04         0₃1         ⁻0.08         0.02
  0.18        ⁻0.2         ⁻0.14         0.16
```

The function *FREDUC1* defined above can also be substituted for *FREDUC* in *GAUSS*

5.8   ```A←YR,UNIT (1↑ρYR)```
```
       A
    1    2    3    4    1    0    0    0
    5    6    7    1    0    1    0    0
    9   10   ⁻5    0    0    0    1    0
   13   14    1    3    0    0    0    1
```

```
       C←1
```
Pivot number

```
       X←ι1↑ρYR
       X
 1 2 3 4
```
Row numbers of A

```
       S←⁻1↓ 0⌽X
       S
 2 3 4
```
Rows for which a zero element must be generated in
column C

```
       A[C;]←A[C;]÷A[C;C]
       A[S;]←A[S;] - A[S;C]∘.×A[C;]
```

```
        A
1     2     3    ‾4    1     0     0     0
0    ‾4    ‾8    19   ‾5    1     0     0
0    ‾8   ‾32   ‾36   ‾9    0     1     0
0    ‾12  ‾38   ‾49   ‾13   0     0     1
```

        $C \leftarrow 2$ In column 2, rows 3, 4 and 1 must be made zero by
        $S \leftarrow 1 \downarrow 1 \phi X$ appropriate operations
        S
3 4 1

        $A[C;] \leftarrow A[C;] \div A[C;C]$
        $A[S;] \leftarrow A[S;] - A[S;C] \circ . \times A[C;]$

```
A
1     0    ‾1    ‾5.5   ‾1.5    0.5     0     0
0     1     2     4.75   1.25  ‾0.25    0     0
0     0   ‾16     2      1     ‾2       1     0
0     0   ‾14     8      2     ‾3       0     1
```

        $C \leftarrow 3$
        $S \leftarrow 1 \downarrow 2 \phi X$
        $A[C;] \leftarrow A[C;] \div A[C;C]$
        $A[S;] \leftarrow A[S;] - A[S;C] \circ . \times A[C;]$ Repetition for columns 3 and 4

        $C \leftarrow 4$
        $S \leftarrow 1 \downarrow 3 \phi X$
        $A[C;] \leftarrow A[C;] \div A[C;C]$
        $A[S;] \leftarrow A[S;] - A[S;C] \circ . \times A[C;]$

```
A
1     0     0     0    ‾0.55   ‾0.5    ‾0.85    0.9
0     1     0     0     0.475   0.5     0.825  ‾0.8
0     0     1     0    ‾0.04    0.1    ‾0.08    0.02
0     0     0     1     0.18   ‾0.2    ‾0.14    0.16
```

                                       The right half of A is the required inverse
        $(-\rho YR) \uparrow A$
```
‾0.55     ‾0.5     ‾0.85     0.9
 0.475     0.5      0.825   ‾0.8
‾0.04      0.1     ‾0.08     0.02
 0.18     ‾0.2     ‾0.14     0.16
```

```
     ∇ R←INV M;N;C;V;S
[1]    R←M,UNIT N←1↑ρM
[2]    C←0
[3]    V←ιN
[4]  L: C←C+1                    Previous steps embodied in a function
[5]    S←1↓(‾1+C)φV
[6]    R[C;]←R[C;]÷R[C;C]
[7]    R[S;]←R[S;]-R[S;C]∘.×R[C;]
[8]    →(C≠N)/L
[9]    R←(-ρM)↑R
     ∇
```

```
      YR
  1   2   3   4
  5   6   7   1
  9  10  ¯5   0
 13  14   1   3
      INV YR
¯0.55         ¯0.5        ¯0.85        0.9
 0.475         0.5         0.825       ¯0.8
¯0.04          0.1        ¯0.08        0.02
 0.18         ¯0.2        ¯0.14        0.16
```

5.9   We begin with the three-dimensional array YR

```
       YR
 12    ¯2   ¯10
 ¯2     6     0
¯10     0    10

  0     0     0
  0    ¯8     0
  0     0     1
```

whose first and second planes correspond, respectively, to the real and imaginary parts of the reduced nodal admittance matrix at the frequency of 1000 rad/s. Using complex APL we generate a two-dimensional complex reduced nodal admittance matrix

```
      YR←¯9 ¯11+.○ YR
      YR
 12    ¯2      ¯10
 ¯2   6J¯8       0
¯10     0     10J1
```

The Gaussian elimination function used is

```
      ∇VR←YR GAUSS IR
[1]    YRI←YR FREDUC IR
[2]    VR←(0 ¯1↓YRI) BKSUB,YRI[;1+¯1↑ρYR]∇
```

and the functions $FREDUC$ and $BKSUB$ are exactly as defined in Chapter 5. Definition of the (complex) current excitation

```
      IR←2.7 0 0
```

and execution of $GAUSS$

```
      VR←YR GAUSS IR
```

yields the same complex nodal voltages as obtained in Section 4.4 of Chapter 4 for the circuit of Fig. 4.4

```
      VR
1.285J¯0.4633  0.2284J0.1501  1.227J¯0.586
```

Thus, if complex APL is available, no change is required in the functions $FREDUC$, $BKSUB$ and $GAUSS$.

Solution of Linear Equations II

Although Gaussian elimination is a straightforward means of solving the set of equations describing a circuit, an entirely different method, called LU factorization, is usually used in practice. There is more than one reason. First it allows the response of the same circuit to different excitations to be computed at little additional cost. As a consequence (but we shall have to wait until Chapter 9 to see why), it is also a simple matter to compute the sensitivity of a circuit voltage to component changes. Second, it offers potential savings in computational effort and storage if, as is usually the case, most of the elements in the nodal conductance or admittance matrix are zero.

6.1 Solution by *LU* Factorization

With reference to the circuit equations discussed earlier,†

$$I \leftarrow Y +. \times V \tag{6.1}$$

we assume that Y can be expressed as the inner product of two matrices of the same size as Y, called L (for 'lower') and U (for 'upper'):

$$Y \leftarrow L +. \times U \tag{6.2}$$

The matrix L has zeros above its diagonal (it is 'lower triangular') whereas U has zeros below, and ones on, its diagonal (it is 'unit upper triangular'). For the circuit of Fig. 6.1, the matrices Y, L and U are shown in that figure both for illustration and ease of checking (6.2). Thus equation (6.1) can be expressed as

$$I \leftarrow L +. \times U +. \times V \tag{6.3}$$

For the moment we do not concern ourselves with *how* L and U are derived from Y, but assume that this factorization has already been carried out.

The solution V of (6.1) is found in two simple steps. First, we denote the inner product $U +. \times V$ in (6.3) by H, so that (6.3) becomes

$$I \leftarrow L +. \times H \tag{6.4}$$

---

† To simplify some expressions we are soon to encounter, the suffix R has been dropped from IR, YR and VR.

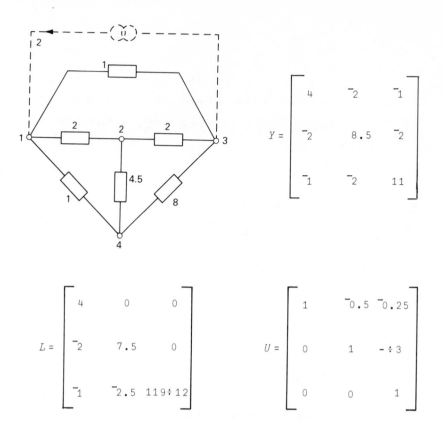

$$Y = \begin{bmatrix} 4 & ^-2 & ^-1 \\ ^-2 & 8.5 & ^-2 \\ ^-1 & ^-2 & 11 \end{bmatrix}$$

$$L = \begin{bmatrix} 4 & 0 & 0 \\ ^-2 & 7.5 & 0 \\ ^-1 & ^-2.5 & 119 \div 12 \end{bmatrix} \qquad U = \begin{bmatrix} 1 & ^-0.5 & ^-0.25 \\ 0 & 1 & ^- \div 3 \\ 0 & 0 & 1 \end{bmatrix}$$

Fig. 6.1
A simple circuit, its reduced nodal conductance matrix and its L and U factors

and calculate the vector H very straightforwardly by *forward substitution*. The reason that this process is straightforward is that L is lower triangular. For the example of Fig. 6.1 the relevant equations are

$$\begin{bmatrix} 4 & 0 & 0 \\ ^-2 & 7.5 & 0 \\ ^-1 & ^-2.5 & 119 \div 12 \end{bmatrix} +.\times \begin{bmatrix} H \end{bmatrix} = \begin{bmatrix} 2 \\ 0 \\ ^-2 \end{bmatrix} \text{ giving } H = \begin{bmatrix} 0.5 \\ 2 \div 15 \\ ^-2 \div 17 \end{bmatrix} \quad (6.5)$$

The first equation is solved for $H[1]$, which then enables the second to be solved for $H[2]$, and so on.

Next we recall that

$$H \leftarrow U +.\times V \qquad\qquad (6.6)$$

Since U is unit upper triangular, V can now be obtained by *back substitution* (recall the second step of Gaussian elimination). For the example of Fig. 6.1 the relevant equations are

$$\begin{bmatrix} 1 & ^-0.5 & ^-0.25 \\ 0 & 1 & -\div 3 \\ 0 & 0 & 1 \end{bmatrix} +.\times \begin{bmatrix} V \end{bmatrix} = \begin{bmatrix} H \end{bmatrix} = \begin{bmatrix} 0.5 \\ 2\div 15 \\ ^-2\div 17 \end{bmatrix} \text{giving } V = \begin{bmatrix} 44\div 85 \\ 8\div 85 \\ ^-2\div 17 \end{bmatrix} (6.7)$$

An algorithm for back substitution was described earlier. Forward substitution is so similar that, to emphasize the similarity, we define an appropriate function in terms of that for back substitution:

$$\nabla R \leftarrow Y \ FSUB \ I$$
$$[1] \quad R \leftarrow \phi (\phi \ominus Y) \ LUBKSUB \ \phi I \qquad\qquad\qquad (6.8)$$
$$\nabla$$

However, because the L matrix involved in the forward substitution (6.5) does not have unity entries on its main diagonal, the back-substitution function called in (6.8) must be altered from its previous form as shown below (the modification is the insertion of $Y[S; S]$ in line 2):

$$\nabla V \leftarrow Y \ LUBKSUB \ I; S$$
$$[1] \quad V \leftarrow (S \leftarrow \rho I) \rho 0$$
$$[2] \quad L1: V[S] \leftarrow (\div Y[S; S]) \times I[S] - Y[S;] +.\times V \qquad (6.9)$$
$$[3] \quad \rightarrow (0 < S \leftarrow S - 1)/L1$$
$$\nabla$$

Using the conventional basis for assignin computational cost, it is found that the combined cost of full forward and backward substitution is $N*2$. Thus, once the L and U factors have been found, the solution V for a new excitation I can be found at a cost $(N*2)$ which is much less than that of a Gaussian elimination $(\approx (N*3) \div 3)$.

6.2 The Factorization

The algorithm for carrying out the LU factorization whose result was assumed in the earlier discussion can be derived by comparing the symbolic result of the inner product $L+.\times U$ with the original matrix Y. Thus, for a 3×3 matrix we compare

$$\begin{bmatrix} L[1;1] & 0 & 0 \\ L[2;1] & L[2;2] & 0 \\ L[3;1] & L[3;2] & L[3;3] \end{bmatrix} +.\times \begin{bmatrix} 1 & U[1;2] & U[1;3] \\ 0 & 1 & U[2;3] \\ 0 & 0 & 1 \end{bmatrix} =$$

$$\begin{bmatrix} L[1;1] & L[1;1] \times U[1;2] & L[1;1] \times U[1;3] \\ L[2;1] & (L[2;1] \times U[1;2]) + L[2;2] & (L[2;1] \times U[1;3]) + (L[2;2] \times U[2;3]) \\ L[3;1] & (L[3;1] \times U[1;2]) + L[3;2] & (L[3;1] \times U[1;3]) + (L[3;2] \times U[2;3]) + L[3;3] \end{bmatrix}$$

with
$$\begin{bmatrix} Y[1\,;\,1] & Y[1\,;\,2] & Y[1\,;\,3] \\ Y[2\,;\,1] & Y[2\,;\,2] & Y[2\,;\,3] \\ Y[3\,;\,1] & Y[3\,;\,2] & Y[3\,;\,3] \end{bmatrix}$$

This simple comparison shows that:

(1) No calculation is involved in finding the first column of L, since

$$\begin{bmatrix} L[1\,;\,1] \\ L[2\,;\,1] \\ L[3\,;\,1] \end{bmatrix} = \begin{bmatrix} Y[1\,;\,1] \\ Y[2\,;\,1] \\ Y[3\,;\,1] \end{bmatrix}$$

(2) Division of the remainder of the first row by $L[1\,;\,1]$ yields $U[1\,;\,2]$ $U[1\,;\,3]$, which is the non-unity part of the first row of U.

(3) Subtraction of the outer product $\begin{bmatrix} L[2\,;\,1] \\ L[3\,;\,1] \end{bmatrix}$ $[U[1\,;\,2] \quad U[1\,;\,3]]$

from the $(N–1) \times (N–1)$ submatrix yields

$$\begin{bmatrix} L[2\,;\,2] & L[2\,;\,2] \times U[2\,;\,3] \\ L[3\,;\,2] & (L[3\,;\,2] \times U[2\,;\,3]) + L[3\,;\,3] \end{bmatrix}$$

The three steps defined above can now be repeated for the $(N–1) \times (N–1)$ submatrix, and subsequently until all elements of L and U are found. It should be noted that, once $L[P\,;\,Q]$ or $U[P\,;\,Q]$ is calculated, $Y[P\,;\,Q]$ is not required again; therefore, $Y[P\,;\,Q]$ can be replaced by $L[P\,;\,Q]$ or $U[P\,;\,Q]$ once it is calculated.

The algorithm for LU factorization can be derived more generally, but again by inspection. We observe (Fig. 6.2) that the value of the element $Y[R\,;\,S]$ is the inner product of the Rth row of L ($L[R\,;\,]$) and the Sth column of U ($U[\,;\,S]$) although the nature of L and U is such that only the first $R\,L\,S$ elements of the row and column are relevant. Once $L[R\,;\,1]$ and $U[1\,;\,S]$ are determined (for all R and S), the contribution of $L[R\,;\,1]$ and $U[1\,;\,S]$ for all R and S except 1 can be subtracted from the $(N–1) \times (N–1)$ submatrix, leaving a new matrix which is also the inner product of a lower triangular and a unit upper triangular matrix (each of size $(N–1) \times (N–1)$). The same comments then apply to the elements of the new submatrix. What enables the iteration to proceed is the nature of L and U: *only* because L is lower triangular and U is unit upper triangular can we write, for the first iteration, that

$$L[R\,;\,1] = Y[R\,;\,1]$$

and

$$U[1\,;\,S] = Y[1\,;\,S] \div Y[1\,;\,1]$$

For example, if there were nonzero elements below the diagonal of U, we could not determine $L[R\,;\,1]$. Similarly, nonzero elements above the diagonal of L would prevent the simple calculation of $U[1\,;\,S]$.

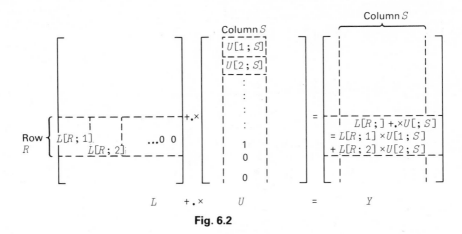

Fig. 6.2

The algorithm can be embodied in a function, as illustrated and annotated below:

	$\nabla LU \leftarrow LUFAC\ Y; C; S$	Row of U and
[1]	$C \leftarrow 1$	column of L to be
		found
		See notes above re
[2]	$LU \leftarrow Y$	$L[P; Q], U[P; Q]$
		and $Y[P; Q]$
		S identifies remain-
[3]	$L1: S \leftarrow C \downarrow \iota\ 1 \uparrow \rho LU$	ing rows and
		columns
		Row C of U is
[4]	$LU[C; S] \leftarrow LU[C; S] \div LU[C; C]$	generated (see 2
		above)
		Preparation of new
[5]	$LU[S; S] \leftarrow LU[S; S] - LU[S; C]$	matrix of reduced
	$\circ . \times LU[C; S]$	size for application
		of line [4] (see 3
		above)
		Repeat for next
[6]	$\rightarrow L1 \times (1 \uparrow \rho LU) > C \leftarrow C + 1\ \ \nabla$	numbered row and
		column if process is
		not complete.

The steps are illustrated diagrammatically in Fig. 6.3, with major steps identified by the line numbers of $LUFAC$.

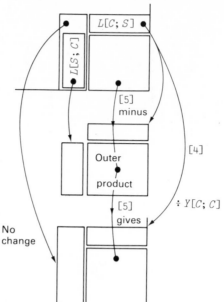

Fig. 6.3
Illustration of the *LU* factorization procedure

The computational cost of *LU* factorization is illustrated in Fig. 6.4 and can be shown to be $((N*3) \div 3) - N \div 3$ for an $N \times N$ matrix. If we add the cost $(N*2)$ of combined forward and backward substitution, the result is identical with the cost of Gaussian elimination. Thus, one advantage of *LU* factorization (inexpensive solution for a new excitation) is obtained at no extra cost.

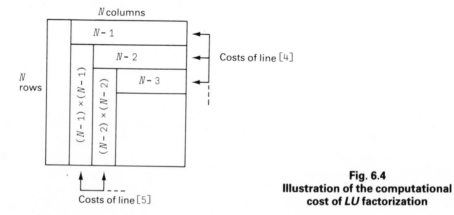

Fig. 6.4
Illustration of the computational cost of *LU* factorization

6.3   A Relation

So far, Gaussian elimination and *LU* factorization have been viewed as two different approaches to the same objective. However, examination of the functions involved and the algorithms they embody reveal similarities.

$$
\begin{array}{cccc}
 & & M & \\
14 & 4 & 8 & 7 \\
10 & 11 & 15 & 3 \\
1 & 5 & 9 & 16 \\
12 & 6 & 13 & 2 \\
\end{array}
$$

```
      U
1    0.28571   0.57143    0.5         1   0.28571   0.57143    0.5
0    1         1.1404    ‾0.24561      0   1         1.1404    ‾0.24561
0    0         1          5.4569       0   0         1          5.4569
0    0         0          1            0   0         0          1


      L
14    0          0          0          14    0          0          0
10    8.1429     0          0          10    8.1429     0          0
 1    4.7143     3.0526     0           1    4.7143     3.0526     0
12    2.5714     3.2105   ‾20.888      12    2.5714     3.2105   ‾20.888

            (a)                 Fig. 6.5              (b)
```

The relation between the two approaches can be made explicit by means of an illustrative example. In Fig. 6.5(a) is shown the result of LU factorization carried out on the 4 × 4 matrix shown. In Fig 6.5(b) are presented two matrices. The first is the result of the forward reduction process involved in Gaussian elimination,[†] and is seen to be identical with U in (a). The second contains, in the appropriate locations, the multiplying factors used to set elements equal to 1 and 0 in the foward reduction process;[‡] it is seen to be identical to L in (a). In the Gaussian elimination scheme these multiplying factors are discarded; by contrast they are retained (as L) in the LU factorization procedure with the result that little extra computation is involved in obtaining a new solution for a new excitation vector IR.

To learn more about the relation illustrated above see Fidler and Nightingale (1978), p. 74, and Chua and Lin (1975), Section 4.3.

6.4 Sparsity

In the course of earlier exercises, and especially those involving a circuit with more than just a few nodes, we have observed that the nodal conductance or admittance matrix usually contains a large number of zeros. We refer to such a matrix as being sparse. For an electronic circuit, sparsity is a consequence of the fact that not more than 2 or 3 components are usually connected to any one node.

Consequently, whether the circuit's response is computed by Gaussian elimination or LU factorization or some other method, there are many occasions when a number is either multiplied by or added to zero. If a result is predictable for these reasons, it may be possible to avoid a redundant calculation. Additionally, if the location of zero-valued elements is known, considerable savings in storage may be possible.

[†] Though with the last, 'current', column removed.
[‡] This matrix can easily be formed (see the function $FREDUC$ of Chapter 5) by setting up an empty matrix $X \leftarrow (-NR,0)\rho 0$ and then performing the catenation $X \leftarrow X,(-NR)\uparrow \overline{YRI[K;K]}$, $YRI[\overline{RR};\overline{K}]$ for each $[K]$.

Some idea of the value of these potential savings in computation and storage can be gained from the fact that, for a circuit of reasonable size (above 20 nodes, for example), the percentage of nonzero elements in the matrix may be as low as 5 or 10%, and even lower for really large circuits. The remainder of this chapter takes a first look at the concept of sparsity in the context of *LU* factorization.

6.4.1 Fills

The *LU* factorization algorithm developed earlier in this chapter comprised three steps. It is the third step, embodied in line [5] of *LUFAC*:

[5] $LU[S;S] \leftarrow LU[S;S] - LU[S;C] \circ . \times LU[C;S]$

that is of interest here. It involves the subtraction of an outer product from a submatrix. Inspection shows that a zero-valued element at location $[K;J]$ will be replaced by a nonzero element if, *and only if*, the elements at locations $[1;J]$ and $[K;1]$ are both nonzero. In such an event a *fill* is said to have occurred at location $[K;J]$. Whereas, before, there was a zero at a certain location in the matrix, there is now a nonzero element. By contrast, no fills can be generated by the first and second steps of *LU* factorization.

A fill is of concern to us because the new (nonzero) element will have to be stored and processed, and the potential savings in computation and storage offered by the previous (zero-valued) element have been discarded. If at all

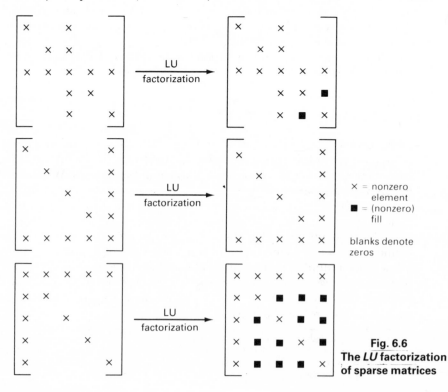

Fig. 6.6
The *LU* factorization of sparse matrices

× = nonzero element
■ = (nonzero) fill
blanks denote zeros

possible, the number of fills that occur in the LU factorization procedure should be kept to a minimum.

Some idea of *how* the number of fills can be minimized can be gained from the examples of Fig. 6.6. This figure shows the result of carrying out LU factorization on three matrices, all of the same size and all equally sparse. No numerical values are shown, since we are only concerned with whether an element value is zero or nonzero. It can be seen from the figure that the disposition of zero elements within the matrix significantly affects the number of fills. But what is really interesting and significant is that all three examples describe the *same* circuit! Only a renumbering of the nodes — or, equivalently, a reordering of rows and columns — has been undertaken to obtain the 'different' matrices on the left-hand side of the figure. Clearly, the numbering of the nodes is of vital interest if we wish to reduce the number of fills occurring during LU factorization.

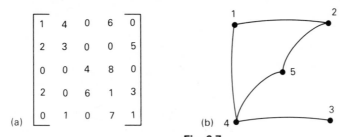

Fig. 6.7
A matrix and the graph representation of the location of its nonzero elements

6.4.2   Graph Representation

The location of nonzero elements in a matrix can be represented by a graph. For the matrix YR of Fig. 6.7(a) the (non-directed) graph G of Fig. 6.7(b) contains a node associated with each circuit node (and hence with a row and column of the nodal conductance matrix YR) and a branch associated with each pair[†] of off-diagonal nonzero elements. Thus, the branch between nodes 4 and 5 represents the nonzero elements $YR[4;5]$ and $YR[5;4]$. Can we deduce, from G, the consequences of choosing any one node to be the first to be processed (i.e. the node to be 'suppressed' or 'eliminated') in the LU factorization, and hence to be renumbered as '1'?

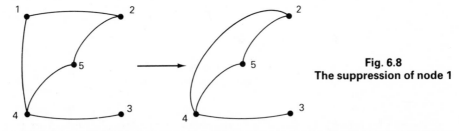

Fig. 6.8
The suppression of node 1

---

† We assume, with insignificant loss of generality, that for each off-diagonal nonzero element at $YR[R;S]$ there is another at $YR[S;R]$. This is not an unreasonable assumption: for a transistor model, voltage-controlled current sources often occur in such pairs, modelling forward and reverse transmission.

Consider first the elimination of node 1. We know from examination of YR that a fill will occur at $[2 ; 4]$ and $[4 ; 2]$, so that the new graph representing the submatrix on which factorization will again be carried out is as shown in Fig. 6.8. But we need a *rule* that will enable us directly to derive the new graph associated with the suppression of any node. Such a rule can be derived from consideration of the third step of LU factorization (as expressed by line $[5]$ of $LUFAC$), and will be illustrated by reference to the suppression of node 4 in our example. First, one identifies the graph nodes to which the node to be suppressed is *directly* connected: for node 4 these are nodes 1, 3 and 5 (Fig. 6.9).

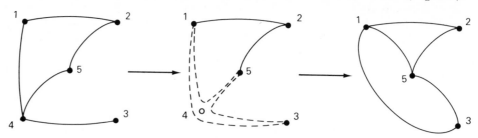

Fig. 6.9
The suppression of node 4

Next, one selects every possible pairing of these nodes; in this case $[1 : 3]$, $[3 ; 5]$ and $[5 ; 1]$. For each pair, a new branch is added to the graph; at the same time the node to be suppressed and its incident branches are removed (Fig. 6.9). If a new branch is found to be in parallel with an existing branch (this does not occur with the present example), the two are merged into a single branch (the interpretation of this is that a nonzero term is subtracted from another nonzero term). If the new branch is not in parallel with an existing branch, then two fills are generated, the location of the fills being given by the nodes at the extremities of that branch. Thus, for our example (Fig. 6.9), we see that six ($= 2 \times 3$) fills are generated at locations $[1 ; 3]$ and $[3 ; 1]$, $[3 ; 5]$ and

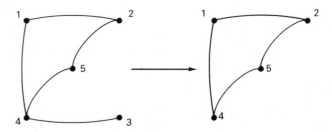

Fig. 6.10
The suppression of node 3

$[5 ; 3]$, and $[5 ; 1]$ and $[1 ; 5]$. For the same example the best node to eliminate first is node 3, since (Fig. 6.10) there are then no fills and the new graph is obtained merely by removing node 3 and the branch connecting it to node 4. The existence of such a node is associated with a row/column of YR having only two nonzero entries.

6.4.3   A 'No-fill' algorithm

To illustrate the above discussion, we first develop an algorithm that will lead to a reordering of nodes if that will ensure an absence of fills during the ensuing stage of LU factorization. As an example (Fig. 6.7) we select the same matrix YR as before

$$
YR = \begin{matrix}
1 & 4 & 0 & 6 & 0 \\
2 & 3 & 0 & 0 & 5 \\
0 & 0 & 4 & 8 & 0 \\
2 & 0 & 6 & 1 & 3 \\
0 & 1 & 0 & 7 & 1
\end{matrix}
$$

whose graph G identifying the nonzero elements is easily generated:

$$
G \leftarrow \sim 0 = YR
$$
$$
G
$$

$$
\begin{matrix}
1 & 1 & 0 & 1 & 0 \\
1 & 1 & 0 & 0 & 1 \\
0 & 0 & 1 & 1 & 0 \\
1 & 0 & 1 & 1 & 1 \\
0 & 1 & 0 & 1 & 1
\end{matrix}
$$

A sum reduction over the columns of G, followed by the addition of $^-1$

$$
^-1 + +/G
$$
$$
\begin{matrix} 2 & 2 & 1 & 3 & 2 \end{matrix}
$$

indicates the number of branches connected to each node of the graph. The '1' in the third position, indicated by a '1' in the vector B

$$
B \leftarrow 1 = {}^-1 + +/G
$$
$$
B
$$
$$
\begin{matrix} 0 & 0 & 1 & 0 & 0 \end{matrix}
$$

identifies node 3 as having only one branch connected to it. Thus, if $^-1 + +/G$ contains at least one '1'

$$
0 \neq +/B
$$
$$
1
$$

which it does in this case, then to generate the node to be eliminated we must locate the position of the first '1' in B:

$$
Q \leftarrow +/0 = \vee \backslash B
$$
$$
Q
$$
$$
2
$$

Here, Q is the number of zeros before the first '1' in B, and hence is one less than the required node number. A simple means of renumbering this node (3) as node 1 is to use Q as the left argument of the rotate function, to generate a permutation vector NN defining the new node numbering:

$$NN \leftarrow Q \phi \iota 1 \uparrow \rho G$$
$$NN$$

$$\begin{array}{ccccc} 3 & 4 & 5 & 1 & 2 \end{array}$$

The permutation vector is then used to index the previous graph G to generate the new graph NG

$$NG \leftarrow 1 \quad 1 \downarrow G[NN; NN]$$
$$NG$$

$$\begin{array}{cccc} 1 & 1 & 1 & 0 \\ 1 & 1 & 0 & 1 \\ 1 & 0 & 1 & 1 \\ 0 & 1 & 1 & 1 \end{array}$$

The dropping of the first row and column of the graph is necessary because, in the next stage of LU factorization, only the smaller matrix represented by NG will be processed.

Based on the above discussion we can now define a function $NOFILLS$ (Fig. 6.10) which identifies a circuit node which, because it corresponds to a node in G connected to only one branch, [†] will, when eliminated, generate no fills. $NOFILLS$ calls a function $RENUMBER$ that reorders the nodes. The application of $NOFILLS$ to our example is illustrated in Fig. 6.11.

6.4.4   Minimum fills

After identifying any nodes in G connected to only one branch, and undertaking the necessary node renumbering and derivation of the new graph, it may then be appropriate to determine the number of fills that would be associated with the elimination of each node. To illustrate a possible approach, we again take the same example as before, for which the matrix G is known:

$$G$$

$$\begin{array}{ccccc} 1 & 1 & 0 & 1 & 0 \\ 1 & 1 & 0 & 0 & 1 \\ 0 & 0 & 1 & 1 & 0 \\ 1 & 0 & 1 & 1 & 1 \\ 0 & 1 & 0 & 1 & 1 \end{array}$$

---

† This is a *sufficient* condition for no fills to be generated, but not a *necessary* condition (see Exercise 6.8).

∇ $NG \leftarrow NOFILLS$ $G; B; Q$ Initialization of NG in case no node is found.
[1] $NG \leftarrow G$ Is there a node connected to only one branch?
[2] $\rightarrow L \times 0 \neq +/B \leftarrow 1 = {}^{-}1 + +/G$ The lowest-numbered such node is found.
[3] $L : Q \leftarrow +/0 = \vee \backslash B$ The nodes are reordered
[4] $NG \leftarrow Q$ $RENUMBER$ G
     ∇

∇ $NG \leftarrow Q$ $RENUMBER$ $G; NN$
[1] $NN \leftarrow Q \phi \iota 1 \uparrow \rho G$
[2] $NG \leftarrow$ 1 1 $\downarrow G[NN; NN]$
     ∇

Fig. 6.10

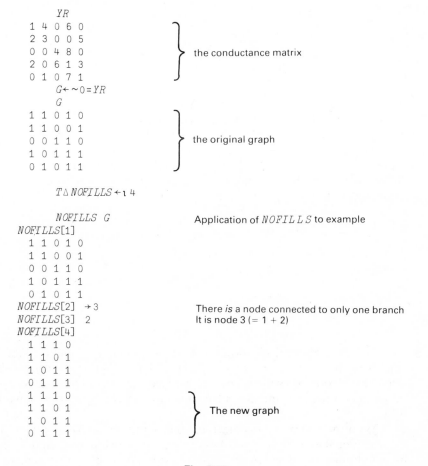

        YR
1 4 0 6 0
2 3 0 0 5
0 0 4 8 0 $\left.\rule{0pt}{3.2em}\right\}$ the conductance matrix
2 0 6 1 3
0 1 0 7 1
        $G \leftarrow \sim 0 = YR$
        G
1 1 0 1 0
1 1 0 0 1
0 0 1 1 0 $\left.\rule{0pt}{3.2em}\right\}$ the original graph
1 0 1 1 1
0 1 0 1 1

        $T \triangle NOFILLS \leftarrow \iota 4$

        $NOFILLS$ G Application of $NOFILLS$ to example
$NOFILLS[1]$
  1 1 0 1 0
  1 1 0 0 1
  0 0 1 1 0
  1 0 1 1 1
  0 1 0 1 1
$NOFILLS[2]$ $\rightarrow 3$ There *is* a node connected to only one branch
$NOFILLS[3]$ 2   It is node 3 ($= 1 + 2$)
$NOFILLS[4]$
  1 1 1 0
  1 1 0 1
  1 0 1 1
  0 1 1 1
  1 1 1 0
  1 1 0 1 $\left.\rule{0pt}{2.2em}\right\}$ The new graph
  1 0 1 1
  0 1 1 1

Fig. 6.11

If node 1 is eliminated, the 'outer-and'

$$W \leftarrow G[; 1] \circ . \wedge G[1 ;]$$
$$W$$

1	1	0	1	0
1	1	0	1	0
0	0	0	0	0
1	1	0	1	0
0	0	0	0	0

identifies, for the 4 × 4 submatrix left after dropping the first row and column, the locations where nonzero elements will be subtracted from existing elements. Thus, the element-by-element 'and' of W and $\sim G$ will give the location of any fills that will occur:

$$LF \leftarrow W \wedge \sim G$$
$$LF$$

0	0	0	0	0
0	0	0	1	0
0	0	0	0	0
0	1	0	0	0
0	0	0	0	0

a result we would expect from our earlier discussion (when we decided that if node 1 were eliminated, fills would be generated at [2 ; 4] and [4 ; 2].).

To examine the effect of suppressing node 2, we must re-order the nodes so that the one previously labelled '2' is now associated with the first row and column of the nodal conductance matrix. Without loss of generality, this can be achieved by simple rotation, such that the new graph is

$$G \leftarrow 1 \ominus 1 \phi G$$
$$G$$

1	0	0	1	1
0	1	1	0	0
0	1	1	1	1
1	0	1	1	0
1	0	1	0	1

From the above discussion it is possible to define a function $MINFILLS$ (Fig. 6.12) that will determine the number of fills associated with the elimination of each node, a result expressed by the vector NF. The result of applying this function to the same illustrative example as before is also shown in Fig. 6.12.

With the objective of achieving a near-optimal ordering of nodes for LU factorization it is possible to employ a combination of functions that identify no-fill conditions (as does $NOFILLS$ for a special case) and others (such as $MINFILLS$) that identify nodes associated with a minimum number of fills. Such a task is left (Exercise 6.11) as an exercise for the reader.

```
          G
    1 1 0 1 0
    1 1 0 0 1
    0 0 1 1 0
    1 0 1 1 1
    0 1 0 1 1
          ∇ NF←MINFILLS G
    [1]     NF←(1↑ρG)ρ0
    [2]     K←1
    [3]   L:W←G[;1]∘.∧G[1;]
    [4]     LF←W∧~G
    [5]     NF[K]←+/1=,LF
    [6]     G←1⊖1⌽G
    [7]     →L×ι(ρNF)≥K←K+1
          ∇

          MINFILLS  G
    2 2 0 6 2
```

Fig. 6.12

MINFILLS **determines the number of fills associated with the elimination of each node**

6.4.5   Sparse data structure

Mention was made earlier of the savings in storage that can be made if a substantial degree of sparsity is present. How can these savings be achieved? Three possible approaches will be illustrated by reference to the matrix M of Fig. 6.13.

$$
\begin{bmatrix}
6.3 & 0 & 0 & 0 & 3.5 & 0 \\
0 & 2.1 & 0 & 4.1 & 0 & 0 \\
0 & 8.7 & 7.2 & 0 & {}^-1.3 & 0 \\
0 & {}^-2.8 & 0 & 5.3 & 0 & 0 \\
4.3 & 0 & 6.2 & 0 & 11.1 & 0 \\
0 & 0 & 0 & 0 & 0 & 2.2
\end{bmatrix}
$$

Fig. 6.13
The sparse matrix M used for illustration

For this matrix the location (row and column positions) and value of each non-zero element must somehow be stored in a data structure.

The table of Fig. 6.14 might first come to mind. However, the unnecessary repetition of row positions in the first column (a repetition that would be far more obvious in an example of realistic size) immediately suggests alternative

Row	Column	Value
1	1	6.3
1	5	3.5
2	2	2.1
2	4	4.1
3	2	8.7
3	3	7.2
3	5	${}^-1.3$
4	2	${}^-2.8$
4	4	5.3
5	1	4.3
5	3	6.2
5	5	11.1
6	6	2.2

Fig. 6.14
A possible data structure representing the matrix M of Fig. 6.13

structures, one of which is shown in Fig. 6.15. Here, for the *n*th row of the matrix, the *n*th element of the vector RP ('row pointer') points to (i.e., indexes) the first entry in a block in the CI ('column index') vector. Thus (see Fig. 6.15) the element 5 in the 3rd position of RP points to the 5th element of CI (which is 2) identifying element $M[3 ; 2]$ as the *first* nonzero element in row 3 of M. Reference to the 5th element of EV ('element value') yields the value (8.7) of $M[3 ; 2]$. Other nonzero elements occur in row 3 of M and their

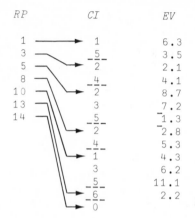

Fig. 6.15
An alternative data structure representing the matrix M **of Fig. 6.13**

column indices (3 5) are listed in the block (within dashed lines) of CI identified by the row pointer. There is no need to separately store information about the location of the block boundaries, since this information is implicit within the vector RP. For example, the block for row 3 contains – $/RP[4 \ 3]$ elements.

Interrogation of the sparse data structure of Fig. 6.15 is quite easy. For example, to find the value of $M[J ; K]$, we simply note that all nonzero elements of row J will appear in vector EV between position $RP[J]$ and $^-1 + RP[J+1]$. If the index K is found within the corresponding elements of CI, then $M[J ; K]$ has a nonzero value which can be found by indexing EV. A suitable function $FIND$ to find the value of an element of M

```
      ∇ R←J FIND K
[1]     X← 0 ¯1 +RP[J+ 0 1]
[2]     CIINDICES←X[1] +0,ιX[2] - X[1]
[3]     Z←(K = CI[CIINDICES])/CIINDICES
[4]     R←+/0,EV [Z]
      ∇
```

can easily be tested by reference to the matrix M of Fig. 6.13.

```
      RP
 1 3 5 8 10 13  14
      CI
 1 5 2 4 2 3 5 2 4 1 3 5 6 0
      EV
 6.3 3.5 2.1 4.1 8.7 7.2 ¯1.3 ¯2.8 5.3 4.3 6.2 11.1 2.2
      1 FIND 1
 6.3
      4 FIND 6
 0
```

The sparse data structure of Fig. 6.15 is, however, not wholly satisfactory for use in the course of LU factorization. We saw earlier that, during LU factorization, *new* nonzero elements (fills) can be — and usually are — generated. Clearly, our sparse data structure must be able to accommodate these additions with minimum inconvenience.

The addition of new elements to the data structure of Fig. 6.15 is not easy. If, for example, a new nonzero element ($M[4\ ;\ 5]$) equal to 1.5 is to be added, then 1 must be added to *all* elements in RP after the 4th, and the correct column index (5) and element value (1.5) must be inserted, after the correct positions for these insertions has been determined, within the block relevant to row 4 of M in columns CI and EV.

RP	CI	NEXT	EV
1	1	2	6.3
3	5	0	3.5
5	2	4(14)	2.1
8	4	0	4.1
10	2	6	8.7
13	3	7	7.2
14	5	0	$\overline{1.3}$
	2	9	$\overline{2.8}$
	4	0	5.3
	1	11	4.3
	3	12	6.2
	5	0	11.1
	6	0	2.2
	0(3)	(4)	($\overline{5}.4$)

Fig. 6.16
A linked list data structure illustrated by the representation of the matrix M of Fig. 6.13

An alternative data structure which is more suited to the addition of new elements is shown in Fig. 6.16, and is called the 'linked-list' data structure (ignore the numbers in brackets: these are referred to later). The most obvious departure from the earlier structures is the vector $NEXT$ which is a list of pointers. For an element whose column index is $CI[X]$, the element $NEXT[X]$ points to the element of CI containing the next nonzero column index in the same row of the matrix. Before any changes are made to the matrix M the vector $NEXT$ is totally redundant. But now consider the addition, to M, of a nonzero element $M[2\ ;\ 3]$ of value $\overline{5}.4$.

To insert the new element the following steps are involved, the results being indicated by the entries within brackets in Fig. 6.16.

(1) Take the row index (2) and, by reference to the corresponding (i.e., 2nd and 3rd) row pointers (3   5) identify the elements (2   4 at locations 3   4) of the corresponding block in the CI vector.

(2) Take the column index (3) of the new element and determine the element in CI that it must follow. Since there is a nonzero element at 2 (representing $M[2\ ;\ 2]$), the new element follows this one. The element at 2 will be termed the immediately preceding element.

(3)   Replace the element in $NEXT$ corresponding to the immediately preceding element in CI by a pointer (14) whose value is $1+\rho NEXT$.

(4)   Replace the last element of CI (a zero only for the virgin state of the structure) by the column index (3) of the new element.

(5)   In the corresponding location in $NEXT$ (by appending to the existing vector $NEXT$) enter the value of the pointer (4) that was replaced (in step 3) by the new pointer (14).

(6)   In the corresponding location in EV (by appending to the existing vector EV) enter the value ($^-5.4$) of the new element.

6.4.6   Comment

It is reasonable to ask whether the computation involved in determining a near-optimal ordering of nodes, followed by the additional effort involved in arranging for the savings in calculation and storage to be exploited, is really worthwhile: would it not be simpler just to proceed with LU factorization? In reply, it should first be remarked that the small examples used for illustration do not give a realistic impression of the savings possible with a large and very sparse matrix. Secondly, any investment made prior to LU factorization is especially valuable if the *same* circuit is analyzed repeatedly with *different* component values. This occurs, for example, when a designer simulates the effect of component tolerances: 300 different analyses may be carried out to see whether a mass-produced circuit will give acceptable performance every time. The same remarks apply to the time-domain analysis of a circuit (see Chapter 8): a typical analysis may involve 400 LU factorizations, so that any initial work designed to save computation is a good investment.

We have attempted here to do no more than provide initial insight into the concept of sparsity, the consequences of node renumbering, the means by which a near-optimal renumbering suited to LU factorization can be achieved and a suitable sparse data structure. Intentionally, we have only scratched the surface of the subject of sparsity.

Exercises

6.1   Explore the effect of the functions LO and UP on a square matrix

```
      ∇R←LO X
[1]   R←((ιN)∘.≥ιN←1↑ρX)×X∇
      ∇R←UP X
[1]   R←((ιN)∘.=ιN)+((ιN)∘.<ιN←1↑ρX)×X∇
```

6.2   The first step in an LU factorization of an $N \times N$ matrix A, and which is repeated with appropriate change of indices, is

```
V←1↓ιN
A[1;V]←A[1;V]÷A[1;1]
A[V;V]←A[V;V]-A[V;1]∘.×A[1;V]
```

Execute this step a sufficient number of times to obtain the L and U factors of a 3×3 reduced nodal conductance matrix (generate this randomly by entering $YR←3\ 3\ ρ\ 9\ ?9$. Test the L and U factors by calculating the inner product $L+.×U$. Carry out a forward and backward substitution to obtain the nodal voltage response, and check that this is correct.

Appropriate algorithms for forward substitution ($FSUB$) and back substitution are shown below. Essentially, $LUBKSUB$ is of the same form as originally defined, but now with $÷Y[S;S]$ in line 2 because it is to be used in the $FSUB$ function.

```
      ∇R←Y FSUB I
[1]   R←Φ(ΦΦ[1] Y) LUBKSUB ΦI∇

      ∇V←Y LUBKSUB I;S
[1]   V←(S←ρI)ρ0
[2]   L:V[S]←(÷Y[S;S])×I[S]-Y[S;]+.×V
[3]   →(0<S←S-1)/L∇
```

6.3   Incorporate the algorithm explored in Exercise 2 in a function $LUFAC$, and test it.

6.4   At the APL terminal define the matrix A to be

```
6   1   8   3
5   2   0   0
1   0   7   0
4   0   0   9
```

Carry out the LU factorization using the function $LUFAC$. How many zero entries are there in the combined LU matrix? By experiment or reasoning see if row or column interchange can lead to more zero entries being retained. Following the generation of an LU matrix having some zero entries, show how forward and backward substitution must be carried out.

6.5   Suppose the voltage response of a four-terminal linear resistive circuit has been calculated by first finding the L and U factors of its 3×3 nodal conductance matrix. Now you are presented with a circuit whose nodal conductance matrix is the *transpose* of that of the original circuit. Show how its voltage response to current excitation can be obtained merely by forward and backward substitution, using the L and U factors of the original matrix.

6.6   Suppose that each node of a circuit is associated with three branches. Under the assumption that the circuit contains only two-terminal components, determine the percentage sparsity associated with a circuit containing (a) 4 nodes; (b) 10 nodes; (c) 50 nodes; (d) 500 nodes.

6.7   For each of the three circuits shown in Fig. E6.1, draw the graph representation of the location of the nonzero elements of the reduced nodal admittance matrix. Also, determine the number of fills that would be associated (with one step of LU factorization) with the suppression of each node.

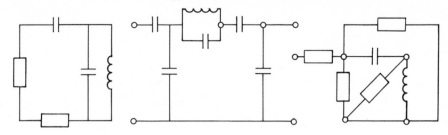

Fig. E6.1

6.8   Find an example of a circuit for which the suppression of at least one of its nodes does not lead to any fills, and where that node is *not* connected to only one branch in the graph representation of the location of nonzero elements of the reduced nodal admittance matrix.

6.9   Define a monadic function whose argument is the graph representation of the location of nonzero elements, and whose returned result is a vector of nodes which, if suppressed in the first stage of LU factorization, will cause no fills.

6.10   For a circuit containing both two-terminal and three-terminal components, derive a systematic method of obtaining the graph indicating the location of nonzero elements in the nodal admittance matrix.

6.11   Define a monadic function whose argument is the graph representation of the location of nonzero elements, and whose result is a vector of node numbers indicating a new node numbering that will be near optimal in the sense of reducing the number of fills that will occur if LU factorization is carried out.

6.12   Define a function that implements the insertion algorithm associated with the linked-list data structure illustrated in Fig. 6.16, and test it.

Solutions

6.2

```
     YR
  4 9 2
  5 3 6
  8 7 1
```
Reduced nodal conductance matrix (not based on an actual circuit).

```
     IR←5 ‾3 10
     VR←IR⊟YR
     VR
1.0672  0.43874  ‾1.6087
```
Nodal current excitation.

Nodal voltage response, for later checks.

```
     X←ι1↑ρYR
     X
  1 2 3
```
Row and column numbering.

```
     S←1↓X
     S
  2 3
```
Rows and columns to be operated upon.

```
     C←1
```
The pivot number.

```
     A←YR
     A[C;S]←A[C;S]÷A[C;C]
     A[S;S]←A[S;S]-A[S;C]∘.×A[C;S]
```
The first stage of the *LU* algorithm.

```
     A
  4            ‾2.25        0.5
  5            ‾8.25        3.5
  8            ‾11          ‾3
```

```
     C←2
     S←1↓S
     S
  3
```

```
     A[C;S]←A[C;S]÷A[C;C]
     A[S;S]←A[S;S]-A[S;C]∘.×A[C;S]
```
The second stage.

The result of *LU* factorization.

```
     LU←A
     LU
  4            2.25         0.5
  5           ‾8.25        ‾0.42424
  8           ‾11          ‾7.6667
```

```
     L←LO LU
     L
  4             0            0
  5            ‾8.25         0
  8           ‾11          ‾7.6667
```
Generation of *L* and *U* matrices.

```
     U←UP LU
     U
  1             2.25         0.5
  0             1           ‾0.42424
  0             0            1
```

```
     L +.×U
  4             9            2
  5             3            6
  8             7            1
```
Product of *L* and *U* yields *YR* .

        $H \leftarrow L \ FSUB \ IR$
        H
  1.25 1.1212 ‾1.6087
        $VR \leftarrow U \ LUBKSUB \ H$
        VR
  1.0672 0.43874 ‾1.6087

Calculation of the nodal voltage response by forward and backward substitution checks with the VR calculated earlier.

6.3   ∇ $R \leftarrow LUFAC \ M$
    [1] $C \leftarrow 1$
    [2] $R \leftarrow M$
    [3] $L: S \leftarrow C \downarrow \iota 1 \uparrow \rho R$
    [4] $R[C; S] \leftarrow R[C; S] \div R[C; C]$
    [5] $R[S; S] \leftarrow R[S; S] - R[S; C] \circ . \times R[C; S]$
    [6] $\rightarrow L \times (1 \uparrow \rho R) > C \leftarrow C + 1$
          ∇

The previously explored steps incorporated in a function.

        $LU \leftarrow LUFAC \ YR$
        LU
    4   2.25   0.5
    5   ‾8.25   ‾0.42424
    8   ‾11   ‾7.6667

Test of the function $LUFAC$. See answer to Exercise 6.2 for a check.

6.4 YR
    6 1 8 3
    5 2 0 0
    1 0 7 0
    4 0 0 9

The sparse matrix.

        $LUFAC \ YR$
    6   0.167   1.33   0.5   Its LU factorization.
    5   1.17   ‾5.71   ‾2.14
    1   ‾0.167   4.71   ‾0.182
    4   ‾0.667   ‾9.14   3.91

        $P \leftarrow 4 \ 3 \ 2 \ 1$
        $YR[P; P]$
    9 0 0 4
    0 7 0 1
    0 0 2 5
    3 8 1 6

A permutation vector.

Original matrix with rows and columns permuted according to P.

        $LUFAC \ YR[P; P]$
    9   0   0   0.444 LU factorization
    0   7   0   0.143   of permuted matrix.
    0   0   2   2.5
    3   8   1   1.02

        $SP \ LUFAC \ YR$
    1 1 1 1
    1 1 1 1
    1 1 1 1
    1 1 1 1
        $SP \ LUFAC \ YR[P; P]$
    1 0 0 1
    0 1 0 1
    0 0 1 1
    1 1 1 1

        $SPV \ LUFAC \ YR$
    16 16

Sparsity of the LU factorization of the original and the permuted matrix (see functions SP and SPV on next page.

        $SPV \ LUFAC \ YR[P; P]$
    10 16

```
        IR←4 2 9 ‾5
        []←VR←IR⊞ YR
‾5.49 14.7  2.08  1.88
```
Current excitation.
Voltage response calculated directly to allow check.

```
        L←LO LUFAC YR[P;P]
        U←UP LUFAC YR[P;P]
```
$\left.\begin{matrix} L \\ U \end{matrix}\right\}$ of permuted matrix.

```
        H←L  FSUB IR[P]
        J←U LUBKSUB H
        J
1.88  2.07 14.7 ‾5.49
        VR←J [P]
        VR
‾5.49 14.7 2.07 1.88
```
Forward substitution requires excitation to be permuted.

Result of back substitution must be permuted to yield correct voltage response.

```
      ∇ R←SP M
 [1]    R←~0=M
      ∇
```
Functions for investigating sparsity.

```
      ∇ R←SPV M
 [1]    R←(+/+/~0=M),×/ρM
      ∇
```

6.5
```
          YR
  4 9 2
  5 3 6
  8 7 1
```
Reduced nodal conductance matrix of original circuit.

```
        YRT← ⍉YR
        IRT←3 16 ‾8
```
Reduced nodal conductance matrix and current excitation of the 'transpose circuit'.

```
        VRT← IRT⊞ YRT
        VRT
  1.91  ‾2.09 0.727
```
Nodal response voltages of 'transpose circuit' calculated directly to allow a later check.

```
        L← LO LUFAC YR
        U← UP LUFAC YR
```
L and U factors of original circuit.

```
        LT← ⍉U
        UT← ⍉L
```
L and U factors of the 'transpose circuit'.

```
              LT+.×UT
```
Check on the above	4	5	8
	9	3	7
	2	6	1

```
        H← LT FSUB IRT
        VRT← UT LUBKSUB H
```
For the given excitation of the 'transpose circuit', forward and backward substitution lead to the correct nodal voltage response.

```
        VRT
  1.91  ‾2.09 0.727
```

6.6 In typical circuits the average number of branches per node is around 2.5, but for simplicity the next higher integer value (3) was chosen to simplify the problem. The choice of such a precise value has one minor drawback, in that it restricts attention to circuits with an even number of nodes (n). This restriction is easily explained. If $n=6$, for example, then there must be 6×3 branch terminations (or ends). Since each branch has two ends, the number of branches (b) for $n=6$ is $(6 \times 3) \div 2 = 9$.

By this argument it follows that, for the problem as stated, n must be even. However, this is not a severe restriction, since the aim of the problem is to gain some feeling for the degree of sparsity to be found in circuits of different sizes.

Consider the complete nodal admittance matrix Y. The number of elements in Y is $N*2$. If a node is connected to three other different nodes, then the corresponding row and column will each contain four nonzero entries. Since there are N rows and columns, the total number of non-zero entries in Y is $4 \times N$. Thus, the following table can be composed in answer to the problem.

N	$N*2$	$4 \times N$	Percentage sparsity
4	16	16	0
10	100	40	60
50	2500	200	92
500	250000	2000	99.2

The above discussion takes no account of the fact that, if each node is associated with three branches, then it may not always be possible for each of these three branches to have a different destination node.

6.7

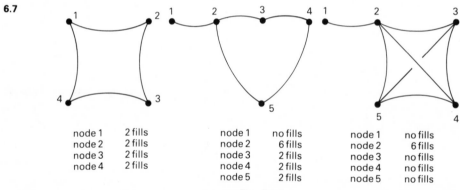

node 1	2 fills
node 2	2 fills
node 3	2 fills
node 4	2 fills

node 1	no fills
node 2	6 fills
node 3	2 fills
node 4	2 fills
node 5	2 fills

node 1	no fills
node 2	6 fills
node 3	no fills
node 4	no fills
node 5	no fills

Fig. E6.2

6.8

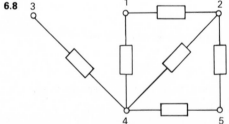

Fig. E6.3

The suppression of node 1, and also of node 5, will cause no fills in the *LU* factorization process. Neither will the suppression of node 3, for reasons discussed in Section 6.4.3.

6.10 Consider a 3-terminal component with the node labelling shown in Fig. E6.4(a). It was shown in Chapter 3 that such a component can be modelled as in Fig. E6.4(b). The two-terminal admittances in the model give rise to branches between a and e, and between c and e, in the graph (Fig. E6.4(c)). Again with reference to Chapter 3, we know that the controlled source $g_{ae}v_{ce}$ gives rise to entries at locations a;c, a;e, e;c, e;e in the nodal admittance matrix (Fig. E6.4(d)). The other controlled source makes similar contributions at c;a, c;e, e;a and e;e, so that the controlled sources together lead to nonzero elements at the locations shown in Fig. E6.4(e). Those that are circled are already handled by the branches in Fig. E6.4(c). The remaining off-diagonal nonzero elements can be

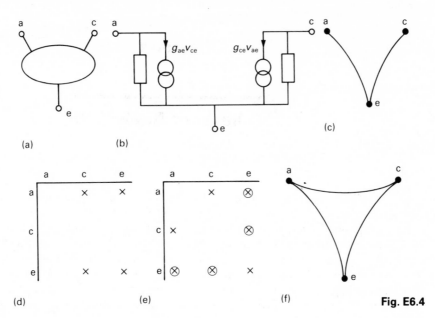

Fig. E6.4

represented by a single branch between a and c, so that the complete graph representing the three-terminal component is as shown in Fig. E6.4(f).

Quiescent Analysis II

In Chapter 2 we developed the Newton–Raphson iterative algorithm by considering the simplest possible example: a *single* nonlinear two-terminal component excited by a *single* current source, and generating a *single* response voltage. The function $ITER$ was defined, tested and refined. This simple approach was adopted for good reason: from one viewpoint, the extension of the algorithm to the analysis of a *circuit* containing a *number* of components (two-, three- or multi-terminal), *more* than one current excitation and *more* than one voltage response (Fig. 7.1) merely involves a straightforward generalization of the function $ITER$.

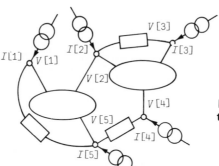

Fig. 7.1
Example of the type of circuit considered in this chapter. The two-terminal components might be diodes, and the three-terminal components transistors, although linear components are allowed

7.1 Generalization of $ITER$

If the circuit contains in addition to the reference node, more than one other node, the variables IR and VR become vectors: so also, therefore, do $ERROR$ and ΔVR. Similarly, whereas in $ITER$ the scalar $SLOPE$ describes the relation between infinitesimal changes ΔIR and ΔVR in a single node current and node voltage, we now require a *matrix* — again called $SLOPE$ — to describe the relation between infinitesimal changes in *all* node currents and voltages. Thus, there is very little modification required to $ITER$, to obtain the new function ($ITERNET$) shown below for the analysis of a nonlinear resistive circuit.

The two modifications are to be found on lines 2 and 4. On line 2, since $ERROR$ is now a vector of imbalance currents at the nodes, we check the largest imbalance current against a tolerance which in this example is 1 μA. On line 4, scalar division ($\div$) is replaced by matrix inversion($\boxminus$),for the reason discussed below, where we comment on the function *ITERNET* line by line. For convenience of reference it is temporarily assumed that the circuit has *N* nodes in addition to its reference node.

```
        ∇VR←IRX ITERNET VRO
[1]    VR←VRO
[2]    L1:→L2×1E¯6<⌈/|ERROR←IRX-NET VR
[3]    L2:SLOPE←DIFNET VR
[4]    VR←VR+ΔVR←ERROR⊟SLOPE
[5]     →L1
        ∇
```

Line 1   The initial guess (the *N*-element vector *VRO*) at the result is assigned to the result in case it is correct.

Line 2   The injected nodal currents which would be associated with the nodal voltage vector *VR* are calculated (details below) by the function *NÈT*, and are then subtracted from the actual injected nodal currents (described by the vector *IRX*) to find the nodal imbalance currents: the latter are described by the *N*-element vector *ERROR.* A check is then carried out to see if the largest imbalance current exceeds 1 μA in magnitude. If so, branching to line 3 (*L2*) occurs; if not, (because the vector *VR* is sufficiently close to the actual response) the function terminates.

Line 3   The function *DIFNET* (details below) generates the matrix *SLOPE* relating infinitesimal changes ΔIR and ΔVR in the nodal currents and voltages:

$$\Delta IR \leftarrow SLOPE +.\times \Delta VR \tag{7.1}$$

The matrix *SLOPE* is called the reduced incremental (or small-signal) nodal conductance matrix.

Line 4   The required changes (ΔIR) in the nodal currents are known: they are the elements of *ERROR.* To achieve these changes the nodal voltages need to be adjusted according tȯ (see 7.1)

$$\Delta VR \leftarrow \Delta IR \boxminus SLOPE$$

or

$$\Delta VR \leftarrow ERROR \boxminus SLOPE \tag{7.2}$$

Even though the actual changes ΔIR and ΔVR may not be infinitesimally small, it is still assumed that (7.1) is sufficiently accurate for our purposes. From (7.2) the adjusted values ($VR + \Delta VR$) of the nodal voltages are obtained.

Line 5   The new *VR* must be tested, on line 2 (*L1*) , to see if these nodal voltages correspond sufficiently closely to those which actually result from the given current excitation.

7.2 The Functions *NET* and *DIFNET*

If the nodal voltages VR are known it is possible to find the branch voltages VB from Kirchhoff's voltage law according to

$$VB \leftarrow (\lozenge A) + . \times V\!R , 0 \tag{7.3}$$

From VB, a knowledge of the linear or nonlinear relations describing the individual components allows the branch current vector IB to be found. Thus, for a circuit composed entirely of linear conductances and exponential diodes (for example), the branch currents might be calculated as the sum ($IBG + IBD$) of a vector associated with the linear conductances

$$IBG \leftarrow G \times VB \tag{7.4}$$

and a vector associated with the exponential diodes:

$$IBD \leftarrow IS \ DIODE \ VB \tag{7.5}$$

where G and IS are vectors of length B (the number of branches): For example, an element of G would give the value of the corresponding conductance, or would be zero if no conductance were present in that branch. Finally, Kirchhoff's current law

$$IR \leftarrow {}^-1 \downarrow A + . \times IB \tag{7.6}$$

allows the vector of nodal currents to be computed. The function NET (whose definition is left as an exercise (Exercise 7.1)) follows this sequence to determine, from its argument VR, the nodal current vector IR; it has the syntax

$$IR \leftarrow NET \ VR$$

The function $DIFNET$ is used to generate the reduced incremental nodal conductance matrix $SLOPE$, which is a function solely of the nodal voltages VR. Thus, the function has the syntax

$$SLOPE \leftarrow DIFNET \ VR$$

To illustrate the basis of $DIFNET$ we show, in the left-hand part of Table 7.1, the derivation, for a linear circuit, of the conductance matrix YR relating *actual* nodal voltages and currents. It invokes (as we saw in Chapter 1) the two Kirchhoff laws and the branch relations. But, as shown in the right-hand part of Table 7.1, *precisely* the same laws and relations are satisfied by *changes* in nodal currents and voltages, although for nonlinear components the linear branch relations are valid only for *small* changes in current and voltage. Thus, *precisely* the same development leads to the matrix relation between ΔIR and ΔVR, a relation which is to be embodied in $DIFNET$. We need only comment on the derivation of the *incremental* branch conductance matrix GBI.

	Linear homogeneous circuits (actual currents and voltages)	Nonlinear circuits (small changes in currents and voltages)
The relations between small changes in the currents and voltages of a nonlinear circuit are compared, below, with the relations between the currents and voltages of a circuit containing only linear homogeneous components.		
Kirchhoff's current law	$I \leftarrow A +.\times IB$ (a)	$\Delta I \leftarrow A +.\times \Delta IB$ (d)
Kirchhoff's voltage law	$VB \leftarrow (\lozenge A) +.\times V$ (b)	$\Delta VB \leftarrow (\lozenge A) +.\times \Delta V$ (e)
Branch relations	$IB \leftarrow YB +.\times VB$ (c)	$\Delta IB \leftarrow YBI +.\times \Delta VB$ (f)
Reduced nodal relations	(a, b, c) yield $IR \leftarrow YR +.\times VR$ where $YR \leftarrow \bar{}1 \quad \bar{}1 \downarrow A +.\times YB +.\times \lozenge A$	(d, e, f) yield $\Delta IR \leftarrow SLOPE +.\times \Delta VR$ where $SLOPE \leftarrow \bar{}1 \quad \bar{}1 \downarrow A +.\times YBI +.\times \lozenge A$

Table 7.1

For a circuit containing only two-terminal components, for example, each diagonal element of GBI is the ratio of small changes in current and voltage for the corresponding component or, in other words, its incremental conductance (Fig. 7.2). The definition of $DIFNET$ is left as an exercise for the student (see Exercise 7.2).

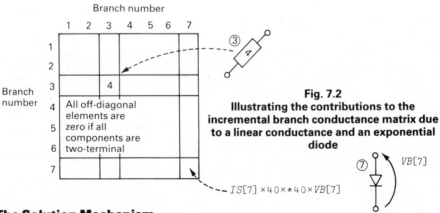

Fig. 7.2
Illustrating the contributions to the incremental branch conductance matrix due to a linear conductance and an exponential diode

7.3 The Solution Mechanism

In line 4 of $INTERNET$ we have used an expression of the form $V \leftarrow I \boxminus Y$ both to *denote* the solution of a set of equations $I \leftarrow Y +.\times V$ and to *calculate* it. The implementation of $\boxminus$ is not discussed here, since it may well vary between APL systems. The actual solution mechanism can, however, be made more explicit by replacing line 4

$VR \leftarrow VR + \Delta VR \leftarrow ERROR \boxminus SLOPE$

with, for example,

$$VR \leftarrow VR + \Delta VR \leftarrow SLOPE\ GAUSS\ ERROR$$

if Gaussian elimination (Chapter 5) is employed as the solution algorithm, or with

$$VR \leftarrow VR + \Delta VR \leftarrow U\ LUBKSUB\ L\ FSUB\ ERROR$$

if the LU factorization approach (Chapter 6) is adopted.

7.4 Circuit Interpretation of the Newton–Raphson Algorithm

It was shown in Chapter 2 that an alternative but equivalent approach to the quiescent analysis of a single nonlinear component could be developed by replacing the component with the parallel connection of a current source and a linear conductance. The same approach is often taken in the quiescent analysis of nonlinear circuits, and is, indeed, adopted in the $BASECAP$ package used as the illustrative example in Part 2 of this book.

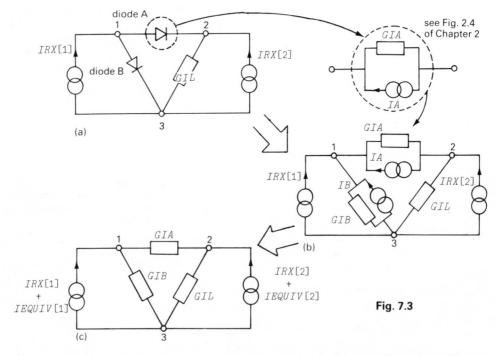

Fig. 7.3

Just as for the single component discussed in Chapter 2, each nonlinear two-terminal component (Fig. 7.3(a)) is replaced by the parallel connection of a linear conductance and a current source (Fig. 7.3(b)) which approximates the component's current/voltage characteristic in the vicinity of the present values of IR and VR.

In the special case of a linear conductance, of course, the current source is zero-valued. Thus, each iteration of the Newton–Raphson algorithm is replaced by the analysis of a linear circuit (containing the conductances GI), but one in which the excitation is now the *sum* of the actual nodal excitation vector IRX and a vector $IEQUIV$ representing the current sources associated with the model of each nonlinear component (Fig. 7.3(c)).

7.5 Convergence and Termination

All the remarks in Chapter 2 concerning convergence of the Newton–Raphson algorithm are generally applicable in the context of nonlinear circuit analysis, as also are the circumstances which can lead to computer overflow ($DOMAIN$ $ERROR$). For the prevention of $DOMAIN$ $ERROR$ the same approaches discussed in Chapter 2 can be applied:

(a) linearization of (e.g. exponential) components beyond a critical voltage;

(b) avoidance of small slope conductances;

and (c) control of the magnitude of the branch voltage correction according to the steepness of the characteristic.

Another method of preventing an unacceptably large change in voltage, the Method of Alternating Bases, will now be described.

Within a nonlinear circuit, consider a single two-terminal exponential diode whose characteristic is illustrated by Fig. 7.4. Let the initially guessed voltage (or the voltage associated with a specific step in the iteration) be VO and the calculated modification ΔVR, so that the new solution is associated with the point A on the linearized characteristic.

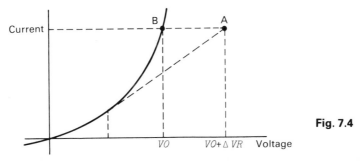

Fig. 7.4

Two choices are now available. The one that is familiar is to select the *voltage* associated with A as the new component voltage ($VR \leftarrow VO + \Delta VR$). However, this may lead to an unacceptably large current and hence a $DOMAIN$ $ERROR$. The other choice is to select the *current* associated with A, by moving horizontally towards the current axis. If the component is an exponential diode, this merely involves computing the current corresponding to the point A and then, by using the *inverse* of the exponential diode relation, finding the corresponding voltage associated with point B. Point B is the new trial point for the next iteration. The nature of this algorithm, which can be used with many types of nonlinearity, requires that the inverse of the current–voltage characteristic (e.g. for the exponential diode, $V \leftarrow 0.025 \times \circledast 1 + I \div IS$) be known.

The use of the method just described for a single component would be artificial, since if the inverse of the component relation is known there is no need for an iterative algorithm. The method is realistic as soon as more than one component is involved.

A variety of criteria can be used to terminate the Newton–Raphson procedure. Termination could occur when (a) the absolute difference between a variable of interest and its previous value is less than some critical value, (b) when the relative error (e.g., $\Delta VR \div VR$) is sufficiently small, or (c) when the maximum magnitude of nodal current imbalance is less than some preset maximum (as in $ITERNET$). In general, no single criterion is superior.

7.6 Bias-dependent Small-signal Behavior

If the circuit being subjected to a quiescent analysis is a wide-band amplifier containing nonlinear devices such as transistors and diodes, its small-signal frequency-domain behavior is of primary interest. However, we have already seen (in Section 7.2) that the relation between the small-signal current and voltage associated with a nonlinear component is a function of its quiescent condition. We must therefore conclude that the frequency-domain analysis of a nonlinear circuit must be preceded by a quiescent analysis of that circuit; only in this way can the small-signal model of the nonlinear components be obtained.

The matrix $SLOPE$ is, in fact, the reduced nodal conductance matrix appropriate to the resistive part (both linear and nonlinear) of the circuit, as expressed in (7.1) and discussed in detail in Section 7.2. Therefore, in a circuit-analysis package (such as $BASECAP$: see Part 2) capable of handling both d.c. and small-signal frequency-domain behavior, it is arranged that the $SLOPE$ matrix generated at the *last* iteration of the quiescent analysis should be 'handed over' to the frequency-domain analysis as the real part of the reduced nodal admittance matrix (see RY in equation (4.6)).

The principle can be illustrated by the example (this example uses a complex arithmetic version of APL (see Section 4.6)) of Fig. 7.5: for simplicity we have chosen a circuit with only two nodes so that quiescent analysis is achieved with $ITER$, and $SLOPE$ is a scalar. In order to make use of the computed value of $SLOPE$, it must be removed, for the purpose of this illustration, from the list of local variables on line $[0]$ of $ITER$. As the illustration shows, the small-signal voltage response VAC is a function of the bias applied to the diode.

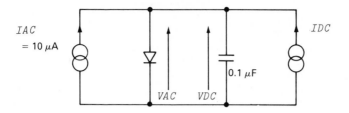

```
   ∇ VR←IRX ITER VR0; ERROR; ΔVR                    ∇ IR←DIODE VR
[1]    VR←VR0                                   [1]    IR←0.00000000001×⁻1+*40×VR
[2]    L1: →L2×0.000001 < | ERROR←IRX - DIODE VR     ∇
[3]    L2: SLOPE←DIFDIODE VR                         ∇ S←DIFDIODE VR
[4]    VR←VR+ΔVR←ERROR÷SLOPE                    [1]    S←0.00000000001×40×*40×VR
[5]    →L1                                           ∇
   ∇
```

The function *ITER* embodying the Newton–Raphson algorithm, with *SLOPE* being a global variable.

Functions, called by *ITER*, describing the exponental diode.

```
        IDC←.00025
        VDC←IDC ITER .4
        VDC
0.4258681493
        SLOPE
0.01026518804
        W←1E5
        C←1E⁻7
        B←W×C
        B
0.01
        Y←⁻9 ⁻11+.○SLOPE,B
        Y
0.01026518804J0.01
        IAC←1E⁻5
        VAC←IAC÷Y
        VAC
0.0004998287888J⁻0.0004869163297
        | VAC
0.0006977939163
```

The quiescent current is set at 250 μA. *ITER* is used to determine the quiescent voltage *VDC*

The last value of *SLOPE* is available ($\approx$0.01 S).

The radian frequency is 10^5 rad/s. The capacitor is assigned a value of 0.1 μF. The admittance *B* of the capacitor is computed.

B is 0.01 S (reactive). The complex admittance of the diode and capacitor in parallel is computed.

The small-signal current *IAC* has an amplitude of 10 μA. The small-signal voltage response *VAC* is computed. Its magnitude is 697 μV.

```
        IDC←.00005
        IDC ITER .4
0.3858503061
        Y←⁻9 ⁻11+.○SLOPE,B
        VAC←IAC÷Y
        VAC
0.0002180448235J⁻0.000949951614
        | VAC
0.0009746546127
```

For a quiescent current of 50 μA the whole calculation is repeated

The magnitude of the small-signal voltage response is now 974 μV.

Fig. 7.5

Exercises

7.1 Consider the class of circuits containing only linear conductances and exponential diodes. For each diode employ the relation between current and voltage discussed in Exercise 2.3. The collection of components can be described by two vectors: G contains, as its elements, the conductances of the branches, while IS defines the saturation currents of any diodes present. Thus, the vectors

$$G \leftarrow 0 \quad 0.1 \quad 2$$
$$IS \leftarrow IE^-11 \quad 0 \quad IE^-10$$

describe the branches shown in Fig. E7.1. The normal incidence matrix A describes the interconnection of the branches.

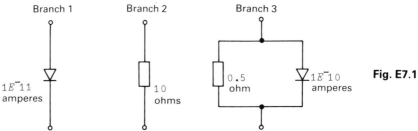

Branch 1   Branch 2   Branch 3

$1E^-11$ 10 0.5 $1E^-10$ **Fig. E7.1**
amperes   ohms   ohm   amperes

Define a monadic function, NET, whose single argument VR is the reduced vector of nodal voltages of the circuit, and which returns the compatible reduced vector of currents that must be injected into the circuit nodes.

7.2 For the class of circuits defined in Exercise 7.1, define a monadic function, $DIFNET$, whose single argument VR is again the reduced vector of nodal voltages of the circuit, and which returns the circuit's reduced nodal incremental conductance matrix.

7.3 For the class of circuits described in Exercise 1, define a dyadic function $ITERNET$ whose arguments are the reduced vector IRX of nodal excitation currents and the initial guess at the reduced vector of nodal voltages. The returned result is the actual reduced vector of nodal voltages. The function $ITERNET$ should call the functions NET and $DIFNET$ defined in response to Exercises 1 and 2 respectively. Test the function $\overline{ITERNET}$ with a number of example circuits, and observe the changes in the vector of nodal imbalance currents (called $ERROR$ in the earlier discussion) and the nodal incremental conductance matrix (earlier called $SLOPE$). Note also the number of iterations involved before termination occurs. In your tests, use a wide range of initial guessed voltages, and explore the effect of variation in the choice of the threshold voltage VT. Make sure that $ITER$ still works if no diodes are present.

7.4 Repeat Exercise 7.3, but employing the circuit interpretation of the Newton–Raphson algorithm discussed in Section 7.4.

7.5 For one of the circuit examples used in Exercise 3, connect a small-signal independent current source between any two nodes, and determine the small-signal voltage response between an arbitrarily selected pair of nodes. See how the voltage response is affected by a change in the quiescent excitation, leading to a change in the quiescent condition of the nonlinear components.

7.6 Repeat Exercise 7.5 with a capacitor connected between an arbitrarily selected pair of terminals.

Solutions

7.1
```
     ∇ IR←NET VR
[1]    VB←(⍉A) +.×VR,0
[2]    IBD←COMLIN VB
[3]    IB←IBD + G×VB
[4]    IR← ¯1↓A+.×IB
     ∇
```

```
     ∇ IR←COMLIN VR
[1]    IR←(DIODE VR⌊VT) +(DIFDIODE VT) ×0⌈ (VR- VT)
     ∇
```

```
     ∇ IR←DIODE VR
[1]    IR←IS× ¯1+*40×VR
     ∇
```

```
     ∇ S←DIFDIODE VR
[1]    S←IS×40×*40×VR
     ∇
```

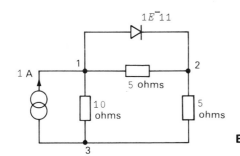

```
IS←0 1E¯11 0
G←.1 .2 .2
A←3 3⍴1 1 0 0 ¯1 ¯1 ¯1 0 1
VT←.7
```

Fig. E7.2
Example circuit and its description

```
     VR←3.744 3.128
       NET VR
0.9996 3.8864E¯5
```

For these values of the nodal voltages at nodes 1 and 2, the injected currents at these nodes are (approx) 1 and 0 A (see Solution 7.3 for confirmation).

7.2
```
     ∇ S←DIFNET VR
[1]    VB←(⍉A) +.×VR,0
[2]    X←VB>0
[3]    S1←DIFDIODE VT⌊ X×VB
[4]    S2←Y\ ((Y/IS) × ¯1+*40×VBR) ÷ VBR←(Y←~X)/VB
[5]    S← ¯1 ¯1 ↓A+.×(DIA S1+S2+G) +.×⍉A
     ∇
```

```
     ∇ R←DIA X
[1]    R←((B,B←⍴X)⍴X) × UNIT ⍴X
     ∇
```

```
     ∇ R←UNIT X
[1]    R←(⍳X) ∘.=⍳X
     ∇
```

```
            VR
3.744 3.128
        DIFNET VR
    20.394        ⁻20.294
   ⁻20.294         20.494
```

For the same circuit example used in Solution 7.1, the same vector of nodal voltages leads to the reduced incremental nodal conductance matrix shown (see Solution 7.3 for confirmation).

7.3

```
        ∇VR←IRX ITERNET VRO
[1]  K←1
[2]  VR←VRO
[3]  L1: →L2×.001 <⌈/| ERROR←IRX - NET VR
[4]  L2: SLOPE←DIFNET VR
[5]  ERROR
[6]  SLOPE
[7]  VR←VR+∆VR←ERROR⊞SLOPE
[8]  K←K+1
[9]  →L1
[10]   ∇
```

The result of applying *ITERNET* to the circuit example shown in Solution 7.1, for different starting voltages and different values of *VT*, is shown opposite.

```
VT←0.7
VRO←5    2
          1  0  ITERNET  5    2
¯1345.1  1.345.2                ¯578.7
    578.8                        578.9
    578.7
¯5.016  5.016                   ¯220.3
    220.4                        220.5
¯220.3
¯1.7213  1.7213                 ¯88.752
    88.852                        88.952
¯88.752
¯0.52203  0.52203               ¯40.988
    41.088                        41.188
¯40.988
¯0.11211  0.11211               ¯24.727
    24.827                        24.927
¯24.727
¯0.0094509  0.0094509           ¯20.669
    20.769                        20.869
¯20.669
3.744  3.128   VR

3.744 - 3.128    <VT
0.616  diode voltage is    <VT

6 iterations
```

```
VT←0.7
VRO←¯10000  ¯345
          1  0  ITERNET  ¯10000  ¯345
2932  ¯1862                     ¯0.2
          ¯0.3                   0.4
          ¯0.2
¯1055.8  1055.8                 ¯578.7
    578.8                        578.9
    578.7
¯5.016  5.016                   ¯220.3
    220.4                        220.5
¯220.3
¯1.7213  1.7213                 ¯88.752
    88.852                        88.952
¯88.752
¯0.52203  0.52203               ¯40.988
    41.088                        41.188
¯40.988
¯0.11211  0.11211               ¯24.727
    24.827                        24.927
¯24.727
¯0.0094509  0.0094509           ¯20.669
    20.769                        20.869
¯20.669
3.744  3.128

7 iterations
```

```
VT←1
VRO←5    2
          1  0  ITERNET  5    2
¯1.9066E8  1.9066E8             ¯9.4154E7
    9.4154E7                     9.4154E7
¯9.4154E7
¯8.6593E5  8.6593E5             ¯3.4637E7
    3.4637E7                     3.4637E7
¯3.4637E7
¯3.1856E5  3.1856E5             ¯1.2742E7
    1.2742E7                     1.2742E7
¯1.2742E7
¯1.1719E5  1.1719E5             ¯4.6877E6
    4.6877E6                     4.6877E6
¯4.6877E6
¯43112  43112                   ¯1.7245E6
    1.7245E6                     1.7245E6
¯1.7245E6
                            .  .
¯0.53544  0.53544               ¯41.521
    41.621                        41.721
¯41.521
¯0.11622  0.11622               ¯24.89
    24.99                         25.09
¯24.89
¯0.010073  0.010073             ¯20.694
    20.794                        20.894
¯20.694
3.744  3.128

18 iterations
```

```
VT←1
VRO←¯10000  ¯345
          1  0  ITERNET  ¯10000  ¯345
2932  ¯1862                     ¯0.2
          ¯0.3                   0.4
          ¯0.2
¯1.4358E8  1.4358E8             ¯9.4154E7
    9.4154E7                     9.4154E7
¯9.4154E7
¯8.6593E5  8.6593E5             ¯3.4637E7
    3.4637E7                     3.4637E7
¯3.4637E7
¯3.1856E5  3.1856E5             ¯1.2742E7
    1.2742E7                     1.2742E7
¯1.2742E7
¯1.1719E5  1.1719E5             ¯1.1719E5
                            .
¯41.521                          41.721
¯0.11622  0.11622               ¯24.89
    24.99                         25.09
¯24.89
¯0.010073  0.010073             ¯20.694
    20.794                        20.894
¯20.694
3.744  3.128

19 iterations
```

Note: the threshold voltage VT has more effect on the convergence than the discrepancy between the 'guessed' and actual voltages

Time-domain Analysis

8.1 The Problem

For many applications, and particularly those referred to as *digital*, a circuit must be designed to exhibit a desired time-domain behavior. In other words, given an input voltage or current of known waveform, a circuit must be designed so that the waveform of its output voltage or current lies within a specification. Typically, the specification might be that the time taken for an output voltage to move from one logic state to the other be less than a given value. The inverter (or NOT-gate) of Fig. 8.1 provides an example.

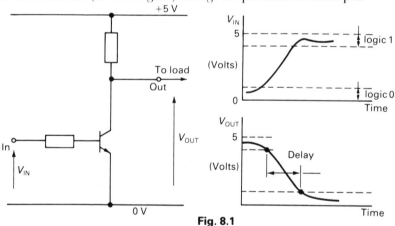

Fig. 8.1

For the great majority of circuits whose time-domain behavior is of interest no *synthesis* procedure exists which will accept details of the input and specifications on the output and, therefrom, generate an acceptable circuit. Manual design must therefore be adopted, and because such a design approach can benefit from frequent analyses of a circuit, the need arises for an efficient and accurate means of time-domain analysis. It is with some fundamental concepts and techniques of time-domain analysis that this chapter is concerned.

A simple linear circuit excited by a current whose waveform is a step function is shown in Fig. 8.2. It is a straightforward matter not only to formulate the *differential* equation describing its behavior

$$I = (G \times V) + C \times dV/dT \tag{8.1}$$

but also to write down an exact solution

$$V \leftarrow (I \div G) - ((I \div G) - V[1]) \times \star - T \div (C \div G) \qquad (8.2)$$

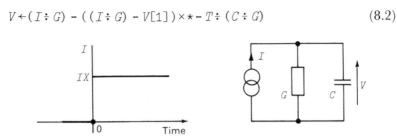

Fig. 8.2
A simple *GC*-circuit with a step current excitation

which is valid for any initial value ($V[1]$) of V at time $T = 0$. Equation (8.2) clearly allows the waveform of V to be determined without approximation.

With more complex and realistic circuits, and especially when some of their constituent components are nonlinear (e.g. when diodes or transistors are present) it is not a straightfoward task to formulate the corresponding differential equation, though it is certainly possible. What does present difficulty is the *solution* of the differential equation: it is usually impossible to write down an explicit expression from which the circuit response can be determined without approximation. In these circumstances it is necessary to compute the circuit's response by making appropriate approximations.

Fortunately, much of the groundwork for the computer calculation of time-domain behavior has been laid in earlier chapters. Briefly, the overall approach to be described is that, within each of a succession of small time periods, the circuit's *differential* equation description is replaced by a set of *algebraic* equations describing a resistive circuit: the smaller the time periods, the better is the approximation to the exact differential equation description. Techniques for the analysis of resistive circuits, both linear (Chapters 1 and 3) and nonlinear (Chapters 2 and 7), have already been described, and will be exploited in the present chapter.

8.2   Discretization of Time: The Backward Euler Algorithm

Every voltage and current in a circuit is a continuous function of time (Fig. 8.3). Since capacitor currents and inductor voltages are expressed as time derivatives of their respective voltages and currents

$$IC = C \times dVC/dT \qquad\qquad VL = L \times dIL/dT \qquad (8.3)$$

it is necessary, in any computation of circuit behavior, to know the values of these derivatives at any instant of time. The approach taken here is, first, to consider only specific instants of time. These instants of time are denoted by elements of the vector T. Thus $T[1]$ is the time reference or origin. Next, the approximation is made that derivatives can be replaced by finite differences in the following way (Fig. 8.3)

$$dVC/dT \big|_{T[N+1]} = (VC[N+1] - VC[N]) \div (T[N+1] - T[N])$$

$$dIL/dT \big|_{T[N+1]} = (IL[N+1] - IL[N]) \div (T[N+1] - T[N])$$

(8.4)

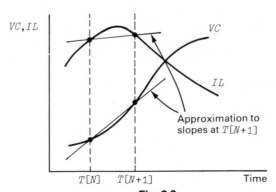

Fig. 8.3
Waveforms of capacitor voltage and inductor current

The indices N and $N+1$ denote values at discrete time instants. The 'formula' of equation (8.4) is known as the *backward Euler* formula because past (i.e. $T[N]$) values are used to approximate present (i.e. $T[N+1]$) derivatives. Clearly, unless the time interval is sufficiently small, the approximations of (8.4), illustrated in Fig. 8.3, can lead to substantial error.

We investigate the consequences of the approximation represented by (8.4) in the context of a simple illustrative circuit (Fig. 8.4) whose behavior is well known. Consider first the capacitor, whose governing equation is

$$IC = C \times dVC/dT$$

Using the approximation of (8.4), the above expression becomes

$$IC[N+1] = C \times (VC[N+1] - VC[N]) \div H$$

(8.5)

where

$$H = T[N+1] - T[N]$$

and is known as the *step size*. Rearrangement of (8.5) in the form

$$IC[N+1] = (C \times VC[N+1] \div H) - (C \times VC[N] \div H)$$

(8.6)

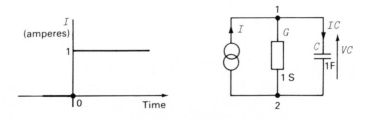

Fig. 8.4
A circuit example chosen for illustration

now leads to the model shown in Fig. 8.5. The first term in parentheses on the right of (8.6) defines the ratio, at time $T[N+1]$, of a current and the voltage $VC[N+1]$, and can therefore be modelled by a conductance. The second term relates a current at $T[N+1]$ to the voltage $VC[N]$ at a different instant of time, and is modelled by a current source. The model of Fig. 8.5 is known as the *companion model* of the capacitor. We note that it is of the same form as the previously encountered companion model of a diode: that was relevant to a small region of a nonlinear characteristic, while the model of Fig. 8.5 is relevant to a small time period.

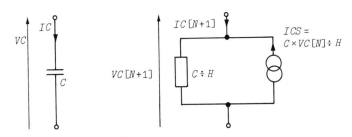

Fig. 8.5
A capacitor, and the companion model approximating its current voltage relationship at time $T[N+1]$

Let us now employ the capacitor companion model in our analysis of the circuit of Fig. 8.4. We describe the circuit by conductance and capacitance vectors:

$GT \leftarrow 1 \quad 2 \quad 1$

$CT \leftarrow 1 \quad 2 \quad 1$

in each of which the first two elements describe the end-nodes and the last element the component value (Siemens or Farads). Reference to Fig. 8.5 allows the circuit of Fig. 8.4 to be approximated, during the time period from $T[N]$ to $T[N+1]$, by the circuit of Fig. 8.6. If we choose a time-step H which is significantly less than the time constant (1 second) of the circuit

$H' \leftarrow 0.05$

then the conductance component of the capacitor's companion model can be described by

$GC \leftarrow CT \div 1 \quad 1 \quad , H$
GC
$\quad 1 \quad 2 \quad 20$

a vector GC whose elements are interpreted in the same way as for GT and CT: that is, a conductance of 20 siemens between nodes 1 and 2. Let us also suppose that the initial value $VC[1]$ of VC is zero:

$VC \leftarrow 0$

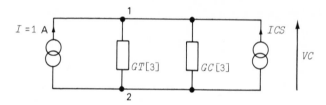

Fig. 8.6
Circuit model associated with a time step H for the circuit of Fig. 8.4

The current source in the capacitor companion model is given by

$$ICS \leftarrow VC \times CT[3] \div H$$

which, of course, has the value of zero

$$\begin{array}{l} ICS \\ 0 \end{array}$$

if VC is zero.

It only requires the application of Ohm's law to the circuit of Fig. 8.6 to determine the value of VC at the completion of the first time step, a period during which the excitation current I is known:

$$\begin{array}{l} I \leftarrow 1 \\ VC \leftarrow (I + ICS) \div GT[3] + GC[3] \\ VC \\ 0.047619 \end{array}$$

This value compares with the theoretical value of

$$\begin{array}{l} VC \leftarrow 1 - *- H \\ VC \\ 0.048771 \end{array}$$

and indicates an error of about 2%.

To extend the calculation to cover a number of time steps we embody the algorithm illustrated above in the function $BACKEULER$. Its left argument X is a two-element vector containing the initial value of VC and the time step H. Its right argument I is the vector of excitation currents at instants separated by H:

```
      ∇  VC←X BACKEULER I;H;K
[1]       VC←X[1],(¯1+ρI)ρ0
[2]       H←X[2]
[3]       K←1
[4]    LOOP: VC[K+1]←(I[K+1]+VC[K]×CT[3]÷H)÷GT[3]+CT[3]÷H
[5]       →LOOP×(ρI)>K←K+1
      ∇
```

For the circuit of Fig. 8.4, described by the vectors GT and CT, application of *BACKEULER* over 9 time steps, using a time step of 0.05 s and an initial voltage of zero, yields

```
     0 .05 BACKEULER 10ρ1
0 0.0476 0.093 0.136 0.177 0.216 0.254 0.289 0.323 0.355
```

The result of applying *BACKEULER* over 60 time steps allows the asymptotic behavior to be checked against expectation, and is shown in Fig. 8.7. Also shown in that figure is the variation, with time, of the difference between computed and actual responses:

$$ERROR \leftarrow (1 - \ast - (H \leftarrow .05) \times 0\lrcorner 160) - 0 .05 \; BACKEULER \; 61\rho1$$

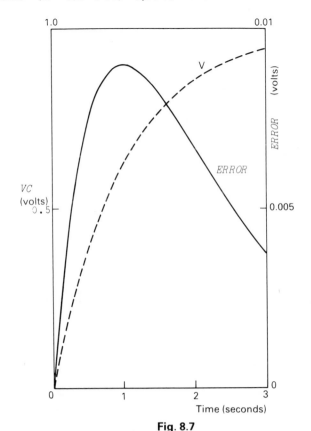

Fig. 8.7
The time-domain response of the circuit of Fig. 8.4, and the error involved in the approximation (for percentage error, see Table 8.1 later in chapter)

One would expect the error associated with the (discrete time) approximation to decrease monotonically with the time step H, though at the expense of increased computation for the same time period. As an experiment, the value of H was increased to three times its earlier value

$$H \leftarrow 0.15$$

and the difference between exact and approximate solutions computed for the same time period (0 to 3 s) as before:

$$ERROR \leftarrow (1 - * - H \times 0, \iota\, 20) - (0, H)\ BACKEULER\ 21\rho 1$$

The results shown in Fig. 8.8 confirm our expectation.

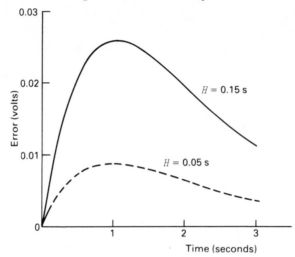

Fig. 8.8
**Showing the effect of time step (H) on the accuracy of the backward Euler
time-domain solution of the circuit of Fig. 8.4**

8.3   The Time-domain Response of a Circuit

For the simple circuit of Fig. 8.4, involving one response voltage and one current excitation, the essential calculation relevant to a time step H is the application of Ohm's law to the model of Fig. 8.6:

$$VC \leftarrow (I + ICS) \div GT[3] + GC[3] \tag{8.7}$$

Little that is conceptually new is involved in extending the above approach to handle more complex circuits containing linear conductances and capacitances. The generalization of Ohm's law, with which we are familiar from Chapter 1, leads to an expression similar to (8.7):

$$VC \leftarrow (I + ICS) \boxplus G + GC \tag{8.8}$$

where VC is now a *vector* of nodal voltages, I a vector of nodal excitation currents, and ICS an equivalent vector of nodal excitation currents arising from the capacitor companion models. G and GC are now reduced nodal conductance *matrices* associated with the conductances and (via the companion models) the capacitances within the circuit.

The development of a function capable of analyzing a circuit containing only linear conductances and capacitances will be illustrated by initial reference to the simple circuit of Fig. 8.9. The circuit is fully characterized by two three-column matrices associated with conductances and capacitances. The first two columns contain the end nodes of each component, and the third its value. For the example circuit of Fig. 8.9,

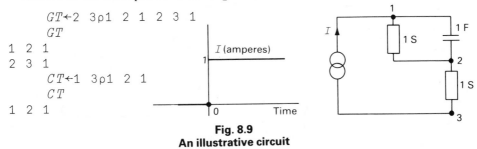

```
GT←2 3ρ1 2 1 2 3 1
GT
1 2 1
2 3 1
      CT←1 3ρ1 2 1
      CT
1 2 1
```

I (amperes)

Fig. 8.9
An illustrative circuit

The conductance component of the circuit is described by a complete nodal conductance matrix which can be generated by a function similar to *FORM* developed in Chapter 1. Here we use a modified version,[†] again called *FORM* (Fig. 8.10), for which the size of the matrix is specified before *FORM* is executed. Thus, having specified the number of circuit nodes

$N←3$

the complete nodal conductance matrix due to the *conductances* in the circuit can be generated

```
      G←FORM GT
      G
 1  ‾1   0
‾1   2  ‾1
 0  ‾1   1
```

and confirmed by inspection of the circuit. In the same way, once the time step is specified, e.g.

$H←0.05$

```
      ∇ Y←FORM GT; B; K; ΔY; NDS
[1]      B←1↑ρGT
[2]      Y←(N,N)ρ0
[3]      K←1
[4]  L:ΔY← 2 2 ρ 1 ‾1 ‾1 1 ×GT[K; 3]
[5]      NDS←GT[K; 1 2]
[6]      Y[NDS; NDS]←Y[NDS; NDS]+ΔY
[7]      →L×(K←K+1) ≤B
      ∇
```

Fig. 8.10
A function which constructs the nodal conductance matrix of a circuit

† For the new function *FORM* to operate correctly, the number of rows of its argument must be non-zero. In other words, with the function *TIMEDOMAIN* developed in this section, an absence of capacitors or conductances is not permitted. This limitation is corrected in Exercise 8.4.

the complete nodal conductance matrix GC associated with the capacitor companion models can be computed:

```
GC←FORM CT[;1 2],CT[;3]÷H
GC
 20          ¯20          0
¯20           20          0
  0            0          0
```

The sum of G and GC, but with the last row and column dropped,

```
GG←¯1 ¯1↓G+GC
GG
 21          ¯21
¯21           22
```

then provides the right argument of ⊞ in the generalized solution (8.8), where it is assumed that the highest-numbered node is selected as the reference node.

We now turn to the current excitation called $(I+ICS)$ in (8.8). For simplicity we have assumed an external excitation which is *constant* for $T>0$, although this restriction is easy to relax:

```
I←1 0 ¯1
```

The other contribution to the nodal current excitation comes from the current source associated with the capacitor's companion model. Its value depends on the capacitor voltage. To provide a numerical example, assume the capacitor's initial voltage to be zero, and select a nodal voltage vector V describing conditions just after $T=0$:

```
V←1 1
```

The expression for the vector of capacitor voltages is then

```
VBC←-/V[CT[;1 2]]
VBC
0
```

and the vector of capacitor companion model current sources is

```
IC←CT[;3]×VBC÷H
IC
0
```

These component currents must now be transformed to equivalent *nodal* currents. If that part (AC) of the incidence matrix corresponding to capacitor branches only were available, the expression

```
AC+.×IC
```

would yield the equivalent *nodal* excitation currents (recall $I \leftarrow A + . \times IB$ from Chapter 1). The alternative approach we adopt here is to employ the function $IFORM$ (Fig. 8.11) which is based on precisely the same principle as $FORM$:

$$IN \leftarrow IFORM \ CT[;1 \ 2],IC$$

```
        ∇ I←IFORM X;K;NN
[1]       I←Nρ0
[2]       K←1
[3]    L:NN←X[K; 1 2]
[4]       I[NN]←I[NN]+ 1 ‾1 ×X[K;3]
[5]      →L×ι(1↑ρX)≥K←K+1
        ∇
```

Fig. 8.11
A function which transforms component currents into equivalent nodal currents

The left argument of $\boxplus$ in the generalized equation (8.8) is then the sum of I and IN, though with the last element dropped to provide the required *reduced* nodal excitation vector:

$$II \leftarrow \ ‾1 \downarrow I + IN$$

Thus, the voltage V after the first time step would be, according to (8.8),

$$V \leftarrow II \boxplus GG$$
$$V$$
$$1.0476 \ 1$$

Based on the above discussion we can now define a function, $TIME$-$DOMAIN$, to undertake the time-domain analysis of linear circuits containing only conductances and capacitances. Its left argument VO is the initial reduced vector of nodal voltages: the right argument NT is the required number of time steps.

```
        ∇ V←VO TIMEDOMAIN NT
[1]       N←⌈/,GT[; 1 2],[1]CT[; 1 2]
[2]       V←((N-1),NT)ρ0
[3]       V[;1]←VO
[4]       G←FORM GT
[5]       GC←FORM CT[; 1 2],CT[;3]÷H
[6]       GG← ‾1 ‾1 ↓G+GC
[7]       K←1
[8]    L:VBC←-/(V[;K])[CT[; 1 2]]
[9]       IC←CT[;3]×VBC÷H
[10]      IN←IFORM CT[; 1 2],IC
[11]      II← ‾1↓I+IN
[12]      V[;K+1]←II⊞GG
[13]     →L×ιNT>K←K+1
        ∇
```

N is number of nodes	
Initialize voltage response matrix to zero	
Initial node voltages	
Conductance matrix due to conductances and capacitances	
K is index of time instant	
Calculation of capacitor voltages	
Capacitor companion model current sources	
Nodal currents due to companion models	
II is total nodal current excitation	
Voltage response at time instant $K+1$	
Increment time instant and repeat if appropriate	

Use of the function $TIMEDOMAIN$ to analyze the behavior of the circuit of Fig. 8.9 requires specification of the circuit

```
        GT
1   2   1
2   3   1
        CT
1   2   1
```

the excitation

```
I←1   0   ⁻1
```

and the time step

```
H←0.1
```

If the initial nodal voltage vector is assumed to be

```
V0←⁻5   1
```

and a trace is put on each line of the function

```
TΔTIMEDOMAIN←ι13
```

then application of the algorithm over 25 time steps yields

```
        V0   TIMEDOMAIN   25
TIMEDOMAIN[1]   3
TIMEDOMAIN[2]
   0 0 0 0 0 0 0 0 0 0 0 0 0 0 0 0 0 0 0 0 0 0 0 0 0
   0 0 0 0 0 0 0 0 0 0 0 0 0 0 0 0 0 0 0 0 0 0 0 0 0
TIMEDOMAIN[3]   ⁻5 1
TIMEDOMAIN[4]
   1   ⁻1   0
  ⁻1   2   ⁻1
   0  ⁻1   1
TIMEDOMAIN[5]
     10         ⁻10          0
    ⁻10          10          0
      0           0          0
TIMEDOMAIN[6]
     11         ⁻11
    ⁻11          12
TIMEDOMAIN[7] 1
TIMEDOMAIN[8]   ⁻6
TIMEDOMAIN[9]   ⁻60
TIMEDOMAIN[10]  ⁻60 60 0
TIMEDOMAIN[11]  ⁻59 60
TIMEDOMAIN[12]  ⁻4 36 1          Voltage at nodes 1 and 2 after the
TIMEDOMAIN[13] →8               first time step
TIMEDOMAIN[8]  ⁻5.36
              ⋮
```

```
        ⋮
TIMEDOMAIN[10]  2.18  ¯2.18  0
TIMEDOMAIN[11]  3.18  ¯2.18
TIMEDOMAIN[12]  1.29  1
TIMEDOMAIN[13]  →14
```

$$
\begin{array}{l}
\begin{array}{cccccc}
¯5 & ¯4.36 & ¯3.79 & ¯3.26 & ¯2.78 & ¯2.35 \\
 & ¯1.95 & ¯1.59 & ¯1.27 & ¯0.969 & ¯0.699 \\
 & ¯0.453 & ¯0.23 & ¯0.0277 & 0.157 & 0.324 \\
 & 0.477 & 0.615 & 0.741 & 0.855 & 0.959 \\
 & 1.05 & 1.14 & 1.22 & 1.29 & \\
1 & 1 & 1 & 1 & 1 & 1 \\
 & 1 & 1 & 1 & 1 & 1 \\
 & 1 & 1 & 1 & 1 & 1 \\
 & 1 & 1 & 1 & 1 & 1 \\
 & 1 & 1 & 1 & 1 &
\end{array}
\end{array}
\quad
\begin{array}{l}
\left.\vphantom{\begin{array}{c}a\\a\\a\\a\\a\end{array}}\right\} V[1] \\[3ex]
\left.\vphantom{\begin{array}{c}a\\a\\a\\a\\a\end{array}}\right\} V[2]
\end{array}
$$

The result of applying $TIMEDOMAIN$ to the same circuit over 100 time steps, each of 0.1 s duration, is shown in Fig. 8.12. The result corresponds with expectation.

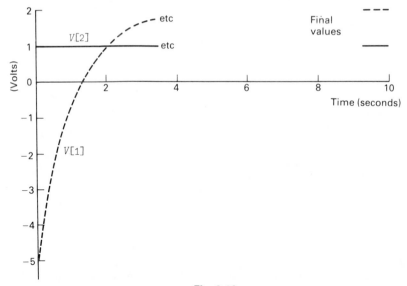

Fig. 8.12
Computed time-domain response of the circuit of Fig. 8.9

One important property of the solution as expressed by (8.8) is that the right argument of ⊟ does not change. The significance of this property is that, rather than execute dyadic ⊟ at each time step, it is possible to carry out an initial LU factorization followed, at each time step, by a forward and backward substitution appropriate to the new vector $I+ICS$ (see (8.8)): such a procedure is far less costly computationally. Only if the value of H were to change for some reason (see Section 8.6) would the LU factorization have to be repeated.

8.4   Nonlinear Components

Many of the practical circuits whose time-domain response is of interest contain nonlinear components such as diodes and transistors. We therefore now remove the earlier restriction to linear circuits although, for simplicity, we still assume that capacitors are the only reactive components in the circuit. As before, we shall develop the necessary procedure by reference to a simple illustrative example, the circuit shown in Fig. 8.13(a).

Even though a circuit may be nonlinear, we at first still follow the procedure developed above in which time discretization results in a totally resistive model of the circuit. For the illustrative circuit of Fig. 8.13(a), this resistive model is shown in Fig. 8.13(b). Unlike previous examples, this model contains a nonlinear conductance due to the presence of the diode. We know, however, how to solve a nonlinear resistive circuit such as that of Fig. 8.13(b): chapters 2 and 7 showed how the iterative Newton–Raphson algorithm can be applied to such a situation. What is necessary is that the current sources (here I and $(CT[3] \times VC[N] \div H)$ shall remain fixed, at the values appropriate to time $T[N]$, and the nonlinear circuit solved by the Newton–Raphson method to find a new value of the diode voltage VD. The final value of VD selected by the Newton–Raphson method is then the value $VC[N+1]$ of VC at the next time instant. Thus, within each time step, a Newton–Raphson solution is carried out.

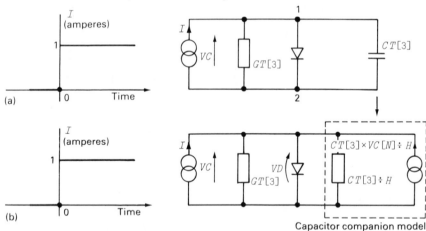

Fig. 8.13
A nonlinear circuit, and the nonlinear resistive model appropriate to a solution at $T[N+1]$

The outline discussion presented above allows the functions of Fig. 8.14 to be defined to allow calculation of the time-domain behavior of the circuit of Fig. 8.13(a). Most of the function $BACKEULERNL$ (the last two characters denoting nonlinearity) is familiar: but line [5] now introduces the iterative solution of the circuit of Fig. 8.13(b). The function $ITER$, whose left argument is the sum of the two (constant) current sources and whose right argument is the value of VC at $T[N]$, embodies the Newton–Raphson algorithm familiar from Chapter 2. The principal departure from the function $ITER$ of Chapter 2 is the consideration of the diode, the conductance $GT[3]$ and the capacitor companion model conductance as a single component. The remaining functions of Fig. 8.14 are familiar from Chapter 2 and its associated exercises.

Use of the functions of Fig. 8.14 requires a description of the circuit's components. The conductance and capacitor are described by

```
      GT
  1  2  1
      CT
  1  2  1
```

where the first and second elements of each vector denote the end-nodes, and the third element the component value. The diode is described by the function *COMT* of Fig. 8.14. The remaining parameter that must be specified is the 'threshold voltage' necessary to ensure satisfactory calculation of the diode current and its incremental conductance:

```
       ∇ VC←X BACKEULERNL I;H;K;ICS
  [1]    VC←X[1],(⁻1+ρI)ρ0
  [2]    H←X[2]
  [3]    K←1
  [4]   LOOP:ICS←VC[K]×CT[3]÷H
  [5]    VC[K+1]←(I[K+1]+ICS) ITER VC[K]
  [6]    →LOOP×(ρI)>K←K+1
       ∇


       ∇ VR←IRX ITER VR0;IR;GC;ERROR;SLOPE;ΔVR
  [1]    VR←VR0
  [2]   A:IR←(COMT VR)+(GT[3]×VR)+(GC←CT[3]÷H)×VR
  [3]   L1:→L2×1E⁻6<|ERROR←IRX-IR
  [4]   L2:SLOPE←(DIFCOM VR)+GT[3]+GC
  [5]    VR←VR+ΔVR←ERROR÷SLOPE
  [6]    →A
       ∇


       ∇ IR←COMT VR
  [1]    IR←(DIODE VR⌊VT)+(DIFDIODE VT)×0⌈(VR-VT)
       ∇


       ∇ IR←DIODE VR
  [1]    IR←1E⁻11×⁻1+*40×VR
       ∇


       ∇ S←DIFCOM VR
  [1]    →(VR≥0)/L
  [2]    S←(COMT VR)÷VR
  [3]    →0
  [4]   L:S←DIFDIODE VR⌊VT
       ∇


       ∇ S←DIFDIODE VR
  [1]    S←1E⁻11×40×*40×VR
       ∇
```

Fig. 8.14
Functions used to compute the time-domain behavior of the circuit of Fig. 8.13(a)

$VT \leftarrow 0.7$

A trace on line 5 of $BACKEULERNL$

$T\triangle BACKEULERNL \leftarrow 5$

allows successive values of VC to be examined, while a trace

$T\triangle ITER \leftarrow 6$

on line 6 of $ITER$ places the number of Newton–Raphson iterations clearly in evidence. Execution of the function $BACKEULERNL$ over 10 time steps of 0.1 s, and with an initial voltage of zero, yields

```
        0 .1 BACKEULERNL 11ρ1
ITER[6]  →2
BACKEULERNL[5]   0.0909
ITER[6]  →2
BACKEULERNL[5]   0.174
ITER[6]  →2
BACKEULERNL[5]   0.249
ITER[6]  →2
ITER[6]  →2
BACKEULERNL[5]   0.317
ITER[6]  →2
ITER[6]  →2
BACKEULERNL[5]   0.379
ITER[6]  →2
ITER[6]  →2
BACKEULERNL[5]   0.435
ITER[6]  →2
ITER[6]  →2
BACKEULERNL[5]   0.487
ITER[6]  →2
ITER[6]  →2
ITER[6]  →2
BACKEULERNL[5]   0.532
ITER[6]  →2
ITER[6]  →2
ITER[6]  →2
BACKEULERNL[5]   0.568
ITER[6]  →2
ITER[6]  →2
ITER[6]  →2
BACKEULERNL[5]   0.59
```

```
0 0.0909 0.174 0.249 0.317 0.379 0.435 0.487 0.532 0.568
    0.59
```

In most time intervals only one Newton–Raphson iteration is called for, though in some others three are required. The result of the same algorithm applied over

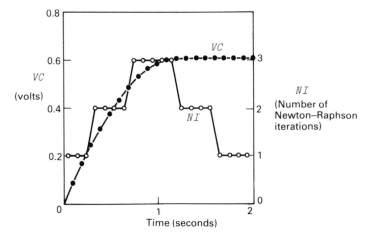

Fig. 8.15
Illustrating the variation in the number of Newton–Raphson iterations at each time step in a time-domain analysis

20 time steps is plotted in Fig. 8.15, which shows successive values of VC_i as well as the number of Newton–Raphson iterations required in each time interval. The final value of VC returned was 0.6096872 V. As a check, we first observe that it is less than the threshold voltage VT. Next, it is wise to compute the steady-state value of the sum of the current in the conductance and the diode current, which should approximate closely the value of the current excitation:

$$.6096872 + DIODE \quad .6096872$$
$$0.99994$$

8.5 Trapezoidal Integration

Trapezoidal integration is an alternative to the backward Euler method, and its name derives from the fact that an integral is approximated by the area of a trapezoid. Thus (see Fig. 8.16), the integral of capacitor current IC over a given time interval H, which is equal to the area under the current plot, is approximated by the area of the trapezoid defined by the values of IC at the extremities of the time interval H.

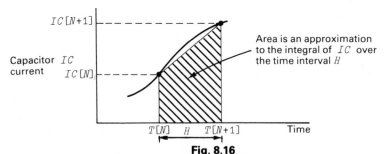

Fig. 8.16
Basis of the trapezoidal integration method of time-domain analysis

As with the backward Euler method, trapezoidal integration can be applied by reference to a companion model of the circuit. To derive the companion model of a capacitor, we observe (Fig. 8.16) that the change in the charge on the capacitor between time instants $T[N]$ and $T[N+1]$ is $C \times VC[N+1] - V[N]$. This can be equated, approximately, to the area of the trapezoid defined by the currents $IC[N]$ and $IC[N+1]$ and the time interval H:

$$(C \times VC[N+1] - VC[N]) = (0.5 \times IC[N] + IC[N+1]) \times H \qquad (8.9)$$

A rearrangement of (8.9)

$$IC[N+1] = -IC[N] + (2 \times C \div H) \times VC[N+1] - VC[N] \qquad (8.10)$$

allows the relation between the capacitor current and voltage at time $T[N+1]$ to be represented by the companion model of Fig. 8.17. Use of the model requires knowledge of both the capacitor current and its voltage at the previous time step. This requirement poses no special problems except at the time origin $T[1]$. We are familiar with the need to specify the initial capacitor voltage $VC[1]$. To determine the initial capacitor current $IC[1]$, the initial quiescent condition of the circuit must be obtained: this will be clarified by the illustration of trapezoidal integration which now follows.

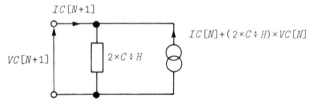

Fig. 8.17
The companion model of a capacitor relevant to trapezoidal integration

To illustrate trapezoidal integration we again employ the circuit of Fig. 8.4 (repeated as Fig. 8.18(a)) to allow comparison with the backward Euler method. As before, we assume the initial capacitor voltage to be zero:

$$VC \leftarrow 0$$

and the time step to be 0.05 s.

$$H \leftarrow 0.05$$

To find the initial capacitor current $IC[1]$ we consider the circuit of Fig. 8.18(b) which is relevant to the initial conditions. Since the initial voltage across the capacitor is zero, it is replaced by a voltage source of this value; that is, by a short circuit. By examination we see that $IC[1] = 1$A:

$$IC \leftarrow 1$$

As before, the circuit is described by the two arrays

$GT \leftarrow 1 \quad 2 \quad 1$
$CT \leftarrow 1 \quad 2 \quad 1$

The circuit model appropriate to the solution of the problem of Fig. 8.18(a) by trapezoidal integration is shown in Fig. 8.19, where

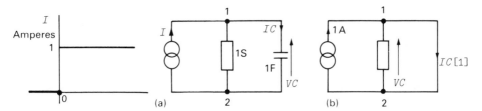

Fig. 8.18
(a) The time-domain response to be computed; (b) circuit relevant to the calculation of the initial capacitor current

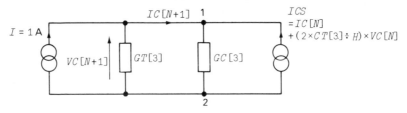

Fig. 8.19
Circuit model appropriate to the analysis of the circuit of Fig. 8.18(a) by trapezoidal integration

$GC \leftarrow CT \div 1 \quad 1, H \div 2$
GC
$1 \quad 2 \quad 40$

The total conductance in the circuit of Fig. 8.19 is

$GG \leftarrow GT[3] + GC[3]$
GG
41

and the total current injected into the conductance GG is the sum of I and the source current ICS associated with the capacitor's companion model:

$ICS \leftarrow IC + (2 \times CT[3] \div H) \times VC$
$II \leftarrow 1 + ICS$
II
2

Thus, the value of VC at the completion of the first time step ($VC[2]$) is

$VC \leftarrow II \div GG$
VC
0.04878

For the same time instant the backward Euler method predicted 0.047619 V, and the exact value $(1-*-H)$ is 0.048771 V. Thus, at least for the first time step for the particular circuit being studied, far better accuracy is obtained by using trapezoidal integration.

In preparation for the next time step (to $T[3]$) we must record, in addition to the capacitor voltage ($VC[2]=0.04879$), the capacitor current $IC[2]$, where the $[2]$ denotes the second instant of time, *not* node 2. Examination of Fig. 8.19 shows that it is equal to $I-VC[2]\times GT[3]$:

```
     I←1
     IC←I-VC×GT[3]
     IC
0.95122
```

The next value of $VC\,(=VC[3])$ can now be found by a new analysis of the circuit of Fig. 8.19:

```
     ICS←IC+(2×CT[3]÷H)×VC
     II←1+ICS
     VC←II÷GG
     VC
0.09518
```

This compares favorably with the true value of 0.095163 V $(=1-*-2\times H)$.

To extend the calculation of the time-domain response of the circuit of Fig. 8.18(a) to any number of time steps, and to allow choice of the time step H and the initial conditions, we define the function $TRAP$. Its left argument X is a three-element vector comprising, in order, the initial capacitor voltage, the initial capacitor current (as obtained separately from an initial quiescent analysis) and the time step H. The right argument I is a vector of the values of the excitation current (I in Fig. 8.18(a)) at successive instants separated by H. For simplicity, the current source in the capacitor companion model is again denoted by ICS.

```
      ∇ VC←X TRAP I;H;ICS;K;IC              Note
[1]      VC←X[1],(⁻1+ρI)ρ0                  X[1] is initial VC
[2]      H←X[3]                             X[2] is initial IC
[3]      ICS←X[2]+(2×CT[3]÷H)×VC[1]         X[3] is time step H
[4]      K←2
[5] LOOP: VC[K]←(I[K]+ICS)÷GT[3]+CT[3]÷H÷2
[6]      IC←I[K]-VC[K]×GT[3]
[7]      ICS←IC+(2×CT[3]÷H)×VC[K]
[8]      →LOOP×(ρI)≥K←K+1
      ∇
```

Thus, for the circuit of Fig. 8.18(a), with an initial voltage VC of zero and a (precalculated) initial capacitor current of 1 A, the voltage response over 24 time steps, each of 0.05 s duration, is obtained by executing the expression

0 1 0.05 *TRAP* 25ρ1

Table 8.1 shows, for comparison, the exact solution for the circuit of Fig.
8.18(a) together with the solutions obtained by the backward Euler and
trapezoidal methods. For convenience, the percentage error associated with the
two methods is also shown. By far the more accurate is the trapezoidal method.
If exploration is carried out (see Exercises 7.3, 7.7) it will be found that, for
comparable accuracy, the trapezoidal method may take larger time steps than
the backward Euler method.

The extension of the trapezoidal method to the time-domain analysis of a
general circuit is straightforward, and follows the same principles enunciated in
the context of the backward Euler method in Sections 8.3 and 8.4. Exercises 7.8
and 7.9 are relevant to such an extension.

8.6 The Choice of Time Step

We have seen that the time step H has a strong influence on the accuracy of
time-domain analysis. If the natural frequencies of a circuit are high, then it is
likely that H will have to be low in order to achieve acceptable accuracy. However, a decrease in H carries with it the penalty of increased computational
effort for a time-domain solution spanning a given time period. Such considerations are taken into account in typical packages for time-domain analysis, in a
manner that can easily be outlined.

First, an initial time step $H1$ is chosen, and the response after one time
step computed. Then, a new time step $H2$ is chosen which is half the previous
one ($H1 \div 2$) and the response at time $H1$ again computed using, necessarily,
two time steps. If the two responses at time $H1$ agree within (say) 5%, the time
step $H1$ is adequate. If not, a further halving of the time step is involved until
reasonable agreement is obtained. A similar technique can be employed to
increase H, and thereby save computational effort. For example, if $H1$ as
described above is found to be adequate then, at the next iteration, a time step
of $2 \times H1$ can be tried.

8.7 Comment

This chapter has introduced some of the key concepts associated with time-
domain analysis. The reader should realize, however, that other topics have
been omitted merely to keep the chapter to a manageable size. To give just one
example we would mention that, with some time-domain analysis techniques,
the potential for numerical instability exists. Such instability is evidenced by a
solution that, at successive time steps, departs more and more from the true
solution. This topic, and other aspects of time-domain analysis, is treated in a
number of references, such as Calahan (1972) and Chua and Lin (1975).

Exact solution	Backward Euler solution	Trapezoidal solution	Backward Euler percentage error	Trapezoidal percentage error
VC	VCB	VCT	$100 \times (VC - VCB) \div VC$	$100 \times (VC - VCT) \div VC$
0	0	0	0	0
0.0488	0.0476	0.0488	2.36	0.0203
0.0952	0.093	0.0952	2.3	0.0198
0.139	0.136	0.139	2.25	0.0193
0.181	0.177	0.181	2.19	0.0188
0.221	0.216	0.221	2.14	0.0183
0.259	0.254	0.259	2.08	0.0179
0.295	0.289	0.295	2.03	0.0174
0.33	0.323	0.33	1.98	0.0169
0.362	0.355	0.362	1.93	0.0165
0.393	0.386	0.394	1.88	0.0161
0.423	0.415	0.423	1.83	0.0156
0.451	0.443	0.451	1.78	0.0152
0.478	0.47	0.478	1.73	0.0148
0.503	0.495	0.503	1.69	0.0144
0.528	0.519	0.528	1.64	0.014
0.551	0.542	0.551	1.59	0.0136
0.573	0.564	0.573	1.55	0.0132
0.593	0.584	0.594	1.51	0.0128
0.613	0.604	0.613	1.47	0.0125
0.632	0.623	0.632	1.43	0.0121
0.65	0.641	0.65	1.39	0.0118
0.667	0.658	0.667	1.35	0.0114
0.683	0.674	0.683	1.31	0.0111
0.699	0.69	0.699	1.27	0.0108

Table 8.1
Comparison of the backward Euler and trapezoidal solutions with the exact solution of the circuit of Fig. 8.4

Exercises

8.1   Derive the companion model of an inductor appropriate to the backward Euler algorithm.

8.2   Modify the function $TIMEDOMAIN$ to use LU factorization instead of dyadic domino (⊞), and test the result.

8.3   By means of the $BACKEULER$ function, repeat the time-domain analysis of the circuit of Fig. 8.4 for a number of values of the time-step H, and determine the maximum difference between the solution obtained and the true solution over a period of 3.2 s. Plot these maximum differences versus H. You may wish to compare the outcome with that of Exercise 8.7.

8.4   Define a set of functions that will permit the time-domain analysis, by the backward Euler method, of a circuit comprising linear resistance, capacitance and inductance, and excited by a current source. Test the set of functions on a number of circuits, including those shown in Fig. E8.1.

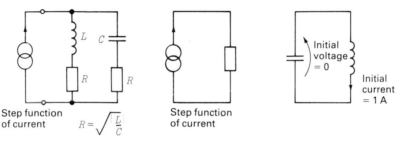

Step function   Step function
of current   of current   Initial current = 1 A

$$R = \sqrt{\frac{L}{C}}$$

Fig. E8.1
Test examples

8.5   By reference to the backward Euler method, define a set of functions that will permit the time-domain analysis of a circuit comprising linear resistance, capacitance, inductance and exponential diodes, and excited by a current source. Test the functions.

8.6   Derive the companion model of an inductor appropriate to the trapezoidal integration algorithm.

8.7   Repeat Exercise 8.3 using the trapezoidal integration algorithm, and compare the outcome with that of Exercise 8.3.

8.8   Define a set of functions that will permit the time-domain analysis, by trapezoidal integration, of circuits comprising linear resistance, capacitance and inductance, and excited by a current source. Test the functions on a number of circuits, including those of Fig. E8.1.

8.9   Modify the functions developed in Exercise 8.8 to handle, in addition, exponential diodes and voltage-controlled current sources.

Solutions

8.1

$$VL = L \times \mathrm{d}\, IL/\mathrm{d}\, T$$
$$VL[N+1] = L \times (IL[N+1] - I[N]) \div H$$
$$IL[N+1] = (VL[N+1] \times H \div L) + I[N]$$

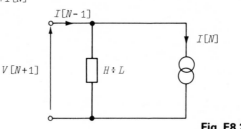

Fig. E8.2

8.2

```
     ∇ V←V0 TIMEDOMAIN NT
[1]    N←⌈/,GT[; 1 2],[1]CT[; 1 2]
[2]    V←((N-1),NT)ρ0
[3]    V[;1]←V0
[4]    G←FORM GT
[5]    GC←FORM CT[; 1 2],CT[;3]÷ H
[6]    GG← ‾1 ‾1 ↓G+GC
[7]    LUFACTOR GG                    LU factorization
[8]    K←1
[9]  LOOP:VBC←-/(V[;K])[CT[; 1 2]]
[10]   IC←CT[;3]×VBC÷H
[11]   IN←IFORM CT[; 1 2],IC
[12]   II←‾1↓I+IN
[13]   V[;K+1]←II LUSOLV GG           LUSOLV replaces ⊟
[14]   →LOOP×ιNT>K←K+1
     ∇


     ∇ LUFACTOR GG;LU
[1]    LU←LUFAC GG
[2]    L←LO LU
[3]    U←UP LU
     ∇


     ∇ R←LUFAC M;C;S
[1]    C←1
[2]    R←M
[3]  L1:S←C↓ι1↑ρR
[4]    R[C;S]←R[C;S]÷R[C;C]
[5]    R[S;S]←R[S;S]-R[S;C]∘.×R[C;S]
[6]    →L1×(1↑ρR)≥C←C+1
     ∇
```

As presented in Section 6.2

```
     ∇ R←LO X;N
[1]    R←((ιN)∘.≥ιN←1↑ρX)×X
     ∇
```

see Exercise 6.1

```
     ∇ R←UP X;N
[1]    R←((ιN)∘.=ιN)+(((ιN)∘.<ιN←1↑ρX)×X
     ∇
```

```
     ∇ VR←II LUSOLV GG
[1]    VR←U LUBKSUB L FSUB II
     ∇
```

```
     ∇ R←Y LUBKSUB I;S
[1]    R←(S←ρI)ρ0
[2] L1:R[S]←(÷Y[S;S])×I[S]-Y[S;]+.×R
[3]    →(0<S←S-1)/L1
     ∇
```
$\left.\right\}$ see Exercise 6.2

```
     ∇ R←Y FSUB I
[1]    R←Φ(ΦΦ[1]Y)LUBKSUB ΦI
     ∇
```

The new function *TIMEDOMAIN* given above was applied to the same problem (Fig. 8.9) as the *TIMEDOMAIN* function in Section 8.3, with identical results.

8.3

Function *TRUE* gives exact response of circuit of Fig. 8.4.

```
     ∇ R←TRUE H
[1]    R←1-*-(H×0,ιNT)
     ∇
```

First, the step function of current is represented by 65 values of 1 A each separated by a time step of 0.05 s.

```
I←65ρ1
NT←64
H←.05
```

We generate the maximum error between the true response and that generated by *BACKEULER* (compare Fig. 8.7).

```
M05←⌈/|(TRUE H)-((0,H) BACKEULER I)
M05
0.00901
```

The calculation is repeated for time steps of 0·1, 0·2 and 0·4 s

```
M1
0.0177
M2
0.034
M4
0.0632
```

The plot (see Solution 8.7) indicates an approximately linear relation between maximum error and time step.

8.4 The circuit description employed is a simple extension of that used in Section 8.3. Thus, arrays *GT* and *CT* are familiar. Array *ILT* (Inverse *L* Table) stores the terminal nodes and the reciprocal inductance of each inductance in the circuit. If the circuit has N nodes and the solution involves $T-1$ time steps, the nodal current excitation is an array of shape N,T; so is the array of nodal voltage response. The vector $V0$ denotes the initial nodal voltages, and $\bar{H}$ is the time step. $\underline{IL}$ is the vector of initial inductor currents.

     The functions developed are displayed below, and concise annotation provided.

```
     ∇ V←V0 TIMEΔRLCΔBE I;N;GG;K;ICX;II;IL
[1]    INITIALIZE
[2]    GG←NODCON
[3]    K←1
[4] LOOP:ICX←COMPANCURRENTS
[5]    II←¯1↓I[;K]+ICX
[6]    V[;K+1]←(II⊞GG),0
[7]    IL←STOREIL V
[8]    →LOOP×(¯1↑ρI)>K←K+1
     ∇
```
$\left.\right\}$ The main function

```
      ∇ INITIALIZE
[1]      N←⌈/,(GT,[1] CT,[1] ILT)[; 1 2]     N is the number of nodes
[2]      V←(ρI)ρ0                             Result V is initialized to zero
[3]      V[; 1]←V0                            The initial nodal voltages are recorded
[4]      IL←IL                               The inductance current vector is initialized.
      ∇
```

```
      ∇ GG←NODCON                            Generation of the reduced nodal conductance
[1]      G←FORM GT                           matrix due to the companion model conductances
[2]      GC←FORM CT[; 1 2],CT[; 3]÷H         as well as the actual conductances.
[3]      GL←FORM ILT[; 1 2],ILT[; 3]×H
[4]      GG← ¯1 ¯1 ↓G+GC+GL
      ∇
```

```
      ∇ Y←FORM GT; B; K; ΔY; NDS
[1]      B←1↑ρGT                             See Section 8.3. The original function is modified so
[2]      Y←(N,N)ρ0                            that, on line [3], execution is terminated if the
[3]      →(B=0)/0                             argument GT is empty.
[4]      K←1
[5]   L:ΔY← 2 2 ρ 1 ¯1 ¯1 1 ×GT[K; 3]
[6]      NDS←GT[K; 1 2]
[7]      Y[NDS; NDS]←Y[NDS; NDS]+ΔY
[8]      →L×(K←K+1) ≤B
      ∇
```

```
      ∇ ICX←COMPANCURRENTS                   Generates the nodal current excitation equivalent
[1]      ICC←CAPCURRENT                      to the companion model source currents associated
[2]      ICL←INDCURRENT                      with capacitors and inductors.
[3]      ICX←ICC+ICL
      ∇
```

```
      ∇ ICC←CAPCURRENT; VBC                  Generates the capacitor companion model source
[1]      ICC←Nρ0                             currents (e.g., ICS in Fig. 8.5).
[2]      →(~((1↑ρCT) >0))/0                  Line [2] terminates execution if no capacitors
[3]      VBC←-/(V[; K])[CT[; 1 2]]           are present.
[4]      IC←CT[; 3]×VBC÷H
[5]      ICC←IFORM CT[; 1 2],IC
      ∇
```

```
      ∇ ICL←INDCURRENT                       Generates the inductor companion model source
[1]      ICL←Nρ0                             currents. Line [2] terminates execution if no
[2]      →(~((1↑ρILT) >0))/0                 inductors are present. On line [3], minus appears
[3]      ICL←IFORM ILT[; 1 2],- IL           before IL because (see Exercise 6) IL flows in
      ∇                                      the opposite direction to the capacitor companion
                                             model source current for which IFORM was defined.
```

```
      ∇ I←IFORM X; K; NN        ⎫
[1]      I←Nρ0                  ⎪
[2]      K←1                    ⎪
[3]   L: NN←X[K; 1 2]          ⎬  See Section 8.3
[4]      I[NN]←I[NN]+ 1 ¯1 ×X[K; 3]  ⎪
[5]      →L×ι (1↑ρX)≥K←K+1      ⎪
      ∇                        ⎭
```

```
      ∇ R←STOREIL V; VL                      The new inductor voltages, and hence the new
[1]      VL←-/(V[; K+1])[ILT[; 1 2]]         inductor currents (see Exercise 6) are calculated
[2]      R←IL+VL×H×ILT[; 3]                  and stored.
      ∇
```

The first circuit selected as a test example is an interesting one. If the values of R, L and C are chosen such that $R'= \sqrt{L/C}$ then the circuit is indistinguishable from a resistor of value R to the excitation.

```
        GT
2     4    0.001
3     4    0.001
        CT
1E0  3E0   1E‾9
        ILT
1E0  2E01 E3
        II
  0  0  1  1  1  1  1  1  1  1  1  1  1  1  1  1  1  1  1  1  1  1
  0  0  0  0  0  0  0  0  0  0  0  0  0  0  0  0  0  0  0  0  0  0
  0  0  0  0  0  0  0  0  0  0  0  0  0  0  0  0  0  0  0  0  0  0
  0  0 ‾1 ‾1 ‾1 ‾1 ‾1 ‾1 ‾1 ‾1 ‾1 ‾1 ‾1 ‾1 ‾1 ‾1 ‾1 ‾1 ‾1 ‾1 ‾1 ‾1
        VO
  0  0  0  0
        IL
  0
```

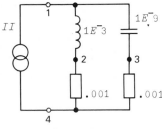

Fig. E8.3

```
      H
0.0000001
```

H is chosen to be small compared with the time constants associated with the circuit.

```
        VM←VO TIMEΔRLCΔBE II
        ⍉VM
0.00E0  0.00E0  0.00E0  0.00E0
0.00E0  0.00E0  0.00E0  0.00E0
0.00E0  0.00E0  0.00E0  0.00E0
1.00E3  9.09E1  9.09E2  0.00E0
1.00E3  1.74E2  8.26E2  0.00E0
1.00E3  2.49E2  7.51E2  0.00E0
1.00E3  3.17E2  6.83E2  0.00E0
1.00E3  3.79E2  6.21E2  0.00E0
1.00E3  4.36E2  5.64E2  0.00E0
1.00E3  4.87E2  5.13E2  0.00E0
1.00E3  5.33E2  4.67E2  0.00E0
1.00E3  5.76E2  4.24E2  0.00E0
1.00E3  6.14E2  3.86E2  0.00E0
1.00E3  6.50E2  3.50E2  0.00E0
1.00E3  6.81E2  3.19E2  0.00E0
1.00E3  7.10E2  2.90E2  0.00E0
1.00E3  7.37E2  2.63E2  0.00E0
1.00E3  7.61E2  2.39E2  0.00E0
1.00E3  7.82E2  2.18E2  0.00E0
1.00E3  8.02E2  1.98E2  0.00E0
1.00E3  8.20E2  1.80E2  0.00E0
1.00E3  8.36E2  1.64E2  0.00E0
```

Columns 1–4 contain nodal voltages 1–4. Time proceeds down the page. The voltage at node 1 is the same as that which would occur if the current excitation were applied to a 1000 ohm resistor. The node 2 voltage increases exponentially towards 1000 V, as expected with a constant 1000 V across a series LR circuit. Similarly, the node 3 voltage decreases approximately exponentially towards zero from 1000 V, as expected.

The second example checks that the set of functions can handle the absence of capacitors and inductors.

```
      GT←1 3ρ1 2 2
      CT←0 3ρ0
      ILT←0 3ρ0
      IG
```

0	0	0	0	0	2.7	2.7	2.7	2.7	2.7
0	0	0	0	0	‾2.7	‾2.7	‾2.7	‾2.7	‾2.7

```
      IL←0
      H←.1
      VO←0 0

      VG←VO TIMEΔRLC ΔBE IG
      ⍊VG
```

0	0
0	0
0	0
0	0
0	0
0	0
1.35	0
1.35	0
1.35	0
1.35	0

The node voltage is as expected

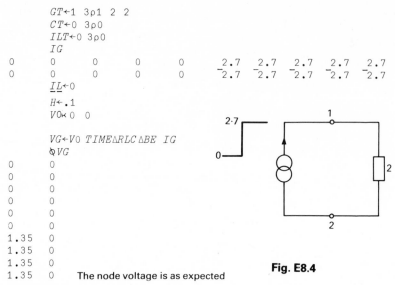

Fig. E8.4

The third example considers the case of a nonzero initial inductance current and a complete absence of current excitation.

```
      H←.1
      I←2 30ρ0
      GT←0 3ρ0
      CT←1 3ρ1 2 1
      ILT←1 3ρ1 2 1
      IL←1
      VO←0 0

      VLC1←VO TIMEΔRLC ΔBE I
      ⍊VLC1
```

A simple check can be made on the first few time steps. We know that, for a capacitor,

$$I = C \, dV/dt.$$

We know the initial value of I, since this is the initial inductor current ($‾1$ A).
Since $C = 1$ F, $dV/dt = ‾1$ volts per second.
This agrees well over the first 2 or 3 time steps.

0	0
‾0.09901	0
‾0.19606	0
‾0.29021	0
‾0.38055	0
‾0.46623	0
‾0.54644	0
‾0.62045	0
‾0.68759	0
‾0.74725	0
‾0.79892	0
‾0.84217	0
‾0.87666	0
‾0.90212	0
‾0.9184	0
‾0.92543	0
‾0.92322	0
‾0.91189	0
‾0.89165	0
‾0.86278	0
‾0.82565	0
‾0.78072	0
‾0.7285	0
‾0.66959	0
‾0.60462	0

‾0.53432	0
‾0.45942	0
‾0.38072	0
‾0.29902	0
‾0.21517	0

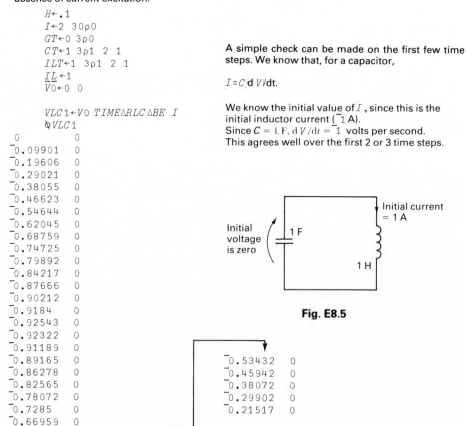

Fig. E8.5

8.6   Increase in inductor current from $T[N]$ to $T[N+1]$ is equal to the integer of VL over H

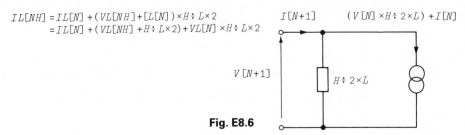

$$IL[NH] = IL[N] + (VL[NH] + [L[N]]) \times H \div L \times 2$$
$$= IL[N] + (VL[NH] + H \div L \times 2) + VL[N] \times H \div L \times 2$$

$I[N+1]$ $(V[N] \times H \div 2 \times L) + I[N]$

$V[N+1]$ $H \div 2 \times L$

Fig. E8.6

8.7   One proceeds in precisely the same way as for Exercise 3. Thus, for a time step of 0.05s and for 64 steps, the maximum error between the true solution and that provided by trapezoidal integration ($M05$) is found as follows:

```
H←.05
NT←64
I←65ρ1

M05←⌈/|(TRUE H)-((0,1,H) TRAP I)
M05
0.0000767
```

Similarly, for time steps of 0·1, 0·2 and 0·4

```
     M1
0.000307
     M2
0.00123
     M4
0.0049
```

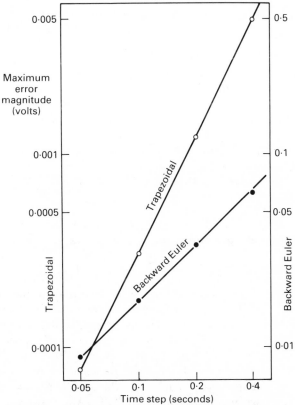

Fig. E8.7
Indicates, for trapezoidal integration, a
quadratic **dependence of maximum error**
on H**, compared with the** *linear* **dependence**
associated with the backward Euler algorithm

CHAPTER

9

Sensitivity and Circuit Modification

The circuit designer is continually concerned with component *change*. For example, a toleranced component can change its value from one sample of a mass-produced circuit to the next, thereby causing a change in the behavior. Or a component value may be changed by the designer either to achieve improved circuit behavior or in the process of exploration to gain insight ('What happens if . . . ?'). Gross changes in a circuit occur when the circuit is modified by the connection of new components in an effort to improve the design or when a fault occurs. Thus, we see the need to be able to do two things: (a) efficiently compute the effect, on circuit behavior, of small or large changes in one or more components within the circuit, and (b) efficiently obtain the description of a circuit modified by the addition of a single component.

In this chapter the topics of sensitivity and circuit modification are considered only in the context of purely resistive circuits, though the algorithms derived can be applied, without limitation, to the frequency-domain performance of circuits containing both reactive and resistive components. The restriction to resistive circuits is adopted merely for simplicity of exposition, and because extension to circuits containing both reactance and resistance involves nothing that is conceptually new.

9.1 Single-component Large-change Sensitivity

Consider a linear non-reciprocal two-port (Fig. 9.1) containing, among other components, a single two-terminal conductance GJ which is subject to large variations. Consider the change ΔGJ in GJ to be represented (Fig. 9.2) by the connection of a separate conductance ΔGJ in parallel with GJ via a newly created port J: in this way the 'boxed' 3-port contains *fixed* components. If the current through ΔGJ is $I[J]$ then, according to the substitution theorem (Desoer and Kuh, 1969) the conductance ΔGJ can be replaced by an independent current source of value $I[J]$ without disturbing any currents or voltages within the circuit (Fig. 9.3). Application of the superposition theorem now allows us to conclude that if $I[1]$ acting alone creates an output voltage $V[2]$, and if the further application of $I[J]$ increases the output voltage by $\Delta V[2]$, then the source $I[J]$ acting alone (Fig. 9.4) will result in a voltage $\Delta V[2]$ at port 2. Thus, from Fig. 9.4,

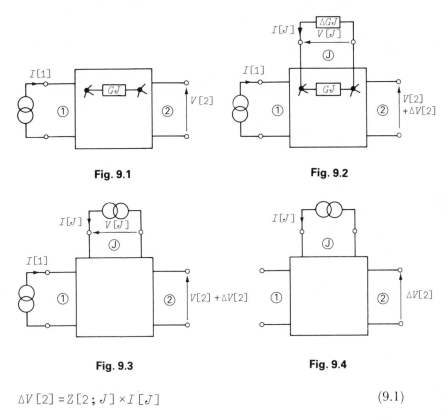

Fig. 9.1   **Fig. 9.2**

Fig. 9.3   **Fig. 9.4**

$$\Delta V[2] = Z[2;J] \times I[J] \tag{9.1}$$

where Z is the port resistance matrix of shape 3,3. Since the source $I[J]$ simulates the conductance ΔGJ we also know that (Fig. 9.2)

$$I[J] = - \Delta GJ \times V[J] \tag{9.2}$$

Furthermore we observe from Fig. 9.3 that

$$V[J] = (Z[J;1] \times I[1]) + Z[J;J] \times I[J] \tag{9.3}$$

The elimination of $I[J]$ and $V[J]$ from the three equations yields

$$\Delta V[2] = - Z[2;J] \times Z[J;1] \times I[1] \div Z[J;J] + \div \Delta GJ \tag{9.4}$$

Thus, knowledge of *three* port resistances is sufficient to allow the *exact* calculation of the effect, on a single response voltage, of a number of values of the conductance change ΔGJ.

Example

For the circuit of Exercise 3.4 of Chapter 3, repeated here as Figure 9.5, the reduced nodal resistance matrix ZR was found to be

```
0.22       0.16       0.066
0.16       0.20       0.079
¯0.016      ¯0.02      0.092
```

The reduced incidence matrix AR is derived from the incidence matrix A by deleting the row corresponding to the reference node

```
        AR← ¯1 0↓ A
        AR
 1  1  0  0  0
¯1  0  1  0  0
 0  0  0  1  1
```

The port resistance matrix ZP (in which a port is associated with each of the five branches) is therefore (see discussion of Section 3.7)

```
ZP←(⍉AR) +.×ZR +.×AR
```

Suppose the conductance whose nominal value is 10 S is subjected to a large change ΔG, and the effect on the voltage of port (branch) 5 is to be found. Then, with reference to equation (9.4), the subscripts are J; 1, 1; 2, 2; 5. If the changes ΔG of interest are

```
ΔG← ¯6 ¯2 2 6
```

we deduce, from (9.4), that

```
        ΔV5←- ( ZP[5 ; 1] ×ZP[1 ; 2] ×5) ÷ ZP[1 ; 1] +÷ ΔG
        ΔV5
0.0118    0.0022  ¯0.0016  ¯0.0036
```

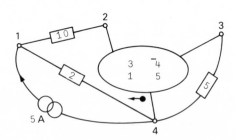

Fig. 9.5

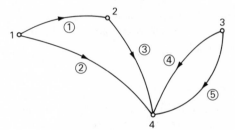

Fig. 9.6

9.2 **Differential Sensitivity**

As a result of infinitesimal changes in the values of one or more components, the currents and voltages in a circuit undergo infinitesimal change. How are these current and voltage changes (ΔIB, ΔVB) related to the component changes?

  First, it is clear that, because the currents and voltages in the circuit obey Kirchhoff's laws both before and after the component changes, then the *changes* in current and voltage *also* obey the same laws. Thus, they can be considered to be associated with a new circuit having the same topology as the original circuit. If this is the case, what are the components of this new circuit?

  The answer is obtained by next considering the individual components of the original circuit. For a two-terminal conductance of *fixed* value ($IB \leftarrow G \times VB$: Fig. 9.7(a)), changes in current and voltage are related by

$$\Delta IB \leftarrow G \times \Delta VB \tag{9.5}$$

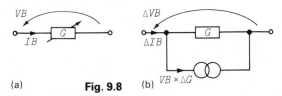

Fig. 9.7

In other words, ΔIB and ΔVB can be considered as the current and voltage associated with a conductance of value G (Fig. 9.7(b)): this conductance models the relation between infinitesimal changes in current and voltage.

  However, if the conductance G of a two-terminal component is variable (Fig. 9.8(a)) then

$$\Delta IB \leftarrow (G \times \Delta VB) + VB \times \Delta G \tag{9.6}$$

where VB is the voltage associated with the conductance and ΔG is the infinitesimal change in conductance. This relation is modelled (Fig. 9.8(b)) by the parallel connection of a fixed conductance G and a current source $VB \times \Delta G$. In other words, as far as infinitesimal changes in current and voltage are concerned, the variable conductance looks like the parallel connection of the original conductance and a suitable current source.

(a)   **Fig. 9.8** (b) $VB \times \Delta G$

  From the above, we can deduce that the infinitesimal changes in current and voltage *are the actual currents and voltages associated with a new circuit* having the same topology as the original circuit: we call this new circuit the *sensitivity model* of the original circuit. For a circuit composed only of fixed and variable two-terminal conductances, the sensitivity model is formed by replacing them, respectively, with the models of Figs. 9.7(b) and 9.8(b). However, before proceeding to an illustrative example, we must establish the sensitivity models of independent current and voltage sources.

Since, for a constant current source (Fig. 9.9(a)) the change in current is constrained to be zero, the corresponding sensitivity model (Fig. 9.9(b)) is an open circuit. Similarly, the sensitivity model of a fixed voltage source is a short circuit (Fig. 9.10).

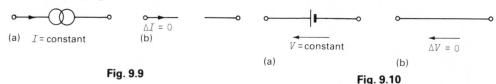

(a) I = constant   (b) $\Delta I = 0$ V = constant $\Delta V = 0$

Fig. 9.9

Fig. 9.10

Example

From the above we can now deduce that if interest lies in the infinitesimal changes in current and voltage arising from a change ΔGA in the conductance GA in the circuit of Fig. 9.11, then the appropriate sensitivity model is as shown in Fig. 9.12. In other words, we would first analyze the circuit of Fig. 9.11 to find the voltage VA associated with the conductance GA. Knowledge of VA completes the description of the sensitivity model of Fig. 9.12, which would then be analyzed. The response ΔV is then the change in the voltage V in Fig. 9.11 due to an infinitesimal change ΔGA in GA.

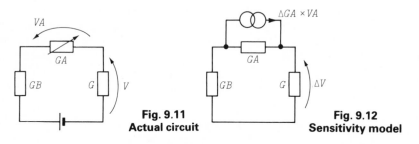

Fig. 9.11
Actual circuit

Fig. 9.12
Sensitivity model

Normally, it is the *sensitivity* $\Delta V \div \Delta GA$ that is of interest, in which case we can (Fig. 9.13) apply a current source of 1 A in the sensitivity model, so that the new response voltage ($\Delta V \div \Delta GA \times VA$) need only be multiplied by VA (from Fig. 9.11) to obtain the sensitivity. More precisely, since changes are infinitesimal, $\Delta V \div \Delta GA$ is known as the *differential sensitivity*.

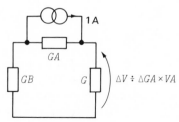

Fig. 9.13
Sensitivity model: new excitation

The above method of calculating differential sensitivity offers a distinct advantage over the more obvious and straightforward 'perturbation' method in which, after the first analysis, the circuit would again be analyzed after changing GA by a very small amount. In the perturbation approach, the change in GA must be kept reasonably small, but not so small as to give problems of accuracy; by contrast, the method we describe is a 'direct' method which does not depend on the difference between two almost identical values of circuit performance.

9.3   Single-response Multiple-parameter Sensitivity

The algorithm developed above can be used to compute the effect of an infinitesimal change in one component on many (or all) response currents and voltages. It is, however, often more useful to be able to compute the effect, on a *single* voltage (e.g., the output voltage) of infinitesimal changes in *all* components in the circuit taken one at a time. The use of perturbation analysis (a new analysis for each component that is changed) would be very expensive.

Assume first that the circuit is *reciprocal*, composed entirely of two-terminal conductances. In this case the excitation and response of the circuit of Fig. 9.13 can be interchanged (Fig. 9.14). As before, the voltage response (now across GA) need only be multiplied by VA (the voltage across GA in the original circuit of Fig. 9.11) to obtain the differential sensitivity $\Delta V \div \Delta GA$. This calculation involves *one* circuit analysis. But the *same* analysis (i.e., at no further expense) yields a voltage across GB which, by analogy, will be equal to $\Delta V \div \Delta GB \times VB$, so that multiplication by VB (the voltage associated with GB in Fig. 9.11) is all that is necessary to find the sensitivity of V to changes in $\overline{GB}$. Thus, we see that the sensitivity of V to changes in each conductance in a circuit calls for only *one* additional circuit analysis and some minor additional multiplications.

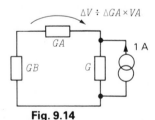

Fig. 9.14
An interchange of excitation and response in a reciprocal circuit leaves their ratio unchanged

This very efficient method of differential sensitivity analysis can be extended to *nonreciprocal* circuits. Conceptually, the extension is straightforward. We simply say that we wish to find a new circuit (Fig. 9.15) having the same ports as the sensitivity model such that, if we apply a 1 A excitation at the same location as before (i.e., at the output port as in Fig. 9.14), then we shall again find, as the response across the port associated with (say) GA, the voltage $\Delta V \div \Delta GA \times VA$. If feasible, this would seem a reasonable request, since the calculation of the required differential sensitivity ($\Delta V \div \Delta GA$) is then as simple as for a reciprocal circuit.

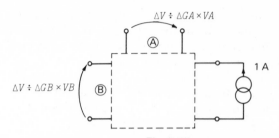

Fig. 9.15
Requirement placed on the sensitivity model associated with the efficient sensitivity analysis of a nonreciprocal circuit

Our objective can, in fact, be achieved if the new circuit, within the dashed lines of Fig. 9.15, is described by a nodal matrix (resistance and conductance) which is the *transpose* of the matrix describing the original sensitivity model (Fig. 9.12).[†] Why should this transpose relation hold? An answer is given in Fig. 9.16: assume that the circuit on the left is the actual circuit, with port K modelling the port across GA and port S representing the port across G. It is shown that if the excitation current is moved from port K to port S, then the response voltage at the other port remains the same if two specific two-port transfer resistances are equal. Generalized to more than two ports, the requirement is satisfied if the circuit is described by a resistance matrix which is the transpose of that of the actual circuit.

To summarize, the relevant relation is

$$\Delta V \div \Delta GA = - VA \times VAT \qquad\qquad (9.7)$$

where VA is the voltage across the conductance in the actual circuit, and $\underline{VAT}$ is the voltage (with the same reference polarity — hence the minus sign) at the same location in the transpose circuit; the latter being excited by a 1 A current source at the port which, in the actual circuit, is associated with the voltage V of interest.

The transpose circuit (sometimes called the adjoint circuit) need not be constructed or explicitly modelled. If we have obtained the nodal resistance matrix or the LU factors of the original circuit, then we can easily find the required response of the transpose circuit (see Exercise 5 of Chapter 6). If, nevertheless, one wished to obtain a model of the transpose circuit, it is only necessary to replace every component in the original sensitivity model (e.g. Fig. 9.12) with a new component whose matrix is the transpose of that describing the original component.

Example

The transpose circuit method of differential sensitivity analysis is illustrated in Fig. 9.17 for a circuit containing two-terminal conductances and a voltage-controlled current source. The example emphasizes the fact that only two circuit analyses — one of the actual circuit and one of the transpose circuit — are needed to find all differential sensitivities associated

---

[†] Since the sensitivity model of Fig. 9.12 is reciprocal, the circuit within the dashed lines of Fig. 9.15 would, in this special case, be identical with that of Fig. 9.12.

with the single response voltage V. From the Exercise 5 it will be seen that the computational effort can be reduced still further to a level just slightly in excess of that associated with the analysis of the actual circuit.

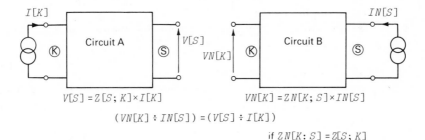

$$V[S] = Z[S; K] \times I[K]$$

$$VN[K] = ZN[K; S] \times IN[S]$$

$$(VN[K] \div IN[S]) = (V[S] \div I[K])$$

if $ZN[K: S] = Z[S; K]$

Fig. 9.16

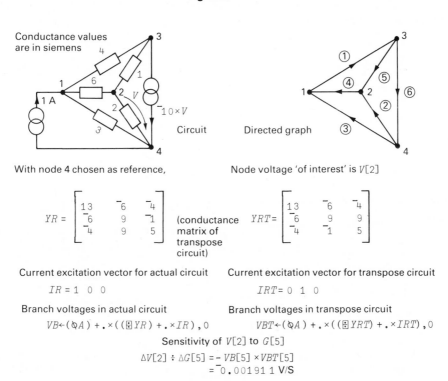

Conductance values are in siemens

With node 4 chosen as reference,

$$YR = \begin{bmatrix} 13 & ^-6 & ^-4 \\ ^-6 & 9 & ^-1 \\ ^-4 & 9 & 5 \end{bmatrix}$$

(conductance matrix of transpose circuit)

Node voltage 'of interest' is $V[2]$

$$YRT = \begin{bmatrix} 13 & ^-6 & ^-4 \\ ^-6 & 9 & 9 \\ ^-4 & ^-1 & 5 \end{bmatrix}$$

Current excitation vector for actual circuit

$$IR = 1 \quad 0 \quad 0$$

Current excitation vector for transpose circuit

$$IRT = 0 \quad 1 \quad 0$$

Branch voltages in actual circuit

$$VB \leftarrow (\lozenge A) + . \times ((\boxplus YR) + . \times IR), 0$$

Branch voltages in transpose circuit

$$VBT \leftarrow (\lozenge A) + . \times ((\boxplus YRT) + . \times IRT), 0$$

Sensitivity of $V[2]$ to $G[5]$

$$\Delta V[2] \div \Delta G[5] = ^- VB[5] \times VBT[5]$$
$$= ^-0.00191\ 1 \text{ V/S}$$

Fig. 9.17

9.4   Circuit Modification: Node Addition

Consider a circuit of N nodes (node 1 is reference) described by a reduced nodal resistance matrix ZR (Fig. 9.18). A new two-terminal component of resistance R is connected to node K, creating a new circuit node labelled $N+1$. We wish to find the reduced nodal resistance matrix ZRN describing the new

$(N+1)$ -node circuit shown within the dashed lines of Fig. 9.19.

To establish ZRN we consider the application of a 1 A current source to each node in turn. We would observe that:

(1) When applied in turn to nodes 2 to N:
    (a) the same voltages as before appear at nodes 2 to N
    (b) At node $N+1$, $V[N+1] = V[K] = ZR[K;]$

(2) When applied to node $N+1$:
    (a) the same voltages appear at nodes 2 to N as for a 1 A excitation applied at node K,
    (b) the voltage at node $N+1$ is $V[K]+R$.

From these observations it follows that ZRN is related to ZR and R as shown and explained in Fig. 9.20 and that, as a consequence, no computation (except for a simple addition) is involved.

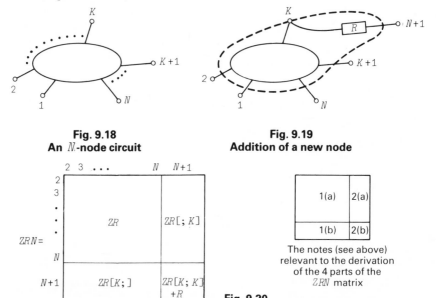

Fig. 9.18
An N-node circuit

Fig. 9.19
Addition of a new node

The notes (see above) relevant to the derivation of the 4 parts of the ZRN matrix

Fig. 9.20

The definition of a function to perform node addition is the subject of Exercise 6. The practical value of the algorithm is not obvious, but is illustrated at the conclusion of the next section.

9.5 Circuit Modification: Branch Addition

Consider (Fig. 9.21(a)) a circuit of N nodes described by a reduced nodal resistance matrix ZR: in this illustrative example, N is equal to 4. A two-terminal component of conductance G is now connected between nodes K and M (Fig.

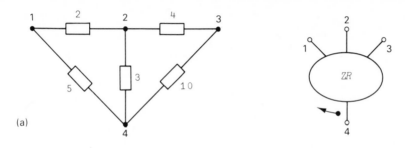

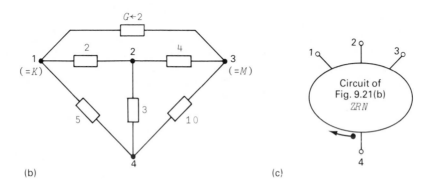

Fig. 9.21

9.21(b)) to form a new N-node circuit (Fig. 9.21(c)). We wish to compute the reduced nodal resistance matrix ZRN describing the new N-node circuit. For the purpose of this discussion we denote, by YR and YRN, the inverses of ZR and ZRN respectively.

The manner in which the conductance G is connected to the original circuit is described by two (identical) connection arrays P and Q:

$$P \leftarrow Q \leftarrow 3 \ 1\rho \ 1 \ 0 \ ^{-}1$$

in which the 1 and $^{-}1$ elements identify the nodes to which G is connected. For the example of Fig. 9.21, the addition ΔYR to YR due to the single conductance G is described by

$$G \leftarrow 1 \ 1 \ \rho \ 2$$
$$\Delta YR \leftarrow P +. \times G +. \times \lozenge Q$$
$$\Delta YR$$

```
   2   0   ‾2
   0   0    0
 ‾2   0    2
```

If YR is known

$$YR$$

```
 7   ‾2   0
‾2    9  ‾4
 0   ‾4   14
```

the new complete nodal conductance matrix YRN is easily obtained by addition:

$$YRN \leftarrow YR + \Delta YR$$
$$YRN$$

```
 9   ‾2   ‾2
‾2    9   ‾4
‾2   ‾4   16
```

But it is not YRN that we wish to obtain. In practice we may have obtained the ZR matrix as the result of matrix inversion and now, *without further matrix inversion*, and as simply as possible, we wish to find ZRN. The algorithm for so doing draws upon a result due to Householder (1957). He showed that if a matrix YR is modified by the addition of another matrix ΔYR which can be expressed in the form $\Delta YR \leftarrow P \,+\,.\times G + .\times \lozenge Q$, then the inverse of this new matrix $YR + \Delta YR$ is related to that of YR by the expression[†]

$$\boxplus(YR+P+.\times G+.\times \lozenge Q)$$
$$=(\boxplus YR)-(\boxplus YR)+.\times P+.\times(\boxplus(\boxplus G)+(\lozenge Q)+.\times(\boxplus YR)+.\times P)+.\times(\lozenge Q)+.\times \boxplus YR \quad (9.8)$$

Despite the profusion of matrix inverses in (9.8), the relation is of considerable value. First, we can replace $\boxplus YR$ by ZR which we assume to be known. Next, we recognize $P+.\times G+.\times \lozenge Q$ as the addition to YR due to the conductance G. The left-hand side is therefore ZRN, the matrix we are seeking. Since, also, $(\lozenge Q)+.\times(\boxplus YR)+.\times P$ is the sum of four appropriately signed elements of ZR, expression (9.8) can be rewritten in the following form, with the interpretation and array dimensions shown (K and M are used to denote the nodes to which G is connected):

```
(N-1),(N-1)  (N-1),(N-1)(N-1),1              1,1                    1,(N-1)
   ZRN   =     ZR   -   (ZR+.×P)+.×  (÷(÷G)+(⍉Q)+.×ZR+.×P)+.×(⍉Q)+.×ZR
```

	difference	reciprocal of ÷ G plus the appropriately	difference
	between columns	signed elements of rows K and M and columns	between rows
	K and M of ZR	K and M of ZR	K and M of ZR

Alternatively, if

$$KM \leftarrow K,M$$

we may write

$$ZRN = ZR - (-/ZR[;KM]) \circ .\times(\div(\div G)+-/-/[1]\; ZR[KM;KM])+-/[1]\; ZR[KM;] \quad (9.9)$$

The outer product in the last expression is needed to maintain the shape of the term subtracted from ZR to give ZRN. This result due to Householder has found useful application in the field of circuit sensitivity and modification.

[†] In conventional mathematical notation, Householder's relation is
$$[ZRN]=[ZR]-[ZR][P]\{[G]^{-1}+[Q][YR]^{-1}[P]\}^{-1}[Q][ZR]$$

Example

For the circuit of Fig. 9.21(a) modified as shown in Fig. 9.21(b), the arrays involved in finding ZRN are:

```
      YR                    ZR                                            G
   7  ¯2   0        0.15406      0.039216    0.011204         2
  ¯2   9  ¯4        0.039216     0.13725     0.039216                  ρ G
   0  ¯4  14        0.011204     0.039216    0.082633         1 1
      ρ YR                   ρ ZR
3  3              3   3                                       
      KM                    P                  Q
1  3                   1                   1
      ρ KM                  0                   0
2                        ¯1                  ¯1
                            ρ P                  ρ Q
                 3   1              3   1
```

First we incorporate equation (9.9) in a function $BRANCHADD$ whose left argument is the original ZR matrix and whose right argument is a three-element vector defining (first and second elements) the nodes to which the new branch is connected and (third element) the conductance of the new branch:

```
          ∇ZRN←ZR BRANCHADD NNG
[1]   KM←NNG[1  2]
[2]   G←NNG[3]
[3]   A←-/ZR[; KM]
[4]   B←÷ (÷ G) +-/-/[1] ZR[KM; KM]
[5]   C←-/[1] ZR[KM; ]
[6]   ZRN←ZR - A ∘ . ×B×C ∇
```

For the example of Fig. 9.21,

```
     ZRN←ZR BRANCHADD 1 3 2
     ZRN
0.12549       0.039216       0.02549
0.039216      0.13725        0.039216
0.02549       0.039216       0.07549
```

a result which can easily be confirmed.

Example

An interesting application (Pinel and Blostein 1967) of the techniques presented for the two methods of circuit modification is the analysis of a circuit. We shall illustrate this application by reference to the simple circuit of Fig. 9.22(a).

Suppose that, by some means, a 'tree' of the circuit's graph has been identified (Fig. 9.22(b)), a tree being defined as a graph connecting all nodes but containing no closed loops. The part of the circuit corresponding to the tree shown in (b) is shown in (c). Starting with the 2 ohm resistor in isolation (d), its reduced nodal resistance matrix is easily found: it has one row and one column, and a single element of 2 ohms. Next, the resistor (3 ohm) that has one node in common with the 2 ohm resistor is connected (e), and the 2×2 reduced nodal resistance matrix of the new three-node circuit found, again at negligible computational cost, by means of the node addition algorithm. The process continues (f) until all three branches have been accounted for.

At this stage each remaining component is connected in turn, and for each such connection the new reduced nodal resistance matrix is found (but not at negligible cost) by the algorithm for branch addition based on Householder's relation. For the 5 ohm resistor the function *BRANCHADD* is employed. However, since the 6 ohm resistor has one terminal connected to the reference terminal, a modified form of the *BRANCHADD* function (shown below) must be employed.

```
        ∇ZRN←ZR  BRANCHTOREFADD  NG
[1]     K←NG[1]
[2]     G←NG[2]
[3]     A←ZR[; K]
[4]     B←÷ (÷ G) +ZR[K; K]
[5]     C←ZR[K; ]
[6]     ZRN←ZR - A ∘ . ×B×C ∇
```

Its form can easily be derived from examination of expression (9.9), bearing in mind that one terminal of the added branch is now the reference terminal. The resulting resistance matrix given in Fig. 9.22(h) can easily be checked by forming and inverting the circuit's reduced nodal conductance matrix.

The method of circuit analysis described above and illustrated in Fig. 9.22 is called the 'link-at-a-time' algorithm, since each branch of a graph that is not part of the chosen tree is known as a 'link'.

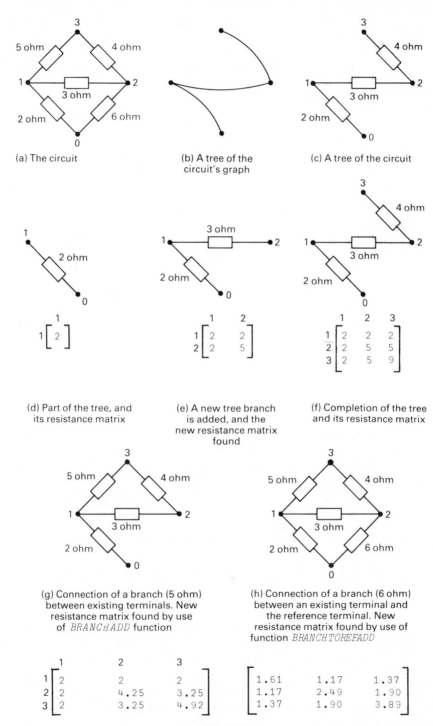

(a) The circuit

(b) A tree of the circuit's graph

(c) A tree of the circuit

(d) Part of the tree, and its resistance matrix

(e) A new tree branch is added, and the new resistance matrix found

(f) Completion of the tree and its resistance matrix

(g) Connection of a branch (5 ohm) between existing terminals. New resistance matrix found by use of *BRANCHADD* function

(h) Connection of a branch (6 ohm) between an existing terminal and the reference terminal. New resistance matrix found by use of function *BRANCHTOREFADD*

Fig. 9.22

Exercises

9.1   Draw a very simple circuit comprising two-terminal linear homogeneous resistive components, and obtain its reduced nodal resistance matrix. For a single excitation current calculate the voltage of a designated output node. Repeat with each conductance parameter increased by a small percentage and hence calculate the non-normalized sensitivity of output voltage to each component parameter.

9.2   For the same circuit as in Exercise 9.1, determine the voltage across each two-terminal component in the nominal circuit. Then, for each of these components in turn, set up the sensitivity model appropriate to small changes in these components, and calculate the 'voltage' at the output node which is, in fact, the *change* in output voltage due to the small change in component conductance. Check your results with those of Exercise 9.1.

9.3   Define the transpose circuit associated with the circuit of Exercise 9.1 and carry out an analysis to find the voltages V across the terminal pairs associated with the two-terminal components for a 1 A excitation applied at the output node. Hence find the non-normalized sensitivity of output voltage to small changes in the two-terminal components. Check your results with those of Exercises 9.1 and 9.2.

9.4   Add a voltage-controlled current source to the circuit of Exercise 9.1 and repeat Exercise 9.3 to find the sensitivity of output voltage to small changes in its mutual conductance. Check your results.

9.5   Repeat the small-change sensitivity calculation of Exercise 9.4 by using the LU factorization of the original circuit's conductance matrix to obtain the LU factors of the transpose circuit.

9.6   A linear two-terminal component of resistance R is connected to node K of a n-node circuit whose complete nodal resistance matrix is OZ, to create a new node numbered $N+1$. Define a function which returns the new complete nodal resistance matrix NZ, and which takes as its arguments the matrix OZ and a three-element vector V: the elements of V are, respectively, the existing node to which R is connected, the newly created node, and the value of the resistance R. Test the function.

9.7   A linear two-terminal component of resistance R is connected between nodes K and M of an n-node circuit whose complete nodal resistance matrix is OZ. Define a function which returns the new complete nodal resistance matrix NZ, and which takes as its arguments the matrix OZ and a three-element vector V: the elements of V are, respectively, the (two) existing nodes to which R is connected, and the value of R. Test the function.

9.8   Extend the function developed in Exercise 9.7 to handle the case where a voltage-controlled current source is connected to existing nodes, or create a new function appropriate to this situation. Again, test the function.

9.9   Devise a function, based on the substitution current source algorithm for large-change sensitivity analysis, whose result is the change in a designated nodal voltage due to a non-infinitesimal change in a conductance connected between two specified nodes. It is suggested that two functions be written; one would carry out the initial analysis needed to find the parameters involved in each execution of the second function for a new value of the conductance change.

9.10 (Prerequisites are the functions developed in response to Exercises 9.6, 9.7 and 9.8). For the circuit shown in Fig. E9.10, obtain a reduced nodal resistance matrix in the following way:

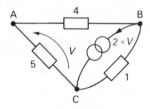

Conductances in siemens

Fig. E9.1

(a) by starting with the nodal resistance matrix of a single two-terminal component, and extending this matrix for each added tree branch;

(b) when no more tree branches are left, by modifying the matrix by the Householder method to account for the connection of each two-terminal link branch and each voltage-controlled current source.

(c) Check your result by writing down the reduced nodal conductance matrix (by inspection) and inverting it.

9.11 (Prerequisite is the function developed in response to Exercise 9.9 above). In the circuit associated with Exercise 9.10 the component of 4 S conductance is changed to a value of 15 S. By using the results of Exercise 9.10 and the substitution current source method of large change sensitivity analysis, compute the change in voltage V_1 (see Figure E9.2) for the excitation shown in the figure. Check the result by another method.

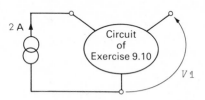

Fig. E9.2

Solutions

9.1

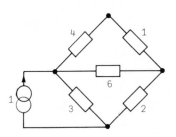

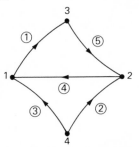

(conductance in siemens)

Fig. E9.3
Circuit of interest

$M\leftarrow\lozenge5\ 2\ \rho\ 1\ 3\ 4\ 2\ 4\ 1\ 2\ 1\ 3\ 2$												Branch connection matrix	
$A\leftarrow INCID\ M$												Incidence matrix	

G

4	2	3	6	1		Branch conductance vector
		YB				Branch conductance matrix
4	0	0	0	0		
0	2	0	0	0		
0	0	3	0	0		
0	0	0	6	0		
0	0	0	0	1		

IR

1	0	0		Nodal current excitation

$VRN\leftarrow(\boxdot^{-}1\ ^{-}1\ \downarrow A+.\times YB+.\times\lozenge A)+.\times IR$ Nominal nodal voltages

$YB[1;1]\leftarrow YB[1;1]\times1.01$
$VR1\leftarrow(\boxdot^{-}1\ ^{-}1\ \downarrow A+.\times YB+.\times\lozenge A)+.\times IR$ Nodal voltages after 1%
change in conductance of
$YB[1;1]\leftarrow G[1]$ branch 1

$YB[2;2]\leftarrow YB[2;2]\times1.01$
$VR2\leftarrow(\boxdot^{-}1\ ^{-}1\ \downarrow A+.\times YB+.\times\lozenge A)+.\times IR$ Nodal voltages after 1%
change in conductance of
$YB[2;2]\leftarrow G[2]$ branch 2

$YB[3;3]\leftarrow YB[3;3]\times1.01$
$VR3\leftarrow(\boxdot^{-}1\ ^{-}1\ \downarrow A+.\times YB+.\times\lozenge A)+.\times IR$ Nodal voltages after 1%
change in conductance of
$YB[3;3]\leftarrow G[3]$ branch 3

Changes in nodal voltages due to 1% change in
conductance of
$\Delta VR1\leftarrow VR1-VRN$ Branch 1
$\Delta VR2\leftarrow VR2-VRN$ Branch 2
$\Delta VR3\leftarrow VR3-VRN$ Branch 3

$\Delta VR1[3]$	$\Delta VR2[3]$	$\Delta VR3[3]$
$7.74E^{-}5$	$^{-}0.000626$	$^{-}0.00138$

Changes in voltage at node 3 due to 1% change in
conductance of branches 1, 2 and 3.

9.2

$VRN\leftarrow(ZRN\leftarrow\boxdot YRN\leftarrow^{-}1\ ^{-}1\ \downarrow YN\leftarrow A+.\times YB+.\times\lozenge A)+.\times IR$ See circuit of Exercise 9.1

VRN

0.22	0.17	0.21		Calculation of nodal voltages and branch voltages of

nominal circuit!

$VBN\leftarrow(\lozenge A)+.\times VRN,0$
VBN

0.01	$^{-}0.17$	$^{-}0.22$	$^{-}0.05$	0.04

```
G←4.04  2  3  6  1          1% change in conductance of branch 1
YB←DIA  G                   Resulting branch conductance matrix

VR1←(ZR1←⊞YR1←¯1 ¯1 ↓ Y1←A+.×YB+.×⍉A)+.×IR    New nodal voltages
ΔVR1←VR1−VRN
ΔVR1                                          Change in nodal voltages
¯3.9675E¯6 5.9512E¯6 7.7366E¯5
```

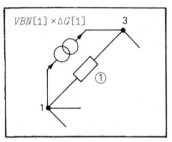

Fig. E9.4
Sensitivity model

Value of current source of sensitivity model

```
IS1←VBN[1]×.04              Nodal current excitation for sensitivity model
IRS1←(-IS1),0,IS1           (- IS1 into node 1, 0 into node 2, IS1 into node 3)

      ZR+.×IRS1             Resulting nodal voltages are the desired changes in
¯4E¯6 6E¯6 7.8E¯5           nodal voltages in the actual circuit.
```

Checks with earlier result

9.3   Reduced nodal conductance matrix of transpose
circuit (Node 3 is output node)

```
YRT←⍉YRN←¯1 ¯1 ↓ A+.×YB+.×⍉A
IRT←0  0  1                          Its current excitation

VBT←(⍉A)+.×(VRT←IRT⊞YRT),0
VBT
¯0.195 ¯0.185 ¯0.21 ¯0.025 0.22      Branch voltages in transpose circuit
VBN
0.01 ¯0.17 ¯0.22 ¯0.05 0.04          Branch voltages in nominal circuit
SENS←-VBN×VBT
SENS                                 Differential sensitivity of node 3 voltage
0.00195 ¯0.0314 ¯0.0462 ¯0.00125 ¯0.0088
SENS[ι3]×.01×G[ι3]
7.8E¯5 ¯0.000629 ¯0.00139            Changes in node 3 voltage due to 1% change in
                                     conductance of branches 1, 2 and 3 (Check with end
                                     of Solution to Exercise 1)
```

9.4

```
        Y
13     ¯6    ¯4
¯6      9    ¯1              Reduced nodal conductance matrix
¯4      9     5

        A
 1      0   ¯1   ¯1    0    0    Incidence matrix
 0      1    0    1   ¯1    0
¯1      0    0    0    1   ¯1
 0      1    1    0    0   ¯1
```

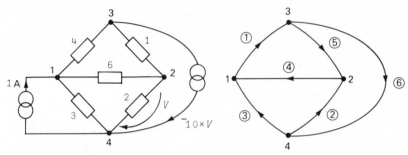

Conductances in siemens

Fig. E9.5
Circuit of Exercise 9.1 with an added voltage-controlled current source, and its directed graph. The differential sensitivity of the voltage of node 2 with respect to all branch conductances is to be found

	M					Branch description
1	2	3	4	5	2	← controlling branch (voltage)
1	2	3	4	5	6	← controlled branch (current)
4	2	3	6	1	¯10	← value in siemens

```
    Z←⊞Y                         Reduced nodal
    Z                            Resistance matrix
0.094737    ¯0.010526    0.073684
0.059649     0.085965    0.064912
¯0.031579   ¯0.16316     0.14211
```

Reduced nodal Resistance matrix

```
      I←1 0 0                    Current excitation
      V←Z+.×I
      V                          Nodal voltages
0.094737 0.059649 ¯0.031579
      VB←(⍉A)+.×V,0              Branch voltages
      VB
0.12632 ¯0.059649 ¯0.094737 ¯0.035088 ¯0.091228 ¯0.031579
```

Actual circuit

```
      ZT←⍉Z                      Reduced nodal resistance matrix
      IT←0 1 0                   Current excitation
      VT←ZT+.×IT                 Nodal voltages
      VT
0.059649 0.085965 0.064912
```

Transpose circuit

```
      VBT←(⍉A)+.×VT,0            Branch voltages
      VBT
¯0.0052632 ¯0.085965 ¯0.059649 0.026316 ¯0.021053 0.064912
```

```
      DIFFSENSITIVITY←-VB[M[2;]]×VBT[M[1;]]
      DIFFSENSITIVITY
0.00066482 ¯0.0051277 ¯0.005651 0.00092336 ¯0.0019206 ¯0.0027147
```

Checks with answer to Question 5   Simple products of branch voltages give differential sensitivities

9.5

```
        Y                        Reduced nodal conductance matrix
 13    ¯6    ¯4
 ¯6     9    ¯1
 ¯4     9     5
```

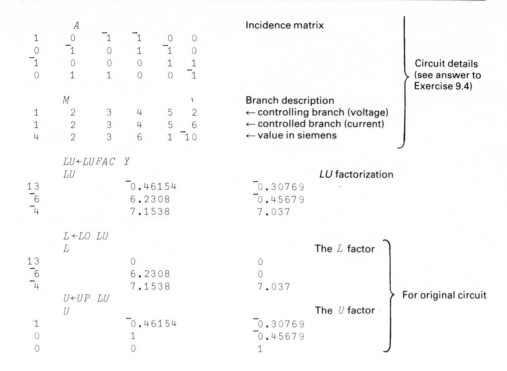

```
         A                                  Incidence matrix
  1      0    ¯1    ¯1     0    0
  0     ¯1     0     1    ¯1    0                                       ⎫
 ¯1      0     0     0     1    1                                       ⎪  Circuit details
  0      1     1     0     0   ¯1                                       ⎬  (see answer to
                                                                       ⎪  Exercise 9.4)
         M                        \        Branch description          ⎭
  1      2     3     4     5    2          ← controlling branch (voltage)
  1      2     3     4     5    6          ← controlled branch (current)
  4      2     3     6     1  ¯10          ← value in siemens

      LU←LUFAC Y
      LU                                        LU factorization
 13                  ¯0.46154           ¯0.30769
 ¯6                   6.2308            ¯0.45679
 ¯4                   7.1538            7.037

      L←LO LU
      L                                         The L factor          ⎫
 13                   0                  0                            ⎪
 ¯6                   6.2308             0                            ⎬
 ¯4                   7.1538             7.037                        ⎪  For original circuit
      U←UP LU                                                        ⎪
      U                                         The U factor          ⎬
 ¯1                  ¯0.46154           ¯0.30769                      ⎪
  0                   1                 ¯0.45679                      ⎪
  0                   0                  1                            ⎭
```

```
      ∇ R←INCID M;N
 [1]   R←((ιN)∘.=M[1;])-(ιN←⌈/,M)∘.=M[2;]
      ∇

      ∇ YB←DIA G;B
 [1]   YB←((B,B)ρG)×UNIT B←ρG
      ∇

      ∇R←LUFAC M
 [1]  C←1
 [2]  R←M
 [3]  L1:S←C↓ι1↑ρR
 [4]  R[C;S]←R[C;S]÷R[C;C]
 [5]  R[S;S]←R[S;S]-R[S;C]∘.×R[C;S]
 [6]  →L1×(1↑ρR)≠C←C+1∇
      ∇R←LO X
 [1]  R←((ιN)∘.≥ιN←1↑ρX)×X∇
      ∇R←UP X
 [1]  R←((ιN)∘.=ιN)+(((ιN)∘.<ιN←1↑ρX)×X∇
      ∇R←Y BKSUB I
 [1]  R←(ρI)ρ0
 [2]  C←0
 [3]  L1:C←C+1
 [4]  S←(1++/ρI)-C
 [5]  R[S]←(÷Y[S;S])×I[S]-+/Y[S;]×,R
 [6]  →(C≠ρI)/L1∇
      ∇R←Y FSUB I
 [1]  R←⌽(⌽⊖Y) BKSUB ⊖I∇
```

Fig. E9.6
Functions used in answers
to Exercise 9.5

```
      I←1  0  0                                      Use of single forward and backward substitution to
      V←U  BKSUB  L  FSUB  I                         find branch voltages of original circuit.
      V
0.094737 0.059649 ‾0.031579
      VB←(⍉A)+.×V,0
      VB
0.12632 ‾0.059649 ‾0.094737 ‾0.035088 ‾0.091228 ‾0.031579

      VT←(⍉L) BKSUB (⍉U) FSUB IT←0  1  0             Use of single forward and backward
      VT                                             substitution to find branch voltages of
0.059649 0.085965 0.064912                           transpose circuit.

      VBT←(⍉A)+.×VT,0
      VBT
‾0.0052632 ‾0.085965 ‾0.059649 0.026316 ‾0.021053 0.064912

      DIFFSENSITIVITY ←- VB[M[2;]]×VBT[M[1;]]        Calculation of differential sensitivities
      DIFFSENSITIVITY                                as simple products of voltages in
                                                     original and transpose circuits.

0.00066482 ‾0.0051277 ‾0.005651 0.00092336 ‾0.0019206 ‾0.0027147
```

Checks with answer to Question 4

PART

2

THE PACKAGE

———————

10

The Design of a Package

10.1 Introduction

Part 2 of this book is concerned with circuit analysis at the package level. The treatment of this subject emphasizes both the internal structure and the design of a circuit analysis package, and the use of such a package as a circuit solution tool.

The material in Part 2 provides an important engineering focal point for the study of the various analysis algorithms undertaken in Part 1, and it permits learning to be extended in several important ways. For example, it provides valuable insight into how a large number of individual algorithms may be integrated into the structure of a useful working package. Similarly, exploring actual uses of the resultant package through its application to the solution of a variety of practical circuits adds realism to the learning process and develops valuable skills. It also results in a much deeper understanding of the capabilities and inevitable practical limitations inherent in the algorithms as well as the package structure.

An exhaustive treatment of the subject of package design for circuit analysis is beyond the scope of this text. Instead, we use one particular package as a vehicle for study which will serve as an example of the concepts involved. *BASECAP* (Burgess and Spence Electronic Circuit Analysis Package) was specifically developed by the authors to incorporate many features which make it an effective choice for this purpose, and it has been extensively tested in this role.

BASECAP is a general-purpose, interactive tool, useful for the d.c. and a.c. solution of a wide variety of electronic circuits. It was designed with a simple modular internal structure that makes it useful as an instructional vehicle. *BASECAP* has much in common with other circuit analysis packages, so much of what is learned is directly applicable to other packages.

A key factor in this choice is that *BASECAP* is written entirely in APL. It should be clear by now that APL is a very powerful, flexible programming language which provides a sophisticated and friendly computing environment. *BASECAP* takes advantage of this to achieve many of its features and capabilities with a minimum of structural complexity.

Also vital in the instructional sense is that *BASECAP* incorporates and extends many of the same basic algorithms developed and presented in Part 1 of this book. Thus, while the material in this part of the book has a different scope and emphasis, it is nevertheless closely related to the material in Part 1, and is a natural application of it.

10.2 Design Objectives

Three broad statements constitute the main overall design objectives that have guided the development of *BASECAP*. These are discussed in the three following subsections, which give an overview of *BASECAP*, and establish how it was intended to be used both as a solution tool and as an instructional vehicle.

10.2.1 Interactive Solutions

BASECAP should serve as a fast, convenient, versatile and highly interactive tool for the solution of a wide variety of electrical and electronic circuits of at least moderate complexity.

From the outset, major emphasis was placed on the importance of the interactive capabilities for the package. In fact, this factor largely dictated the choice of APL as the language for its implementation. Most of the other desirable features, such as fast response time and user convenience, are closely related to this interactive requirement, and are readily obtained with APL.

Unquestionably, the capacity a package has to provide rapid feedback of solution results is very important from the user's point of view. In particular, the overall efficiency of the circuit solution task may be greatly affected by whether immediate processing is used, or whether batch processing is used where the results may not be available for several minutes or hours. Response time also influences the nature and scope of the tasks that may be undertaken conveniently, as well as the approach used in obtaining a solution. For example, it will be seen that a fast response time helps to make *BASECAP* useful for *design* purposes, as it rather effectively puts the 'user in the design feedback loop.' In this case, the user may often modify circuit data based on his or her knowledge of circuit operation, and obtain a solution which meets specified design requirements using a relatively small number of design iterations.

The user should be aware, however, that there may be a penalty associated with such an emphasis on fast interactive response capabilities. Response times deteriorate as circuit size increases, and computing resources such as CPU time and user workspace allocations may be quite limited. These factors will determine the maximum circuit size that can be handled by a package such as *BASECAP*. In practice, a very large amount of design work in industry involves circuits with 50 nodes or less, so our emphasis is not misplaced. It is not, however, directed towards the solution requirements of entire very large-scale integrated circuits that may contain several thousand transistors.

Having chosen a given design emphasis, one must also evaluate its effect on the complexity of the package implementation. In fact, this objective is readily accommodated with APL. Rather than causing any special programming complexities, the opposite is achieved by using the facilities provided within the APL host environment. The resultant package is relatively fast, powerful, and compact.

10.2.2   Instructional Vehicle

BASECAP should serve as an instructional vehicle for students and practicing engineers in courses concerned primarily with the basic methods and algorithms used in current CACD practice.

Essentially this means that it must be relatively easy for students to understand the program structure, to follow the flow of the processing, and to identify and interact with individual algorithms making up the package. In many respects this is a rather novel objective to set for a circuit analysis package. However, our experience in teaching university courses, as well as courses for practicing engineers in industry, has confirmed the desirability of such a feature.

In part this objective is met in *BASECAP* through the use of a simple open highly modularized type of structure for the package design. (For a more extensive treatment of this topic, see Gibson (1982).) All parts of the program structure are accessible to the user. Also, a well-defined specific APL function is used for each processing operation so that its part in the program structure is readily apparent. Students may readily investigate an alternative processing method, for example, by substituting a new routine for an existing one in the program structure. Several such alternate algorithms are provided for this purpose, and are used in various student exercises.

The use of APL to implement *BASECAP* inherently provides other advantages in the educational context. One is that the concise logical nature of the APL notation itself is an aid to understanding the ideas expressed in program form, once some familiarity is gained with the APL language.

Another advantage is that an APL function can readily be tested, using simple test data, in isolation from the main program structure (except for other functions called from within the function under test). Also, the step-by-step processing of data within a function can be followed in detail, and interacted with during the actual execution of that function, by means of the APL *TRACE* and *STOP* system controls. Overall, this results in a very effective means for developing an understanding of the methods and algorithms used in the *BASECAP* package.

An essential requirement in meeting this second overall design objective is good documentation, which is needed at several levels. This should

(a)   describe what variables are used, the type and shape of the data they contain, and where they are used in the package,

(b)   describe what each function does and how it operates,

(c)   explain the structure of the overall package, show where each function fits into that structure, and show how both control and data are transferred throughout the package, and

(d)   provide sample results and test examples to demonstrate what various programs do, and to guide one's investigation of various aspects of the package structure and operation.

These documentation requirements are not totally unique to our circuit analysis package and, fortunately, most APL systems contain facilities, such as workspace documentation aids, which greatly assist with most of these tasks.

10.2.3 Easy Modifications

BASECAP should be capable of being easily modified or extended to incorporate improvements or additional features.

Typically a package such as *BASECAP* is the result of an evolutionary development process carried out by its developers over a period of time. It is also advantageous if knowledgeable users can make modifications, or incorporate special capabilities not designed into the original package. A package can then continue to evolve with the user's changing requirements. An open-ended structure is essential to support this form of package development and permit easy extension.

The same factors discussed in the previous subsection are also important in helping to meet this third objective. Clearly, a simple, open (i.e. accessible), highly modularized structure makes interfacing straightforward. Further, the availability of good documentation is essential if it is to be possible for anyone other than the original designers to make such modifications or extensions to the package.

10.3 General Capabilities

The previous section established the overall objectives that guided the design of the *BASECAP* circuit analysis package. We now consider the actual package that has resulted, beginning with an overview of its circuit solution capabilities. An introductory example is also included to demonstrate broadly how it is used, and to illustrate the nature of the user interaction required. A detailed user's guide is presented in Chapter 11.

BASECAP is classed as a general-purpose d.c. and a.c. interactive electronic circuit analysis package. It can accommodate a wide variety of electrical and electronic circuits. In its present form, it can produce two main categories of solutions; nonlinear d.c. and small-signal a.c. Circuits may contain resistors, capacitors, inductors, semiconductor diodes, bipolar transistors, operational amplifiers, independent d.c. voltage and current sources, independent a.c. voltage and current sources, and linear a.c./d.c. voltage-controlled current sources. A simplified Ebers–Moll model is used to represent the transistors. Diode and transistor model parameter values are readily specified by the user.

The user interface to the package is simple and straightforward. Circuit data is input to the program by means of a circuit description function (such as *EXAMPLE*, used later in this section) which may be kept in the circuit library. The method of nodal analysis is used to obtain solutions for the circuit node voltages. These results are processed and displayed using the output section of the package. The entire solution process is controlled interactively by the user.

As established in the previous section, *BASECAP* is imbedded in APL, from which it derives much of its power and flexibility, so a working knowledge of APL is advantageous. However, most of the user's contact with the program is in the form of user-oriented functions, so an extensive knowledge of APL is not required to make use of *BASECAP*. It is, nevertheless, necessary to have access to an APL computing facility, and be familiar with terminal–computer communication, workspace management, and APL function basics.

   The following example illustrates the use of *BASECAP* for the solution of a simple circuit. Before starting the actual solution it is of course necessary to load a copy of the *BASECAP* package into your active workspace as illustrated below;

```
)LOAD     4900   BASECAP
PARMSET
```

where the *BASECAP* workspace is assumed to be stored in library number 4900 and *PARMSET* initializes program solution parameters prior to first time operation.

Example   Circuit diagram

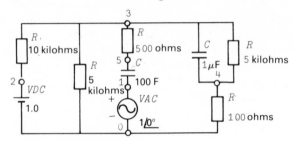

Fig. 10.1
Display already existing function

Circuit description function

```
        ∇EXAMPLE[□]∇
        ∇ EXAMPLE
[1]     STARTCCT
[2]     VAC 1  0 1  0
[3]     R 5  3  500
[4]     C 1  5  100
[5]     VDC 2  0 1
[6]     R  2 3  10000
[7]     R  3 0  5000
[8]     C  3 4  1E¯6
[9]     R  3 4  5000
[10]    R  4 0  100
[11]    END
        ∇
```

```
        EXAMPLE
      ˙ DCSOLVE
CPUSECONDS =0.08
ITERATIONS =3
```

Set up the circuit for solution
Execute a d.c. solution

`↓↓DC NODE VOLTAGES  1  TO 5↓↓`

`2.02E¯16  1  0.202  0.00395  0.202`   d.c. results (volts)

`FLOG 1 1E6  12`

Choose frequencies for a.c. solution
(12 points/decade)

```
        ACSOLVE                          Execute a.c. solution
SOLVING FOR 73 FREQUENCY POINTS
CPUSECONDS=0.723
DONE
```

                `20 40 PLOT LOGF, MAGDB ACVOUT 3`   Request log–log plot
                                                               for voltage at
                                                                node 3 (dB)

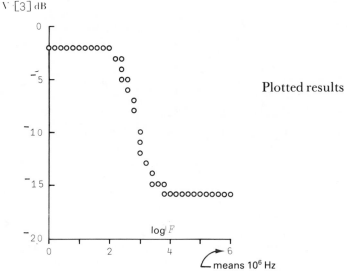

Plotted results

10.4 Structure of *BASECAP*

The overall structure of the *BASECAP* package is presented in this section. This structure is illustrated in the form of interconnected functional blocks in Fig. 10.2. The four central blocks labelled INPUT, DC, AC, and OUT perform all of the actual circuit processing work in the package. Each is further discussed in Sections 10.4.2–10.4.5. The CIRCUITS, MODELS and CONTROL blocks are considered in Sections 10.4.6–10.4.8.

    The functional blocks and the data and control lines depicted in Fig. 10.2 are organizational in nature, and do not represent rigid boundaries or fixed connecting links. The *BASECAP* structure is actually made up of loose associations or groups of functions based only on functional purpose. It is therefore completely open-sided, and all functions are readily accessible.

10.4.1   Functional Groups

Many of the blocks in Fig. 10.2 are also defined for convenience as APL groups.[†]

---

† Some APL implementations do not support the *GROUP* system command feature. Some APL implementations support similar or more sophisticated features using system functions and variables. The group concept is used in this discussion for convenience, and as a pedagogical aid in characterizing and explaining the organization of the *BASECAP* package. APL system support for groups is not vital, however, either to make effective use of this text or the *BASECAP* package.

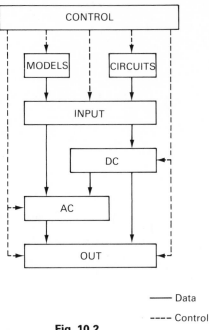

Fig. 10.2
Global structure of *BASECAP*

Each group name has been chosen to reflect its functional purpose in the *BASECAP* package, and each is recognized by the APL system used here. The names of all of the groups defined within *BASECAP*, for example, can be displayed from within the *BASECAP* workspace using the *GRPS* system command as follows:

```
        )GRPS
AC      CIRCUITS      DC        INPUT       OUT
```

Each group contains a number of functions which share a closely related purpose. This helps to fit the member functions into the overall operational structure of the package. It is also easy to display the names of all the functions in a group or to copy or erase all of the functions in a group using a single command. Thus,

```
)COPY BASECAP INPUT
```

results in all of the functions in the *INPUT* group being transferred to the active workspace from the *BASECAP* workspace in the user's library. Similarly,

```
)ERASE INPUT
```

eliminates from the active workspace all of the functions that are part of the *INPUT* group.

Sometimes, these commands are useful for managing the size of the active workspace. For example, when a very large circuit must be solved, the user can copy, execute, and erase each group of functions in turn, in order to minimize the amount of workspace consumed by the functions themselves. This discussion illustrates another reason why a knowledge of the internal structure is beneficial to the package user.

We now briefly consider the composition and overall functional purpose of each of the groups comprising the *BASECAP* package.

10.4.2 The *INPUT* Functions

The *INPUT* group of functions provides the main input interface between the user and the solution routines. Specifically, the *INPUT* functions process the designer's description of a circuit as illustrated in Fig. 10.3, and generate circuit data in a form compatible with the d.c. and a.c. solution functions. Functions contained in this group are readily displayed with the *GRP* system command as follows:

```
)GRP INPUT

ALPHA      C         D         END       IAC       IDC       INCID
L          OA        QCALC     QNPN      QPNP      R
REFNODE    STARTCCT  VAC       VCIS      VDC
```

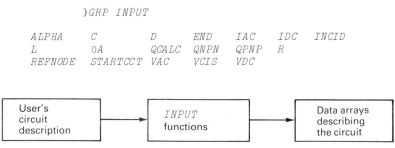

Fig. 10.3
Overall operation of the *INPUT* **functions**

The main functions of interest in this group are the ones called by the user to specify components, devices and sources, along with their interconnections in a circuit. These include the functions R, L, C, D, OA, IAC, IDC, $QNPN$, $QPNP$, VAC, $VCIS$ and VDC. Other functions called directly by the user are *STARTCCT* and *END*, which respectively initialize and format the data arrays for the circuit in question, prior to the solution step. The syntax and use of the main functions in this group are described and illustrated in Section 11.2. Details of the internal structure and operation of the *INPUT* functions are presented in Section 12.2. The *INPUT* functions occupy about 5300 bytes in the workspace.

In practice the data array block in Fig. 10.3 is composed of a number of global variables (matrices and vectors) which completely describe the composition and the interconnections for the specific circuit and also contain parameter values for the components, sources, and devices used.

Once the data array block in Fig. 10.3 has been produced, the user's description of the circuit (circuit description function) and the *INPUT* functions are normally no longer required and can be erased from the workspace until needed for a new or modified circuit. The detailed structure and composition of this data are given in Section 12.1.

10.4.3 The *DC* Functions

The *DC* group contains the functions necessary to accept data prepared by the *INPUT* group and actually carry out the nonlinear d.c. Newton–Raphson solution for one or more sets of parameter values (see Fig. 10.4). This group contains the

Fig. 10.4
Overall operation of the *DC* functions

following functions, including a number of alternative processing functions, and occupies approximately 9700 bytes of storage:

```
)GRP DC

ALTBASE 1              DCSOLVE                   DIA
DIFDIODE   DIFNETL   DIODE   GTRANS   IEQUIV   INCID
INITIALIZE         ITRANS  NEWTALT NEWTALT1      NEWTDC
NEWTDC 1 PARMSET PARMTELL   TRANSDC    UNIT     VBCHKCAN
VBSTART VBSTART 2       VCFORM
```

The main user-called routine in this group is *DCSOLVE* which produces one set of d.c. node voltage solutions. These results are stored in a global vector called *VANS*. *TRANSDC* produces a sequence of nonlinear solutions, one for each value of a specified independent parameter such as a resistance value. These results are stored in the global matrix *VALL*. The linearized nodal conductance matrix (*SLOPE*) for the circuit is an additional important result of this solution since it is used by the a.c. solution functions. Complete details on how to use *DCSOLVE* and *TRANSDC* for d.c. solutions are given in Chapter 11. Further information on the internal structure and detailed operation of the various *DC* group functions is found in Chapter 13.

10.4.4 The *AC* Functions

This group interfaces with the nodal conductance matrix (*SLOPE*) generated from the d.c. solution, and also uses the a.c. component and source data prepared by the *INPUT* group. This is illustrated in Fig. 10.5. The functions in the *AC* group occupy about 3100 bytes in the workspace and are listed below:

```
)GRP AC

ACPREP   ACSOL    ACSOLVE BB     FLIN    FLOG    FORM
FSTEP IFORM    IREIM    REALSOLV    YNET
```

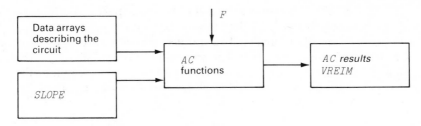

Fig. 10.5
Overall operation of the AC **functions**

A small-signal a.c. solution is set up and carried out for each of the frequencies specified in the global variable F. Real and imaginary voltage results are stored in the global array $VREIM$. The main solution routine in this group is $ACSOLVE$. Chapter 11 explains and illustrates the use of these functions, while Chapter 14 provides a detailed discussion of their internal structure and operation.

10.4.5   The OUT Functions

The main functions in this group are $PLOT$ and $PRINT$, used to display results either graphically or in tabular form. Supporting functions facilitate the selection and formatting of the output data. Results may be displayed as a function of frequency or, where relevant, as a function of the values of an independent parameter. Depending on the complexity of the actual $PLOT$ routine used, these routines occupy up to about 6800 bytes of storage and are listed below:

```
)GRP OUT

ACVOUT     ALL      AND      ANGLE      LOG      LOGF      MAG
MAGDB  PLOT      PRINT     VS       ZATN
```

The method of using these functions is explained in Chapter 11. Details of their internal structure and operation are provided in Chapter 15.

10.4.6   The $CIRCUITS$ Library

The $CIRCUITS$ block in Fig. 10.2 represents part of the support structure of the package. This feature allows complete circuit descriptions to be stored and readily accessed for future use.

The $CIRCUITS$ block may be realized in $BASECAP$ as an APL group. It consists of a collection of APL functions known as circuit description functions, each of which fully describes a separate circuit. More specifically, each circuit description function consists of a set of executable statements, including one for each component and device in the circuit. Any number of circuit description functions may reside in the active workspace at any given time. A specific circuit is set up for solution by executing the corresponding circuit description function, thus yielding the data arrays describing the circuit as illustrated in Fig. 10.3.

The *CIRCUITS* library may be stored as part of the main *BASECAP* workspace, or in a separate library workspace at the user's option (file storage might also be used in some package implementations). It should be noted that it is not mandatory for user-defined circuit description functions to be assigned to the *CIRCUITS* group specifically. All functions and variables will automatically be preserved in the named workspace by the system command

)SAVE

If it is desired to use the group facility, the user must assign a newly created circuit to the *CIRCUITS* group. Assuming, for example, that *NEWCCT* is the name of a new circuit description function, and also that the group is present in the active workspace, *NEWCCT* may be added to the *CIRCUITS* group by the following command.

)GROUP CIRCUITS CIRCUITS NEWCCT

By its nature, the *CIRCUITS* group does not have a fixed composition. The following is a representative display.

)GRP CIRCUITS

```
CACDCCT          CEBIAS          CEBIASAC          CEBIASOA
CEBIASOAPAR      COLPITTS        DIFFAMP          DIFFAMPLOOP
FBTRIPLE          FILTER FRANKLIN          HFIGAIN HOLT73
LPF    MH133   MH133IF NOTRANS   PROB647   SCHMITT   TC22
TC22IF    TC6    TESTCCT    TTLGATE    DIC    FRANKOSC
LOGOPR
```

Some of these circuits, such as *FBTRIPLE,FRANKOSC* and *LPF,* will be featured as examples in Chapter 11.

10.4.7   The *MODELS* Library

The *MODELS* block in Fig. 10.2 also represents part of the support structure of the package. The concept of a *MODELS* library is that component values and device parameter data may be stored and readily accessed for future use. As used in *BASECAP*, this essentially refers to a collection of global variables. The names and contents of these variables follow APL conventions and are readily assigned by the user. The following examples illustrate this process:

```
RIN ←100E3
Q100←100  .1  4E¯9 1E¯9 20E¯11  2E¯12
```

If desired, such data variables may be assigned by the user to an APL group, which can then be used in basically the same ways as explained previously for the *CIRCUITS* group. The global variables in the *MODELS* library can reside permanently in the main *BASECAP* workspace, or optionally in a separate stored workspace and copied into the active workspace as required.

10.4.8 Solution Control Structure

The solution control structure has a significant impact on the nature of any circuit analysis package. A Solution CONTROL block is shown at the top of Fig. 10.2 for *BASECAP*. It does not exist, however, as an identifiable group of control programs, and is therefore a purely conceptual part of the *BASECAP* package. In this case, the control function exists in the person of the user, with the support and assistance of the APL host system. The dashed control lines in Fig. 10.2 serve to emphasize that the user can access and interact with all parts of the package while managing the entire solution process.

This structure is very simple and completely open-ended. It is also the most flexible as there are no inherent rigid pre-programmed sequences or boundaries to restrict the user. Control is achieved by executing various desired *BASECAP* solution routines or APL system commands interactively. Choices are based on the user's particular objectives, combined with information obtained from previous solution steps.

Considerably more extensive and complex are closed package structures, such as menu-based systems, which both control the operation and use of a package and guide the user. Clearly the arrangement used in *BASECAP* places more responsibility for a successful circuit solution in the hands of the user. A knowledge of system commands for workspace management is essential, as explained earlier. A working knowledge of APL variables and functions is also necessary. Beyond this, the more knowledgeable user can often obtain advantages from the many additional advanced tools and features available with the APL system. Examples include use of the *TRACE* and *STOP* system commands for error diagnostics.

In Section 11.5, guidelines are given to assist the user in carrying out various interactive solution control tasks, and this process is illustrated extensively in the examples in Section 11.6. As an alternative mode of operation, however, a particular solution control sequence can be pre-programmed and stored for use as a control function. This approach is faster and more convenient, for example, if a particular solution sequence is needed over and over again.

As an illustration of these concepts, consider the following user-written APL function called *DO*, which uses a number of *BASECAP* functions:

Example   User-written Control Function

```
        ∇ DO                              Display function 'DO
[1]     EXAMPLE                           Set up the circuit called EXAMPLE.
[2]     DCSOLVE                           Obtain the d.c. solution.
[3]     FLOG 1 1000000 12                 Specify frequencies logarithmically.
[4]     ACSOLVE                           Obtain the a.c. solution.
[5]     20 40 PLOT LOGF,MAGDB ACVOUT 3    Plot a graph of the results at node 3.
        ∇
```

This is a simple, if somewhat restrictive, example of a user-written control function for *BASECAP* . A comparison with the example in Section 10.3 will show that this function contains the same essential processing steps. Hence, by executing this function, the reader will confirm that the two sets of solution results are identical.

11

Using the CACD Package

An essential requirement for any circuit analysis package is the provision of easily followed documentation explaining how the package is used. In this chapter we present representative user documentation for the *BASECAP* package. This material emphasizes the input–output features of *BASECAP*, and is designed to show how to use the package as an effective circuit solution tool. Chapter 10, notably Section 10.3, contains relevant background material on the design and use of *BASECAP*. Appendix 4 presents a summary of workspace management details which also may be helpful to *BASECAP* users.

The first two sections in Chapter 11 focus on *BASECAP*'s overall solution capabilities and on the use of valid circuit components, devices, and sources. The requirements for specifying complete circuits are given in Section 11.3. Guidelines for setting solution control parameters and for computing and displaying various types of results are presented in Sections 11.4 and 11.5. The material in Section 11.6 consists of annotated sample circuit solutions illustrating applications of *BASECAP*. Finally, Section 11.7 deals with the diagnosis and correction of errors and problems which can occur when using the *BASECAP* package.

11.1 Capabilities

BASECAP is a general-purpose circuit analysis package designed to carry out linear and nonlinear d.c. solutions and linear small-signal a.c. solutions of electrical and electronic circuits. Circuit solutions in the package are produced using the method of nodal analysis.

As discussed in Chapter 10, *BASECAP* is embedded in APL, and features a simple, flexible, and highly interactive user interface. Circuit elements are specified by means of easy-to-use element description functions, and each circuit is contained in a circuit description function for convenience. The overall solution process is directed interactively by the user with simple *BASECAP* commands and statements. Circuits, model parameters, and solution data can be stored and retrieved conveniently for subsequent use. Error checking and diagnostic capabilities are provided through the APL system.

The various circuit components, devices, and sources available in *BASECAP* are summarized in Table 11.1. Each entry in this table includes a symbol, for example *VDC*, which is the name of the corresponding element

Table 11.1   BASECAP components, devices and sources

	Components and Devices		Sources
R	resistor	VDC	d.c. independent voltage source
L	inductor	VAC	a.c. independent voltage source
C	capacitor	IDC	d.c. independent current source
D	diode	IAC	a.c. independent current source
$QNPN$	NPN transistor	$VCIS$	linear voltage-controlled current source (a.c. & d.c.)
$QPNP$	PNP transistor	OA	operational amplifier

description function. The syntax of each function is presented in Section 11.2. Details of the internal design and operation of these functions are found in Section 12.2. Combinations of basic elements from Table 11.1 can also be used to develop representations for additional devices in $BASECAP$ (the operational amplifier is an existing example).

There is no structural limitation on the number of elements of a given type that are allowed in the $BASECAP$ package. Similarly, there is no inherent limit on the size of the circuits, i.e. the number of nodes, that can be accommodated. In an APL-based package such as this, the internal data arrays are automatically adjusted by the APL system to the size required for the particular circuit being solved.

Each user must, however, work within the constraints of a certain workspace size which ultimately determines the maximum circuit size $BASECAP$ can accommodate. The maximum allowed workspace size may be adjusted by the APL administrator on some large time-sharing systems. In other cases, such as with stand-alone work stations or personal computers, it may be possible to increase the user's workspace size only if additional memory (RAM) can be installed in the machine. Computing charges and solution/wait times may constitute other practical constraints involving the solution of large circuits with any CACD package.

For guidance, all of the circuit examples in this text may be solved with $BASECAP$ in an APL workspace size of 128 Kbytes or more. This figure may be reduced to approximately 64 Kbytes by applying the workspace management techniques outlined in Section 10.4.1. Representative computing times are given for the various solution examples presented in Section 11.6.

11.2 Specifying Components, Devices and Sources

This section contains details of the various elements defined in $BASECAP$. These are the only elements normally permitted in a circuit description function intended for solution by $BASECAP$. Each element is described and illustrated in one of three tables: Table 11.2 contains the basic two-terminal elements; Table 11.3 contains the bipolar transistors; and Table 11.4 presents the voltage-controlled source and the operational amplifier.

All items in these tables are linear except for the diode and the transistor.

Table 11.2 Two-terminal components, devices and sources

Item/Symbol	General Format	Examples
Resistor $N1 \circ\!-\!\square\!-\!\circ N2$ $+ \quad -$	$R\ N1\ N2$ ohms	$R\ 11\ 4\ 1E3$ (5)
Inductor $N1 \circ\!-\!\curvearrowright\!-\!\circ N2$ $+ \quad -$	$L\ N1\ N2$ henrys	$L\ 5\ 9\ 1E^-3$
Capacitor $N1 \circ\!-\!\dashv\!\vdash\!-\!\circ N2$ $+ \quad -$	$C\ N1\ N2$ farads	$C\ 4\ 7, C10$ (6)
Diode $NA \circ\!-\!\triangleright\!\!\vert\!-\!\circ NC$ $+ \quad -$	$D\ NA\ NC\ IS\ CJ$ (1 2) (for ETA and $TEMP$, see $PARMSET$)	$D\ 5\ 2, D1$ (7) $D\ 5\ 2\ 1E^-12\ 3E^-11$
d.c. voltage source $N1 \circ\!-\!\dashv\!\vdash\!-\!\circ N2$ $+ \quad -$	$VDC\ N1\ N2$ volts	$VDC\ 5\ 6\ 10$ $VDC\ 6\ 5\ ^-10$ (8)
d.c. current source $N1 \circ\!-\!\oplus\!-\!\circ N2$	$IDC\ N1\ N2$ amperes	$IDC\ 1\ 3\ 0.1$ (9)
a.c. voltage source $N1 \circ\!-\!\sim\!-\!\circ N2$ $+ \quad -$	$VAC\ N1\ N2$ volts degrees (3 4)	$VAC\ 5\ 0\ 1\ 90$ (10)
a.c. current source $N1 \circ\!-\!\ominus\!-\!\circ N2$	$IAC\ N1\ N2$ amperes degrees (3 4)	$IAC\ 2\ 0, IMAG, IPH$ (11)

Notes: (refer to numbers in parenthesis above)
1. IS is the diode reverse saturation current in amperes.
2. CJ is the diode a.c. capacitance in farads (assumed constant).
3. $N2$ for VAC and IAC must always be ground (node 0).
4. The a.c. sources require a magnitude and phase angle (degrees).
5. A 1 kilohm resistor between nodes 11 and 4.
6. $C10$ is a global variable used as a model name. $C10$ must be assigned a value later e.g. $C10 \leftarrow .0001$. Note that a comma is always required in the parameter list before a model name. Each model name must be unique.
7. The model $D1$ normally requires two parameter values.
8. These are equivalent specifications for VDC.
9. This specifies 0.1 amperes out of node 1.
10. An a.c. source with a magnitude of 1.0 volt and a phase angle of 90 degrees.
11. Two model names ($IMAG$ and IPM) require two commas (see 6 above).

Table 11.3 Bipolar Transistors

Symbols	NPN transistor	PNP transistor

E o‒\/‒o C (NPN), o B

E o‒\/‒o C (PNP), o B

General Format

QNPN E   B   C   BF   BR   IES   ICS   CEB   CCB

QPNP E   B   C   BF   BR   IES   ICS   CEB   CCB

where the following parameters pertain to the Ebers–Moll transistor model (see Section 13.4.2):

BF, BR	— forward and reverse Betas
IES, ICS	— emitter-base and collector-base junction reverse saturation currents in amperes
CEB, CCB	— emitter-base and collector-base small signal capacitances in farads. These are assumed to be constant.
Note	— ETA and Temperature ($TEMP$) are global transistor parameters specified in $PARMSET$.

Examples

```
QNPN   5   6    10   100 .1   1E¯14    1E¯13    80E¯12  3E¯12
QPNP   5  7  11,  Q1
```

where $Q1$ contains the model parameters for the PNP transistor.
   $Q1$ must be specified separately, for example:

```
     Q1 ← 100    0.1    1E¯14    1E¯13    80E¯12   3E¯12
or

     BET← 100
     Q1 ← BET,   0.1    1E¯14    1E¯13    80E¯12  3E¯12


         Q1
  100  0.1    1E¯14    1E¯13   80E¯12  3E¯12
```

The resistor and the d.c. and a.c. current sources are represented internally in *BASECAP* as ideal elements. Each d.c. and a.c. voltage source is represented by a Norton equivalent circuit consisting of a 1-ohm resistor in parallel with a current source. The inductor implementation is ideal under a.c. conditions, but a small place-holding resistor (0.00001 ohms) is included in the circuit at each inductor location under d.c. conditions. Each capacitance is represented by a large resistor ($1E12$ ohms) for d.c. and this resistor is in parallel with the capacitor during the a.c. solutions.

  The d.c. diode behavior is represented by an ideal exponential relationship (Section 13.4), and the corresponding transistor implementation is based on the simplified Ebers–Moll model (Section 13.4). A capacitance is used to account for the charge-storage effects of each diode and transistor junction. Model parameters for these devices, defined in Tables 11.2 and 11.3, are

Table 11.4   Controlled source and operational amplifier

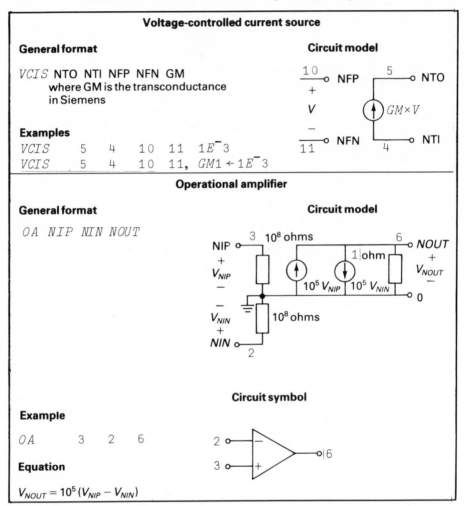

Voltage-controlled current source

General format

$VCIS$ NTO NTI NFP NFN GM
     where GM is the transconductance
     in Siemens

Examples

$VCIS$ 5   4   10   11 $1E^-3$
$VCIS$ 5   4   10   11, $GM1 \leftarrow 1E^-3$

Circuit model

Operational amplifier

General format

OA NIP NIN $NOUT$

Circuit model

Circuit symbol

Example

OA 3   2   6

Equation

$V_{NOUT} = 10^5 (V_{NIP} - V_{NIN})$

assumed to be constant. To completely specify the above exponential relationship requires two additional parameters, namely eta and temperature (designated as ETA and $TEMP$ in $BASECAP$). These are global parameters which
apply for all diodes and transistors in a circuit, and are specified as explained in
Section 11.3. In the $BASECAP$ implementation of these devices, a large resistor
($1E12$ ohms) is placed in parallel with each diode and transistor junction to
minimize convergence problems during the Newton–Raphson d.c. solution
process.

　　Each element description function, for example D , is designed to accept a
vector argument consisting of node numbers and one or more parameter
values. Arguments must be specified according to the rules for regular APL
functions. An example of this requirement is the comma before each model
name in the input parameter list (see Table 11.2 note 6).

The *BASECAP* model facility is very useful for storing commonly-used parameters for components and devices. Each model name is actually a global variable which resides in the workspace. A unique name must be chosen for each model in accordance with the conventions for naming APL variables. Parameter values are also assigned and displayed usirg the rules for variables (see the examples in Table 11.3).

11.3 Describing the Circuit

Prior to analyzing a circuit it is necessary to describe it in a manner compatible with the requirements of the circuit analysis package. The requirements for specifying a circuit for solution with *BASECAP* are discussed in this section.

11.3.1   Circuit Preparation

The circuit description process normally begins with a conventional circuit diagram. All nodes in the circuit must be numbered upwards in sequence from zero. The ground or reference node must be assigned the number zero. This is significant since *BASECAP* computes all other node voltages with respect to the reference node. Otherwise, the position of a given node number is entirely arbitrary, as long as no gaps exist in the numbering sequence. See Section 11.6 for sample circuit diagrams.

11.3.2   The Circuit Description Function

A circuit is described in *BASECAP* by means of an APL function, called a circuit description function, which must be given a unique name for a given circuit. Figure 11.1 shows the general structure and definition of a *BASECAP* circuit description function. Rules for defining, editing, manipulating, and storing circuit description functions are the same rules used for regular APL functions. Virtually any number of different circuit functions may reside in the *BASECAP* workspace.

Each circuit description function normally includes all the necessary circuit information to support both d.c. and a.c. solutions with *BASECAP* . Prior to obtaining a solution, values must be assigned to any models used. Model values may be assigned inside the circuit description function, or separately in the immediate execution mode. Considerable simplification often results from the use of models in cases where circuits contain two or more identical devices. *FBTRIPLE* in Section 11.6 is an example of a *BASECAP* circuit description function which contains several models and is useful for both d.c. and a.c. solutions.

Having generated a circuit description function, it is easy to display its

```
        ∇CCTNAME                    ...1)       CCTNAME
[1]     STARTCCT                    ...2)
[2]     VDC 1 0 10                  ...3)
[3]     R 1 0 1000                  ...3)
[4]     ⍝ THIS IS A COMMENT LINE    ...4)
[5]     END∇                        ...5)
```

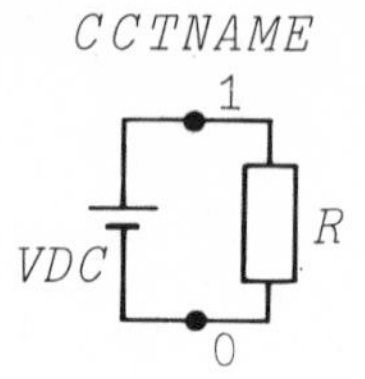

Comments (refer to the numbers on the right above)

(1)   Enter the APL function definition mode by typing the del operator, and then type the function name ($CCTNAME$). The APL system automatically supplies the subsequent line numbers.

(2)   Enter the word $STARTCCT$ on line 1. $STARTCCT$ initializes the input data arrays and normally must be the first executable statement in each circuit description function.

(3)   Describe the circuit elements by entering, for each element in turn, the element description function name, the node numbers, and parameter value(s). Elements may be entered in any order, one element per line.

(4)   Comment lines may be placed anywhere in the body of a circuit description function. These statements typically contain notes for explanation or documentation purposes which can be read by the user when the function is displayed or printed. Comment lines are non-executable, and do not provide information to other $BASECAP$ functions.

(5)   Enter the word END followed by the del operator to terminate the APL function definition mode. END is a $BASECAP$ statement which processes the data generated by the element description statements. END must normally be the last executable statement in each $BASECAP$ circuit description function.

Fig. 11.1
The general structure of a $BASECAP$ circuit description function

contents or modify it, as we illustrate below using standard APL function editing procedures:

(a)   To display the contents of $CCTNAME$, type
      ∇CCTNAME [☐] ∇

(b)   To display and replace line 2 of $CCTNAME$, type

      ∇CCTNAME [2☐]

and when the system prompts for input, enter the new contents of line 2 and terminate the entry as illustrated below:

```
[2]     VDC 1 0 10
[2]     VDC 1 0 15∇
```

(c)   A permanent copy of the new circuit description function can be created by typing

)SAVE

which causes a current copy of the entire workspace contents to be saved in the user's workspace library under the workspace name.

11.4 Solution Control Parameters

A number of global variables which contain $\overline{BASECAP}$ solution control parameters are described here. These parameters are used primarily during the Newton–Raphson d.c. solution process. A brief description of each parameter is included in the function $PARMTELL$ which may be displayed 'on line' by the user as shown in Fig. 11.2. These variables are listed also in Appendix 7, and are discussed more fully in subsequent chapters.

In the function $PARMSET$, a default value is assigned to each control parameter. Thus, all control parameters are automatically reset to their default values by executing $PARMSET$. The default values used in this function may be seen by displaying the contents of $PARMSET$ as illustrated in Fig. 11.3. Any individual parameter may be displayed or changed by the user as desired. Most users of $BASECAP$ will need to change only parameters such as ETA and $TEMP$ used in the diode and transistor model implementations.

$PARMTELL$

$ALTREF$ is a control parameter for the method of alternating bases used by the function $ALTBASE1$ with the version of the d.c. Newton–Raphson process using the functions $NEWTALT$ and $NEWTALT1$.

ETA is normally between 1 and 2.
It is used by functions $DIODE$ and $DIFDIODE$.

$TEMP$ is the global temperature in degrees C of the device junctions. It is used by the functions $DIODE$ and $DIFDIODE$.

$M1$, $M2$ and VT are used by $VBCHKCAN$ to control the junction voltage increments during the d.c. Newton–Raphson process by the functions $NEWTDC$ and $NEWTDC1$.

$SMIN$ and $VBMAX$ are used by $DIODE$ and $DIFDIODE$ to place limits on the minimum slope of the diode junction I–V relationship, and the maximum forward junction voltage, which are permitted during the d.c. solution process.

TOL sets the convergence accuracy for the d.c. Newton–Raphson solution in percent.

Fig. 11.2
Descriptions of solution control parameters in $PARMTELL$

11.5 Solution Guidelines

We now focus on the management of the overall circuit solution process. A general solution sequence is presented to guide the user in obtaining and displaying the desired circuit analysis results with $BASECAP$. Since this process is entirely interactive, actual solution sequences may vary significantly in accordance with the objectives of the user (see the examples in Section 11.6).

```
            ∇PARMSET[□] ∇
            ∇ PARMSET
  [1]       ' '
  [2]       'VARIOUS PROGRAM PARAMETERS ARE'
  [3]       'SET TO THEIR REFERENCE VALUES.'
  [4]       'TYPE PARMTELL FOR DETAILS.'
  [5]       ' '
  [6]       ALTREF←0.26
  [7]       ETA ←1
  [8]       TEMP ← 20
  [9]       M1 ←2
  [10]      M2 ←10
  [11]      SMIN←1E⁻12
  [12]      VBMAX ←1
  [13]      VT←0.026
  [14]      TOL ← 0.1
            ∇
```

Fig. 11.3
Solution control parameter values in *PARMSET*

It is important to note the following overall requirement which must be met throughout this process. As explained in Chapter 13, the d.c. solution of a circuit generates data which is required by the a.c. solution functions in *BASECAP* . Thus, a valid d.c. solution must be obtained for all linear and non-linear circuits, before obtaining one or more a.c. solutions.

A general *BASECAP* solution sequence is illustrated in the flow chart of Fig. 11.4, and described in detail in the remainder of this section.

(1) *Circuit Description*   Each circuit to be solved with *BASECAP* must normally be described in a circuit description function. The methods used to define, display, and edit circuit description functions are given in Section 11.3.

(2) *Circuit Selection*   Several different circuit description functions typically reside in the workspace simultaneously. A specific circuit must therefore be selected for solution. This requires executing the circuit description function by entering its name, for example

   CCTNAME

Prior to this step, it is necessary to ensure that values are assigned to all models used in the circuit description function as discussed in Sections 11.2 and 11.3.

The effect of this circuit selection step is to particularize the *BASECAP* input data arrays for the specified circuit. Consequently, if any changes are subsequently made to circuit connections or to circuit or device parameter values, the circuit description function must be executed again in order to update the data arrays before a subsequent solution is obtained.

(3) *D.C. Solution*   The d.c. node voltage results for the selected circuit

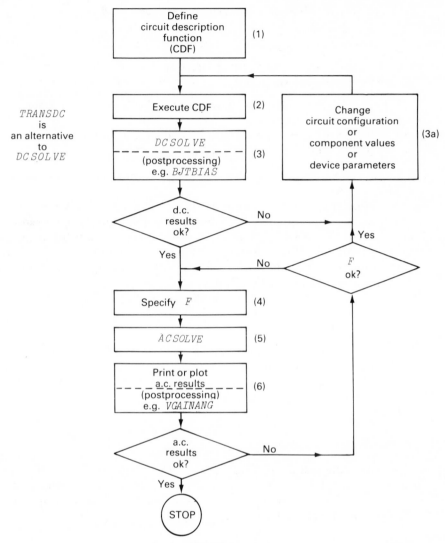

Fig. 11.4
Flow chart of the general $BASECAP$ solution sequence

are computed and displayed (in volts) by entering

    DC SOLVE

In addition, the node voltages are stored in the vector *VANS*, and the d.c. branch voltages are stored in the vector *VBN* (both in volts).

    BASECAP automatically assigns branch numbers to circuit elements in the order in which they appear in the circuit description function. Each element is given a single branch number, except for the transistor which requires two branches (one for the base–emitter junction and one for the base–collector junction), and the operational amplifier which is composed of five branches. To see the voltage across branch 2 in a circuit, for example, type

    VBN[2]

Various postprocessing functions may be used following a d.c. solution to obtain additional information on circuit behavior. For example,

(a) *BJTBIAS* conveniently summarizes the d.c. bias conditions for all bipolar transistors in the circuit,
(b) *IRDC* displays the d.c. current in a given resistor, and
(c) *IDIODE* displays the d.c. current in a given diode.

The use of these postprocessing functions is described and illustrated in Section 15.5.

Before proceeding, it is important to confirm the validity of the d.c. results obtained. In many amplifier and filter circuits, for example, it is essential for the active devices to operate within specified d.c. biassing limits before an a.c. solution is attempted. If necessary, additional d.c. solutions must be carried out under different conditions to ensure that these biassing requirements are met. Note the requirement in Fig. 11.4 to return to box 2 in order to execute the circuit description function before obtaining each new d.c. solution. The function *TRANSDC* (described later) may be a useful alternative to *DCSOLVE* in this context.

(4) *Frequency Specification BASECAP* produces an a.c. solution at each frequency value contained in the global variable *F*. Frequency values in hertz may be assigned to *F* in several ways. A direct approach is illustrated by the following two simple examples:

    F←1000
    F←100 200 300

The support functions *FLIN* and *FSTEP* are designed to help the user specify *F* over a range of frequencies from *FMIN* to *FMAX* with equal spacing between frequency points. *FLOG* is a similar function which produces logarithmically spaced frequencies. The syntax of these functions is

    FLIN   *FMIN*   *FMAX*   *NF*
    FSTEP   *FMIN*   *FMAX*   *FI*
    FLOG   *FMIN*   *FMAX*   *FPD*

where *NF* is the number of frequency points,
      FI is the desired frequency interval between points,
      FPD is the number of frequency points per decade.
The following example illustrates the use of *FLOG*:

    FLOG 1.5*E*2 1.5*E*4 3
      F
    15. 323.2 696.2 1500 3232 6962 1.5*E*4

Several other examples of the use of these functions are found in Section 11.6. For details of their internal design and operation, consult Section 14.4.

(5) *A.C. Solution*   The small-signal a.c. node voltage solutions are computed by entering

> *ACSOLVE*

No results are displayed directly by *ACSOLVE*. Instead the complex node voltages are stored (in volts) in the array *VREIM* for subsequent selection and display. Special functions provided for this purpose are discussed later.

For reference, *VREIM* contains the node voltage results in two planes (in real and imaginary parts respectively), in rows 1 to *NR*, with a separate column for each frequency contained in the variable *F* (*NR* is the number of circuit nodes not counting the reference node). The shape of *VREIM* is therefore (2, *NR*, (ρF)).

The following examples illustrate how the user can access the results in *VREIM* for special requirements:

(a) To display only the real part (plane 1) of all node voltages for the fifth frequency value, enter

> *VREIM*[1;;5]

(b) To display both real and imaginary parts of the voltage at node 3 for all frequencies, enter

> *VREIM*[;3;]

Consult Sections 14.2 and 15.1 for further details.

(6) *Results Selection and Display*   Following the a.c. solution of a circuit, selected a.c. node voltages may be processed and displayed by entering a sequence of choices. One choice must be entered in sequence from each set of square brackets, following the general format shown below:

$$
\left[\begin{matrix} CV\ CH\ PLOT \\ \text{or} \qquad\ PRINT \end{matrix} \right]
\left[\begin{matrix} F \\ \text{or} \quad LOGF \end{matrix} \right]
\left[\begin{matrix} , \\ \text{or}\ AND \end{matrix} \right]
\left[\begin{matrix} MAG \\ \text{or}\ MAGDB \\ \text{or}\ ANGLE \end{matrix} \right]
\left[ACVOUT\ NOUT \right]
$$

where

(a) the square brackets are shown to clarify the various choice categories, and are not entered by the user,
(b) *CV* and *CH* are integers which specify the size of the resulting plot (approximately *CV* characters vertically by *CH* characters horizontally).
(c) *NOUT* represents the node number or numbers for which the a.c. voltages are to be processed and displayed, and
(d) the ',' symbol in the middle set of brackets is an essential APL operator (lamination). Either this operator or the function *AND* must be used here.

ACVOUT is a function which selects the designated node voltages from the array *VREIM* for processing by one of the functions *MAG*, *MAGDB*, or *ANGLE* to produce, respectively, the voltage magnitudes in volts, their magnitudes in dB, or their phase angles in degrees. Further details concerning the design and operation of these functions are contained in Chapter 15, Sections 15.1 to 15.4. Sample choices are illustrated below:

(a)   `30 40 PLOT LOGF, MAGDB ACVOUT 5`

produces a plot (approximately 30 characters by 40 characters in size) of the magnitude in dB of the voltage at node 5 versus frequency on a logarithmic scale.

(b)   `PRINT F,ANGLE ACVOUT 1 8`

produces a printed list of frequencies (in column 1) together with the phase angles (in degrees) of the a.c. voltages at nodes 1 and 8 (in columns 2 and 3 respectively).

(c)   `PRINT F,(MAGDB ACVOUT 3),(ANGLE ACVOUT 3)`

shows how multiple choices may be made in accordance with the standard rules of APL (see the low-pass filter example (*LPF*) in Section 11.6 for an actual printout produced by this statement).

Postprocessing functions, such as *VGAINMAG* and *VGAINANG*, which produce the magnitude and phase angle of the a.c. voltage gain respectively, also may be used following an a.c. solution. See Section 15.5.2 for details of these functions.

Note from Fig. 11.4 that a new a.c. solution may be obtained directly after changing frequencies in the vector *F*. However, if changes are made to the circuit configuration or to component or device parameter values, it is necessary to loop back to box 2 in the flow chart before subsequently producing a new a.c. solution.

(7) *Alternative Solutions (D.C. Transfer Function)* The *BASECAP* function *DCSOLVE* (item 3 above) produces a single d.c. node voltage solution of a circuit for a given set of conditions. As an alternative, the function *TRANSDC* may be used to obtain a d.c. transfer function solution, which consists of a series of separate solutions with one set of node voltages for each value of a circuit or device parameter. *TRANSDC* is useful as a design aid, for example, to determine an optimum value for a parameter which affects the d.c. conditions in a circuit.

The results returned by *TRANSDC* are compatible with *PRINT* or *PLOT* (values of the independent parameter are in column 1 and the corresponding d.c. node voltages for one or more designated nodes are in one or more succeeding columns). The syntax for *TRANSDC* is typically

   `30 40 PLOT TRANSDC NOUT`

where *NOUT* consists of one or more node numbers for which the d.c. node voltages are to be displayed, for example,

   `NOUT←2 5`

The use of *PLOT* (or *PRINT*) is optional, but for the case shown, the designated node voltages will be plotted conveniently as a function of the independent parameter.

During execution, *TRANSDC* prompts the user to enter the following information:

(a)   the name of the circuit description function for the circuit to be solved.
(b)   the name of the independent parameter to be varied.
(c)   the values to be used for the independent parameter.

The independent parameter specified above must normally be either

(a)   a model name used to reference a parameter in an element description statement in the specified *BASECAP* circuit description function (for an example, see *BET* in the transistor parameter list in Table 11.3), or
(b)   a global variable used as a parameter in *BASECAP*, such as *TEMP* (temperature).

An illustration of the use of *TRANSDC* is given in the example solution for the circuit *FBTRIPLE* in Section 11.6. Further details concerning the internal design and operation of *TRANSDC* are found in Sections 13.1 and 13.2.

For reference, all results computed by *TRANSDC* (that is, all node voltages in volts for all values of the independent parameter) are automatically stored in the matrix *VALL* (see Table 15.2 for an illustration). Each set of node voltages is contained in a row of *VALL*, with each row corresponding to a different value of the independent parameter. For example,

   VALL[2;]

displays all node voltages for the second value of the independent parameter, and

   VALL[;5]

displays the voltage at node 5 for all values of the independent parameter.

11.6 Annotated Circuit Examples

Three *BASECAP* circuit examples of varying complexity are presented in this section. Each example includes a printed record of the terminal session used to carry out the solutions on an interactive computer with *BASECAP*. Explanatory comments are added to help the reader interpret the computer print-out. For simplicity, these examples focus primarily on the user's interactions with *BASECAP*, and little or no emphasis is placed on the use of APL system commands associated with loading *BASECAP* or managing the active workspace in the APL environment. Guidance on these matters is provided in Appendix 4.

The *CIRCUITS* library supplied with *BASECAP* includes the circuit description function for each of these examples. Readers are encouraged to obtain appropriate 'hands on' experience by similarly working these and other circuit examples while using *BASECAP* on an interactive computer system.

BASECAP automatically prints the amount of computer time used to produce both the d.c. and a.c. solutions. For reference, all the example solutions in this section were produced by *BASECAP* on an IBM 3081 computer with VS-APL, release 4.0.

11.6.1   **Low-pass Filter** (*LPF*)

The a.c. performance of the simple low-pass filter circuit (*LPF*) is featured in this example. The specific voltage-controlled current source and load resistance combination clearly results in a transfer gain of unity (0 dB) at low frequencies. A simple calculation will confirm that the voltage at node 3 (or 2) is expected to be 3 dB down at 1 kHz.

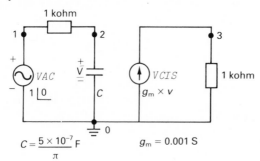

$$C = \frac{5 \times 10^{-7}}{\pi} F \qquad\qquad g_m = 0.001 \text{ S}$$

**Fig. 11.5
The circuit LPF.**

```
        ∇LPF[□] ∇           The listing of the BASECAP circuit
        ∇ LPF               description function LPF. The a.c.
[1]     STARTCCT            voltage source at node 1 has a magnitude
[2]     VAC 1 0 1 0         of 1.0 V for convenience.
[3]     R 1 2 1000
[4]     C 2 0 ,5E ‾7÷O1
[5]     VCIS 3 0 2 0 0.001
[6]     R 3 0 1000
[7]     END
        ∇
```

LPF   Select (execute) the circuit description
                            function *LPF*.

DC SOLVE   Obtain a d.c. solution (it is mandatory).
CPUSE CONDS = 0.02   The Newton–Raphson solution requires
ITERATIONS = 2   two iterations and takes 0.02 s on the
                            IBM 3081 computer.

↓↓*DC NODE VOLTAGES* 1 *TO* 3↓↓

0 0 0   The resultant d.c. voltages are 0 V.

FLOG 100   100000   10   Set up a range of frequencies.

ACSOLVE   Obtain the a.c. solution.
SOLVING FOR 31 *FREQUENCY POINTS*
CPUSECONDS = 0.094   The CPU time for this solution on an
DONE   IBM 3081 computer is 0.094 s.

30 40 *PLOT LOGF, MAGDB ACVOUT* 3

Obtain a plot of the magnitude in dB of
the voltage at node 3 versus the log of
frequency (Fig. 11.6).

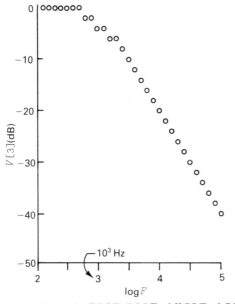

The low frequency result is 0 dB.

The output decreases by 20 dB per
decade at high frequencies as
expected.

Fig. 11.6

30 40 *PLOT LOGF, ANGLE ACVOUT* 3   Request a plot of $\underline{/V[3]}$ versus
log of frequency (Fig. 11.7)

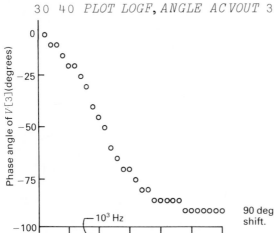

Fig. 11.7

90 degrees is the maximum phase
shift.

```
PRINT F,(MAGDB ACVOUT 3),(ANGLE ACVOUT 3)
1.000E 2    ‾4.330E‾2    ‾5.716E 0
1.259E 2    ‾6.843E‾2    ‾7.182E 0
1.585E 2    ‾1.080E‾1    ‾9.015E 0
1.995E 2    ‾1.699E‾1    ‾1.129E 1
2.512E 2    ‾2.662E‾1    ‾1.411E 1
3.162E 2    ‾4.147E‾1    ‾1.756E 1
3.981E 2    ‾6.401E‾1    ‾2.173E 1
5.012E 2    ‾9.750E‾1    ‾2.664E 1
6.310E 2    ‾1.458E 0    ‾3.228E 1
7.943E 2    ‾2.128E 0    ‾3.849E 1
1.000E 3    ‾3.015E 0    ‾4.503E 1
1.259E 3    ‾4.130E 0    ‾5.157E 1
1.585E 3    ‾5.462E 0    ‾5.778E 1
1.995E 3    ‾6.980E 0    ‾6.340E 1
2.512E 3    ‾8.646E 0    ‾6.831E 1
3.162E 3    ‾1.042E 1    ‾7.247E 1
3.981E 3    ‾1.227E 1    ‾7.591E 1
5.012E 3    ‾1.418E 1    ‾7.873E 1
```

Request the magnitude (dB) and phase for V[3] vs frequency.

Note that F is in column 1, and $V[3]$ is in column 3.

The response is 3 dB down and the phase shift is 45° at 1 kHz (approx.) for the components used.

(Partial printout shown)

11.6.2   Feedback Triple (*FBTRIPLE*)

The d.c. and a.c. operation of the circuit *FBTRIPLE* are investigated in this example. This amplifier circuit features direct-coupled biassing circuitry, and extensive use is made of both d.c. and a.c. negative feedback.

As a check on the a.c. node voltage results computed with *BASECAP*, we calculate the approximate mid-band voltage gain in the circuit between nodes 10 and 8. In the mid-frequency band, the impedance of *C*1 is negligibly small, and that of *C*2 is large enough to have negligible effect. Consequently the mid-band gain is determined essentially by the overall negative feedback connection of *RI* and the two *R*1 resistors in series. Assuming the gain of the three transistors inside the feedback loop is sufficiently high, we can write the following

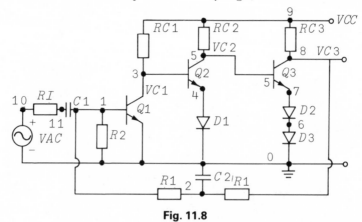

Fig. 11.8

expression

$$|VOUT| = (|VIN| \times 2 \times R1) \div RI$$

For example, if $R1 = 2 \times 10^4$ kohms, and $RI = 10^3$ kohms, the ratio ($|VOUT \div VIN|$) is approximately 40.0, or 32.04 dB. The circuit *FBTRIPLE* is shown in Fig. 11.8.

	$\nabla FBTRIPLE[\Box] \nabla$	The circuit description function is
	∇ FBTRIPLE	requested *FBTRIPLE* is displayed.
[1]	STARTCCT	
[2]	R 1 0 ,R2	
[3]	QNPN 0 1 3 ,Q1	The symbols *Q1*, *QN*, *R1*, etc. are
[4]	R 9 3 14600	user-defined models.
[5]	R 9 5 14600	
[6]	QNPN 4 3 5 ,QN	
[7]	D 4 0 ,D1	
[8]	R 9 8 14600	
[9]	QNPN 7 5 8 ,QN	The a.c. voltage used for the source at
[10]	D 7 6 ,D1	node 10 is 1.0 V for convenience.
[11]	D 6 0 ,D1	Note that any value can be used since
[12]	VDC 9 0 20	the a.c. solution is linear.
[13]	R 8 2 ,R1	
[14]	R 2 1 ,R1	
[15]	R 10 11 ,RI	
[16]	C 11 1 ,C1	
[17]	C 2 0 ,C2	
[18]	VAC 10 0 1 0	
[19]	END	
	∇	

We first assign parameter values and carry out a d.c. transfer function solution for *FBTRIPLE* as shown below.

```
Q1←QN←50 .1 1E¯9 1E¯9 8E¯11 3E¯12
R2←8.2E3                          Values are assigned to circuit and device
C1←100E¯6                         models used in FBTRIPLE
C2←1E¯12
D1←1E¯9 3E¯12
RI←1E3                            R1VAL contains values to be used for the
                                  independent parameter R1
R1VAL
2E5 1.5E5 1E5 7.5E4 5E4 4E4 3E4 2.5E4 2E4 1.8E4 1.6E4 1.4E4 1.2E4 1E4
   8000 6000 4000 2000 1000 500
                                  Executing PARMSET assigns default
PARMSET                           values to various BASECAP solution
                                  control parameters.

VARIOUS PROGRAM PARAMETERS ARE
SET TO THEIR REFERENCE VALUES.
TYPE PARMTELL FOR DETAILS.        Obtain a d.c. transfer function plot of the
   40 60 PLOT TRANSDC 8           voltage at node 8 of the circuit FBTRIPLE
```

```
TYPE CIRCUIT NAME
FBTRIPLE
TYPE NAME OF PARAMETER TO BE VARIED
R1
TYPE PARAMETER VALUES   R1  is the independent parameter.
□:
            R1VAL                    The parameter values in R1VAL are
IT= 13                               used for R1
IT= 4
IT= 6
IT= 3
IT= 3
IT= 3
IT= 3
IT= 2                                One d.c. solution is obtained for each
IT= 2                                value of R1 in R1VAL, and the
IT= 2                                number of iterations printed out for each
IT= 2                                separate solution obtained. Note that IT
IT= 2                                is generally very small since each new
IT= 2                                solution starts from the previous solution.
IT= 2
IT= 2                                (The requested plot appears as Fig. 11.9)
IT= 2
IT= 2
IT= 12
IT= 4
IT= 12
IT= 4
IT= 3
```

We now carry out conventional d.c. and a.c. solutions:

`R1←20E3`	$R1$ is chosen to be 20 kohm
`FBTRIPLE`	Set up the circuit *FBTRIPLE*
`DCSOLVE`	Request a d.c. solution
`CPUSECONDS = 0.86`	The Newton–Raphson d.c. solution
`ITERATIONS = 14`	requires 14 iterations and uses 0.86 seconds of CPU time on an IBM 3081 computer
`↓↓DC NODE VOLTAGES 1 TO 11↓↓`	The d.c. node voltages are listed below

`0.356 1.74 0.712 0.356 1.06 0.352 0.703 3.13 20 3.56E⁻13 3.57E⁻10`

`FLOG 10 1E8 8`	Select the solution frequencies
`ACSOLVE`	Carry out the a.c. solution
`SOLVING FOR 57 FREQUENCY POINTS`	
`CPUSECONDS = 1.06`	This a.c. solution uses 1.06 s of CPU time
`DONE`	on an IBM 3081 computer
`40 60 PLOT LOGF AND MAGDB ACVOUT 8`	
	Obtain a log–log plot (Fig. 11.10)

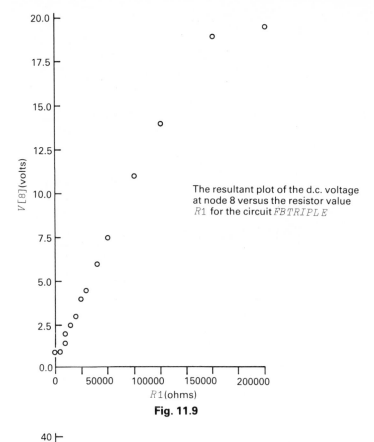

The resultant plot of the d.c. voltage at node 8 versus the resistor value $R1$ for the circuit $FBTRIPLE$

Fig. 11.9

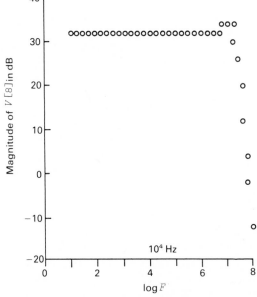

The mid-band gain is $\simeq 32$ dB as expected.

Fig. 11.10

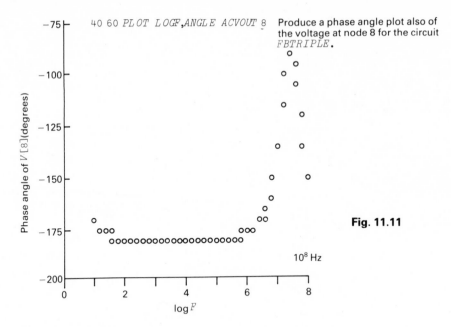

40 60 *PLOT LOGF,ANGLE ACVOUT* 8   Produce a phase angle plot also of the voltage at node 8 for the circuit *FBTRIPLE*.

Fig. 11.11

We now obtain a new solution of *FBTRIPLE* with a different value for the capacitor $C2$ (1000μF).

$C2$	It is interesting to note from the circuit
$1E^-12$	diagram that a very large value for $C2$
$C2\leftarrow1000E^-6$	effectively 'kills' the a.c. negative feed-back from output to input
FBTRIPLE	Note that *FBTRIPLE* must be executed
	again but a new d.c. solution is not
FLOG 1 1E8 8	mandatory here
	A wider frequency range is selected
ACSOLVE	

SOLVING FOR 65 *FREQUENCY POINTS*
CPUSECONDS = 1.21   The a.c. solution is repeated for the new
DONE   frequency values using the new value of $C2$

40 60 *PLOT LOGF, MAGDB ACVOUT* 8
Plot the same node voltage result as before (results appear in Fig. 11.12)

11.6.3 Franklin Oscillator (*FRANKOSC*)

Two operational amplifiers, a transistor, and a tuned circuit are used in a basic Franklin oscillator configuration in the circuit *FRANKOSC*. In the arrangement used here, however, the main feedback loop has been broken between nodes 2 and 9, and an a.c. voltage source is connected to node 9 for testing purposes.

Of interest in this example are the circuit conditions required to produce oscillation (when the loop is closed), as determined by open-loop testing. When

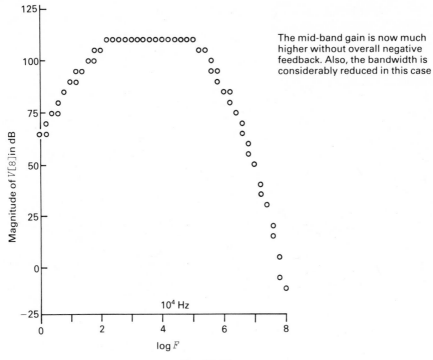

The mid-band gain is now much higher without overall negative feedback. Also, the bandwidth is considerably reduced in this case

Fig. 11.12

connected for oscillation, the output will be sinusoidal due to the selective response of the parallel tuned circuit in the collector of the transistor. The general oscillation conditions of interest are

(a) when the gain (around the loop) from node 9 to node 2 is slightly greater than unity, and
(b) the phase shift in the signal between nodes 2 and 9 is zero degrees.

Consult Exercise 11.2 for a detailed analysis of this circuit.
The circuit *FRANKOSC* is shown in Fig. 11.13.

```
           ∇ FRANKOSC
[1]      ⍝ BASED ON ELECTRONICS II PROJECTS
[2]      ⍝ BY MICHAEL POTHIER AND BERNARD PLOURDE
[3]        STARTCCT
[4]        R  6  0  76000
[5]        R  6  7  10000
[6]        R  10 7  1000
[7]        R  8  9  22000
[8]        R  1  2  ,RF
[9]        R  3  1  20000
[10]       R  5  0  5.5
[11]       R  0  4  ,RP
```

STARTCCT The circuit listing for *FRANKOSC*.

The symbols *RF*, *RP*, *C*1 and *Q*1 are model names used in this circuit description function for convenience. Note that the magnitude of the a.c. voltage source at node 9 is 1.0 V, so the voltage at node 2 equals the gain directly.

```
[12]      L 4  5  0.069
[13]      C 4  0  ,C1
[14]      C 6  8  1E‾6
[15]      VDC 0 7  15
[16]      VAC 9 0  1  0
[17]      OA 4  3  3
[18]      OA 0  1  2
[19]      QNPN 10 6  4  ,Q1
[20]      END
          ∇
```

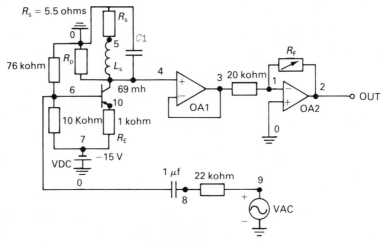

Fig. 11.13

In this example, we first assign values to various solution control parameters by executing *PARMSET* and then we specify values for the device and component models used in *FRANKOSC.* Following this, we compute and display selected d.c. and a.c. node voltage results.

PARMSET	Execute *PARMSET*

VARIOUS PROGRAM PARAMETERS ARE
SET TO THEIR REFERENCE VALUES.
TYPE PARMTELL FOR DETAILS.

SMIN←1E‾16	Assign a new value to *SMIN*
RF←91	Assign values to the models used. A large
RP←1E 20	value of *RP* results in a high *Q* for the
C1←1.47E‾8	tuned circuit
Q1←100 .1 1E‾16 1E‾16 0 0	
□*PW ←70*	Specify a 70-character maximum print width
FRANKOSC	Execute the circuit description function *FRANKOSC*

DC SOLVE	Produce the d.c. solution

CPUSECONDS = 0.362
ITERATIONS = 6

$\downarrow\downarrow$ *DC NODE VOLTAGES 1 TO 10* $\downarrow\downarrow$

$^-$2.28E^-10 0.0000226 $^-$0.00496 $^-$0.00496 $^-$0.00496 $^-$13.3 $^-$15 $^-$2.93E^-7
    1.33E^-11 $^-$14.1

	The voltage drop across *RE* is 0.9 V giving an emitter current of 0.9 mA (approx.)
BJTBIAS	The function *BJTBIAS* produces a convenient summary of the d.c. biassing conditions for each bipolar transistor in the circuit.

TYPE	QNO	ENODE	BNODE	CNODE	VBE	VCE	IE(MA.)	IC(MA.)
NPN	1	10	6	4	.754	14.084	$^-$.910	.901

FLIN 4970 5020 50	Choose frequencies near the expected resonance
AC SOLVE	Obtain the a.c. solution.

SOLVING FOR 51 *FREQUENCY POINTS*
CPUSECONDS = 0.728
DONE

Figure 11.14 shows a linear plot of the magnitude of the a.c. voltage at node 2 versus frequency for the circuit *FRANKOSC*.

$\square PP \leftarrow 5$	The printing precision is set to five decimal places.

PRINT F,(MAG ACVOUT 2*) AND ANGLE ACVOUT* 2

4970	0.2278	76.769
|	|	|
|	|	|
|	|	|
4994	0.88915	27.332
4995	0.94175	19.832
4996	0.98107	11.553
4997	1.0004	2.7552
4998	0.99601	$^-$6.1753
4999	0.96872	$^-$14.819
5000	0.92377	$^-$22.83
5001	0.86818	$^-$30.003
|	|	|
|	|	|
|	|	|

The magnitude and phase angle of the voltage at node 2 are printed along with frequency.

Resonance appears to be between 4997 and 4998 Hz.

Only partial results are shown here.

40 60 *PLOT F', MAG ACVOUT 2*

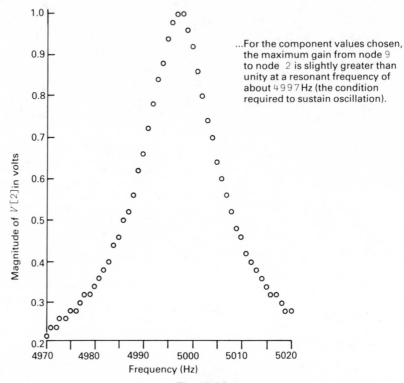

...For the component values chosen, the maximum gain from node 9 to node 2 is slightly greater than unity at a resonant frequency of about 4997 Hz (the condition required to sustain oscillation).

Fig. 11.14

40 60 *PLOT F AND ANGLE ACVOUT 2*

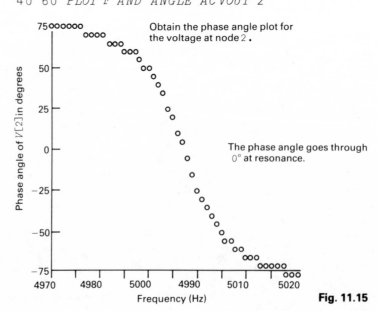

Obtain the phase angle plot for the voltage at node 2 .

The phase angle goes through 0° at resonance.

Fig. 11.15

11.7 When Things go Wrong

It is essential for a CACD package to produce results which are meaningful and sufficiently accurate to be accepted and used with confidence for their intended purposes. The combination of the computer system, the CACD package and the user is a very sophisticated one, and sometimes things go wrong. Occasionally, for example, no results may be obtained at all on a first attempt. In all cases, it is important to remember that the computer and CACD package can only simulate the operation of an electronic circuit, and there are many possible sources of error which the user should be aware of. In the end, it falls to the user to make the CACD package work in his intended application (if that is possible), and it is the user who must ultimately accept responsibility for the validity of the results.

In practice, a credible CACD package will normally produce results that are acceptable within reasonable accuracy limits, when used as intended by the package designers. Sometimes, however, the successful use of a CACD package requires, or is greatly facilitated by, a more extensive knowledge of CACD and the package internals than is normally provided in the user documentation. The material in Chapters 1–9 on circuit analysis methods and algorithms and the subsequent material in Chapters 12–15 on package structure and design are devoted in part to satisfying this requirement for *BASECAP* . In the event of difficulties, or when contemplating an unusual or critical application, *BASECAP* users should refer to these sections of the book for more detailed information and assistance.

In addition to familiarity with the CACD package, the process of checking for errors and diagnosing problems requires a thorough understanding of the specifications and operation of the proposed circuit. *Even in routine applications, the results produced by a CACD package should be accepted with confidence only after carefully checking both the circuit data supplied to the package, and the output generated.*

The following material focusses on potential problems and errors, and is particularly intended to guide the user in checking results obtained with the *BASECAP* package. For purposes of this discussion, errors and problems are assumed to fall into one or more of the following categories: input errors, program-related errors, and system-related errors. While not all errors are directly caused by the user, virtually all can be diagnosed, and corrected or overcome, so that acceptable results are ultimately obtained.

11.7.1   Input Errors

The majority of the errors in this category are made by the user while coding the input data or describing the circuit to *BASECAP* . Examples include typographical mistakes, node numbering errors, using incorrect device or component parameter values, and forgetting to execute the circuit description function prior to obtaining a solution.

A significant amount of error checking is automatically carried out by the APL system. Examples include tests for incorrect element description function syntax, missing data, and certain types of invalid data entries. By means of additional programming within *BASECAP* , other automated tests could also be carried out on the input data. For an example, see Exercise 12.3 which deals with an algorithm to check for gaps in the node numbering sequence. Most numerical input errors, however, are such that they can only be discovered by the user.

The validity of *BASECAP* input data can be directly checked by the user in two main ways. The most common is to display the contents of the circuit description function and compare each entry with the corresponding element in the circuit diagram. See the example displays in Section 11.6.

The second approach is to display and check various *BASECAP* internal data variables, after executing the circuit description function. An understanding of the material in Chapter 12 concerning *BASECAP* data structure is required to do this effectively, however. Representative input data variables are: G, IS, IX, M, Q, $VACL$, $IACL$, LL, RL, and CL. Example 11.1 illustrates selected input data variables for $FBTRIPLE$ using a utility function called $DISPLAY$. Consult Chapter 12 for detailed information and additional examples.

Example 11.1

The contents of three internal data variables, IX, M, and CL, are illustrated below for the circuit $FBTRIPLE$ using the utility function $DISPLAY$:

```
FBTRIPLE
DISPLAY 'IX M CL'                          IX is the d.c. source current vector
```

ITEM		*CONTENTS*

```
IX    ↔   0  0  0   0   0  0   0  0 20  0  0
M     ↔   1  1  1   9   9  3   3  4  9  5   5  7  6  9  8  2 10 11  2 10
         12 12  3   3   5  4   5 12  8  7   8  6 12 12  2  1 11  1 12 12
CL    ↔   1.20E1   1.00E0   8.00E⁻11
         3.00E0   1.00E0   3.00E⁻12    M contains the pair of node
         4.00E0   3.00E0   8.00E⁻11    numbers for each circuit branch.
         5.00E0   3.00E0   3.00E⁻12    Internally, the reference node
         4.00E0   1.20E1   3.00E⁻12    number 0 is replaced by the
         7.00E0   5.00E0   8.00E⁻11    number 12 which is one more than
         8.00E0   5.00E0   3.00E⁻12    the highest user-assigned node
         7.00E0   6.00E0   3.00E⁻12    number
         6.00E0   1.20E1   3.00E⁻12
         1.10E1   1.00E0   1.00E⁻4
         2.00E0   1.20E1   1.00E⁻6    CL contains the pair of nodes and
                                      the value for each
                                      capacitance in the circuit.
```

Fortunately, many types of input errors result in 'gross' solution errors. The resulting APL system error messages greatly assist in identifying and correcting such errors. In other cases the symptoms are more obscure, such as the failure of the d.c. solution process to converge. Examples 11.2 to 11.4 illustrate system responses for deliberately induced errors using circuits from Section 11.6.

Example 11.2

The use of a zero-valued resistance is not allowed, as illustrated below:

	$R2\leftarrow0$	$R2$ is set to zero.
	$FBTRIPLE$	The execution of $FBTRIPLE$ in this
$DOMAIN\ ERROR$		case produces a domain error in the
$R[4]$	$GIN\leftarrow GIN, \div P[3]$	resistor function as $1 \div 0$ is infinite.
	$\wedge$	

	$P[3]$	$P[3]$ is clearly zero.
0		

	$)SI$	Checking the state indicator reveals
$R[4]$	$\star$	that the problem occurred on line 2
$FBTRIPLE[2]$		of $FBTRIPLE$.

	$\nabla FBTRIPLE[2\square]\ \nabla$	We display line 2 of $FBTRIPLE$ and note
$[2]$	$R\ 1\ 0\ ,R2$	the use of the resistor element description
	$R2$	function R with the model name $R2$.
0		The subsequent display of $R2$ confirms
		the value of zero ohms.

Example 11.3

A problem with the input data causes a domain error when the d.c. solution is attempted in this example with the circuit $FBTRIPLE$.

	$\nabla FBTRIPLE[2\square]$	
$[2]$	$R\ 1\ 0\ ,R2$	Here we deliberately set up the
$[2]$	$R\ 21\ 0,R2\nabla$	problem with an invalid node
		number.

	$R2\leftarrow8.2E3$	
	$FBTRIPLE$	$FBTRIPLE$ executes without any
		indication of the problem.

	$DCSOLVE$	
$DOMAIN\ ERROR$		The d.c. solution terminates with
$NEWTALT[6]$	$V\leftarrow ICOM\boxdot SLOPE$	a domain error – a problem with
	$\wedge$	$ICOM$ or $SLOPE$.

	$\rho SLOPE$	We begin checking the data and find
$21\ 21$		some strange numbers here and for
		the contents of M.

	M	

```
21  1 1 9 9 3 3   4 9 5 5 7   6   9 8 2 10 11   2 10
22 22 3 3 5 4 5 22 8 7 8 6 22 22 2  1 11   1 22 22
```

	NH	The highest user-assigned node
22		number in this circuit is 11. There-
		fore NH should contain 12,
		suggesting a node numbering
		error exists.

```
      ∇FBTRIPLE[□]∇
      ∇ FBTRIPLE
[1]    STARTCCT
[2]    R 21 0 ,R2
[3]    QNPN 0 1 3 ,Q1
[4]    R 9 3 14600
[5]    R 9 5 14600
[6]    QNPN 4 3 5 ,QN
[7]    D 4 0 ,D1
[8]    R 9 8 14600
[9]    QNPN 7 5 8 ,QN
[10]   D 7 6 ,D1
[11]   D 6 0 ,D1
[12]   VDC 9 0 20
[13]   R 8 2 ,R1
[14]   R 2 1 ,R1
[15]   R 10 11 ,RI
[16]   C 11 1 ,C1
[17]   C 2 0 ,C2
[18]   VAC 10 0 1 0
[19]   END
      ∇
```

Upon checking the contents of *FBTRIPLE* with the circuit diagram, we discover a typographical error has occurred on line 2. Thus the 21 should be 1. Note that gaps in the node numbering sequence are not allowed.

Example 11.4

In this example we explore the effects of using a zero-valued inductor in the circuit *FRANKOSC*, and a zero-valued capacitance in the circuit *FBTRIPLE*.

```
      ∇FRANKOSC[12□ ]
[12]   L 4 5 0.069
[12]   L 4 5 0∇
```

The inductance is given a value of zero.

```
      FRANKOSC

      DCSOLVE
CPUSECONDS = 8.2
ITERATIONS = 6
```

A d.c. solution is obtained without difficulty.

```
↓↓DC NODE VOLTAGES 1 TO 10↓↓
```

```
¯2.28E¯10 0.0000226 ¯0.00496 ¯0.00496 ¯0.00496 ¯13.3 ¯15 ¯2.93E¯7
     ¯1.33E¯11 ¯14.1
```

```
      F←5000
      ACSOLVE
DOMAIN ERROR
ACPREP[2] BL ←FORM LL[; 1 2],÷LL[;3]
                       ∧
```

A domain error occurs when the zero-valued inductance is encountered during the formation of the reciprocal inductance matrix.

```
      LL
  4 5 0
```

We next obtain d.c. and a.c. solutions for *FBTRIPLE* under the condition of having the capacitance *C2* set to zero.

```
      C2 ←0
      FBTRIPLE
      DCSOLVE
CPUSECONDS =0.858
ITERATIONS=14
```
A valid d.c. solution is obtained with $C2$ set to 0.

```
↓↓DC NODE VOLTAGES 1 TO 11 ↓↓
```

```
0.356 1.74 0.712 0.356 1.06 0.352 0.703 3.13 20 3.56E¯13 3.57E¯10
      F←1000
      ACSOLVE
SOLVING FOR 1 FREQUENCY POINTS
CPUSECONDS =0.023
DONE
```
No difficulty is encountered during the a.c. solution (a display of the a.c. results will confirm that they are as expected for a zero value for $C2$).

```
         CL
   1.20E1       1.00E0       8.00E¯11
   3.00E0       1.00E0       3.00E¯12
   4.00E0       3.00E0       8.00E¯11
   5.00E0       3.00E0       3.00E¯12
   4.00E0       1.20E1       3.00E¯12
   7.00E0       5.00E0       8.00E¯11
   8.00E0       5.00E0       3.00E¯12
   7.00E0       6.00E0       3.00E¯12
   6.00E0       1.20E1       3.00E¯12
   1.10E1       1.00E0       1.00E¯4
   2.00E0       1.20E1       0.00E0
```
$C2$ is definitely zero.

We therefore conclude that a zero-valued capacitance is allowed in *BASECAP*.

Once a solution is obtained for a new circuit, the numbers should be checked and compared with expected values. Any problems with the d.c. results should be corrected before producing an a.c. solution. It is very important to make sure all transistors are biassed correctly. The function *BJTBIAS* is very convenient for this purpose as illustrated in Section 11.6.3. Finally, the a.c. results should be checked, preferably by carrying out a nominal solution at a convenient test frequency where conditions are familiar to the user.

11.7.2   Program-related Errors

Implementation factors.   Potential sources of error exist in the methods and algorithms used in the implementation of the *BASECAP* package. Such errors are not normally significant, but they may become so under certain circumstances. Of concern mainly are the internal representations used for some of the components, sources, and devices in the package (see Section 11.2) as noted below:

(1)   The 1 ohm source resistance used in the Norton equivalent representation of the d.c. and a.c. voltage sources may distort the potentials in circuits with very low impedance levels. See Exercise 14.4 for information on how to remove the effects of these source resistances.

(2)   The 10^{12} ohm resistor used in parallel with each capacitor and each diode and transistor junction may disturb impedance levels at very low

frequencies or in very high-impedance circuits. If it becomes necessary or desirable to remove the effects of these resistors, see Exercise 14.5 for assistance.

(3)   The simplified bipolar transistor model used in *BASECAP* does not include such secondary effects as base-width modulation (Early effect), parasitic resistance in the base, emitter or collector regions, high-level injection, and dependence of β on d.c. current levels. Also, all diodes and transistors in the circuit must have the same junction temperature, and no provision is made for automatically adjusting junction capacitance values to account for their dependence on d.c. operating conditions.

(4)   The operational amplifier model is relatively simple and may prove restrictive in some applications. For example, the existing model has no provision for changing the gain, and there is no built-in frequency compensation. Such features can be incorporated easily in a new model as considered in Exercise 12.5.

(5)   Models may be required for other devices not included in *BASECAP* . In some cases, these can be implemented directly with existing elements as in the case of the op amp model. See Exercise 12.4 for an a.c. model of the MOSFET and JFET.

Convergence problems. *BASECAP* and many other CACD packages use a modified Newton–Raphson iterative d.c. solution process which converges satisfactorily for most circuits. However, convergence cannot be guaranteed in all cases. An understanding of the material in Chapters 2, 7, and 13 (particularly Sections 2.4, 7.5, 13.3, and 13.5) will greatly assist in analyzing and overcoming such difficulties when using *BASECAP* .

Currently in *BASECAP*, d.c. convergence is attained (see *NEWTALT* line 9 in Section 13.3.1) when the largest branch voltage change from one iteration to the next is less than or equal to an absolute error (typically $1E^-12$) plus a relative error determined by the tolerance parameter *TOL* (typically 0.1%). These error parameters are chosen to give a reasonable compromise between solution efficiency and solution accuracy for most cases. Both values may be changed by the user to accommodate specific requirements (see Section 11.4).

Circuit configuration, starting conditions, device parameters, and solution control parameters significantly affect solution efficiency. Generally, convergence with *DCSOLVE* is obtained in less than 50 iterations, and typically in less than 20 for small or medium circuits. In some cases, however, well over 100 iterations may be required. The solution can always be interrupted by the user (with the ATTN key). Also, a system interrupt will occur on some systems if a system time-out limit is exceeded. The number of iterations can then be checked by displaying the iteration counter IT. If desired, the solution may be continued from that point by entering

$\rightarrow\square LC$

In extreme cases, convergence may not be obtained at all without the use of special procedures. Included in *BASECAP* are a number of alternative d.c. solution functions which are sometimes useful as discussed and illustrated in

Chapter 13. For example, the function *NEWTDC* incorporates a brute-force type of Newton–Raphson control algorithm which is sometimes more effective than the corresponding algorithm (*NEWTALT*) employed in *DCSOLVE*.

A procedure known as source-stepping is also sometimes useful in extreme cases. This involves obtaining a series of steady-state d.c. solutions, using a function such as *TRANSDC*, with the d.c. supply voltage as the independent parameter. The solution at each new value of supply voltage is obtained by iteration using the previous solution voltages as starting values. Thus, the d.c. voltage is effectively adjusted in steps from zero (or some convenient value at which convergence can be obtained) to the desired final value (at which convergence is impossible to obtain directly). This procedure is explored further in Exercise 11.3.

11.7.3   System-Related Errors

Number representation. The finite number of binary digits used to represent numbers internally in the computer limits both the size of numbers which can be represented, and their precision. These factors are dependent on both the machine architecture and the APL language implementation used. Consequently the user should consult system reference manuals for specific information when required.

Following are representative guidelines which pertain to computers having the IBM 370 series architecture and an APL implementation which utilizes a double word of 64 bits for floating-point numbers (see LePage 1978):

(a) Integers in the range from -2147483648 to 2147483647 can be accommodated in a 32-bit word.

(b) A single binary digit is stored in one bit.

(c) Floating-point numbers are stored in 64 bits, and may include positive and negative numbers with a magnitude in the range from approximately $7.2E^{-}75$ to $7.2E75$. Internal accuracy of the numbers is limited to approximately 16 decimal digits.

Note that such an APL floating-point number representation corresponds to the use of double precision which often requires special programming in equivalent CACD packages implemented in the FORTRAN language.

Round-off errors. Numerical errors tend to increase during many types of calculations. A well-known example is the subtraction of two large, nearly equal, limited-precision numbers. The majority of the computations in *BASECAP*, for both d.c. and a.c. solutions, involve the simultaneous solution of a large number of linear equations. As a guide, it can be expected that round-off errors in such situations will decrease precision by

$$1 + 2 \times \mathrm{Log}\, N$$

decimal digits, where *N* is the number of equations (see Ralston 1965). Thus for 100 equations (in *BASECAP* this corresponds to 50 nodes under a.c. conditions), the precision of the results will be reduced to approximately 16-5, or 11 decimal digits. For engineering purposes, even for 1000 equations, such results are therefore normally quite acceptable.

Domain errors.   Electronic circuits typically involve a very wide range of numbers for voltage levels, impedance levels, and frequencies. Moreover, both the semiconductor diode and the bipolar transistor contain exponential I–V relationships which generate extremely large numbers under certain conditions. During a circuit solution, therefore, it is possible for the result of a calculation to exceed the number size capacity of the computer. In this case the computation will be halted, and the APL system will report a domain error.

Such domain errors can generally be avoided by scaling the electrical units used in order to keep the resultant numbers within acceptable size limits. *BASECAP* assumes units of volts, amperes, ohms, henrys, farads, and hertz as specified in Section 11.2, and no internal scaling is applied to the calculations. However, the user can easily scale the input data, as long as a consistent set of units is used. For example, millivolts, milliamperes, ohms, microhenrys, microfarads, and megahertz constitute a valid alternative set.

In practice, we find that the 64-bit floating-point number representation used by APL enables *BASECAP* to accommodate a wide variety of circuits and solution conditions without scaling. Example 11.5 illustrates how some of these factors may be interpreted in circuit terms. Shown in this example are a.c. results for a simple R–L test circuit ($L\,TEST$) obtained over a wide range of frequencies using *BASECAP* with no scaling. Valid node voltages ranging from 60 to about 350 dB are indicated for a frequency range from 0.1 Hz to $1E18$ Hz. A domain error did result, however, during the matrix solution of the same circuit at a higher frequency of $1E20$ Hz.

Example 11.5

The results shown below for the simple test circuit $L\,TEST$ illustrate, in circuit terms, the range of numbers and solution conditions that *BASECAP* can accommodate without scaling.

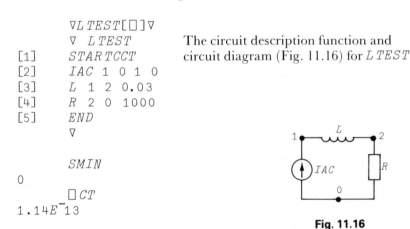

```
        ∇LTEST[□]∇
        ∇  LTEST
[1]     STARTCCT
[2]     IAC 1 0 1 0
[3]     L  1  2  0.03
[4]     R  2  0  1000
[5]     END
        ∇

        SMIN
0
        □CT
1.14E‾13
```

The circuit description function and circuit diagram (Fig. 11.16) for $L\,TEST$

Fig. 11.16

$\square CT \leftarrow 1E^-50$ Solution control parameters are
$TOL \leftarrow 1E^-20$ specified.
$LTEST$
$DCSOLVE$ Obtain d.c. and a.c. solutions. Note
$CPUSECONDS = 0.018$ the very wide frequency range used.
$ITERATIONS = 2$

$\downarrow\downarrow DC\ NODE\ VOLTAGES\ 1\ TO\ 2\ \downarrow\downarrow$

$0\ 0$

$FLOG\ .1\ 1E18\ 5$
$ACSOLVE$
$SOLVING\ FOR\ 95\ FREQUENCY\ POINTS$
$CPUSECONDS = 0.25$
$DONE$

$30\ 50\ PLOT\ LOGF,MAGDB\ ACVOUT\ 1$ The results are in Fig. 11.17.

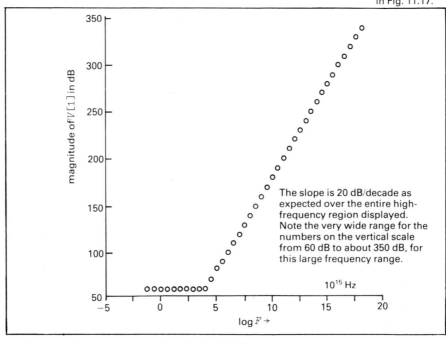

Fig. 11.17

Next we try an a.c. solution at a much higher frequency.

$F \leftarrow 1E20$
$ACSOLVE$
$SOLVING\ FOR\ 1\ FREQUENCY\ POINTS$
$DOMAIN\ ERROR$
$REALSOLV[3]\quad V \leftarrow (\rho I)\rho\ (,I)\boxdot YT,[1]YB$
$\qquad\qquad\qquad\qquad\qquad\wedge$

A domain error occurs during the matrix solution indicating an out-of-range number condition at this high frequency.

Simultaneous equations in *BASECAP* are solved using the APL matrix divide operator (see *NEWTALT* and *REALSOLV*). Other methods of matrix solution manipulate numbers differently, and each may result in a different limit on the effective range of numbers which can be accommodated in a circuit solution (see the discussions on Gaussian elimination, pivotal condensation, and *LU* factorization in Chapters 5 and 6). For example, a Gaussian elimination algorithm called *GAUSS* (complete with pivoting) was implemented in *BASECAP* in place in the regular matrix divide operator. With *GAUSS* in *BASECAP*, tests were again carried out using the circuit *LTEST* in Example 11.5. In this case, no domain errors resulted during the matrix solutions over the much wider frequency range from 0.1 Hz to 10^{60} Hz! It became necessary to scale the current source value only in order to compute and display the voltage magnitudes in dB (this calculation uses a squaring operation). The following rather remarkable results were obtained for the voltage at node 1 of *LTEST*, printed with a precision of 14 digits:

678.83935579489 dB	at	4.6416E59 Hz
$-$ $^-$321.16064420511 dB	at	4.6416E09 Hz
1000.00000000000 dB	over	50 decades!

Alas, *GAUSS* is about 5.5 times slower than the regular matrix divide operator. Still, if you need to use 10^{60} Hz (or some other special feature), it is good to know that a different matrix solution algorithm can be substituted very simply in *BASECAP*

Suspended functions.   Whenever the solution process is interrupted, either by the user (with the ATTN key) or by the APL system as a result of an error condition, the contents of local variables from suspended functions are preserved in the workspace for diagnostic purposes. However, a local variable in a suspended function may have the same name as a global variable or a function elsewhere in the package. In such cases, access to that global variable or function may not be permitted during a new solution, until all suspensions are removed. This requires clearing the state indicator by entering the right arrow operator ($\rightarrow$) as many times as necessary to make

```
)SI
```

yield a null response. For an illustration, see Example 11.6.

Example 11.6

Upon attempting to execute the circuit description function *LTEST* we encounter a syntax error associated with the function *L* . The problem relates to suspended functions as shown below:

```
          LTEST
SYNTAX ERROR
LTEST[3]    L  1  2  0.03        There appears to be a problem in
             ∧                   recognizing and accessing L
          ∇LTEST[□]∇
          ∇ LTEST
[1]       STARTCCT
[2]       IAC  1  0  1  0        The contents of the circuit
[3]       L  1  2  0.03          description function LTEST are
[4]       R  2  0  1000000       displayed and found to be in order
[5]       END
          ∇                      Display the contents of the state
          )SI                    indicator
LTEST[3]         *
REALSOLV[3]          *           Suspended functions exist in the
ACSOL[8]                         workspace from an earlier solution
ACSOLVE[3]                       attempt
          →
          →                      Clearing the state indicator
          →
          →
          )SI                    The state indicator is now clear
          LTEST                  We proceed with the execution of
          DCSOLVE                LTEST and encounter no further
CPUSECONDS=0.018                 problems
ITERATIONS=2

↓↓DC  NODE  VOLTAGES  1  TO  2  ↓↓

  0  0
```

Exercises

11.1 Investigate the operation of the circuit *FRANKOSC* with *BASECAP* under the condition when R_p is set to 1 kilohm. Determine the value of R_f required to produce a loop gain ≥ 1.0, and compare results with those obtained for *FRANKOSC* in Section 11.6.3.

11.2 An approximate calculation is often sufficient for checking the computed results obtained from a circuit analysis program.
   (a) Carry out such an analysis for the circuit *FRANKOSC* in Section 11.6.3 to determine the minimum value of R_f needed to support oscillation at a frequency of approximately 5 kHz with $L_s = 69$ mH and $R_p = 1E20$ ohms.
   (b) Repeat using $R_p = 1$ kilohm. See Exercise 11.1.

11.3 It has been found that d.c. convergence is very slow (or impossible to obtain) for the circuit *FBTRIPLE* in Section 11.6.2 using *DCSOLVE* under the conditions given in Table E11.1.

Parameter	Value(s)
R2	8.2 kilohms
ETA	1
TEMP	20 °C
R1	20 kilohms
RI	1 kilohm
D1	$1E^-16$ $5E^-11$
Q1, QN	100 0.1
	$1E^-16$ $1E^-16$

Table E11.1

Convergence in this case is readily obtained with the alternative d.c. solution function *NEWTDC*. However, we wish to investigate the method of source stepping here, so *NEWTDC* will be used only for checking purposes.
   (a) Change line 12 of *FBTRIPLE* to the following to enable different d.c. supply voltages to be specified conveniently:

   [12] VDC 9 0, VSCE

   Using the above circuit and device parameters, confirm that convergence with *DCSOLVE* is not readily obtained with *VSCE* set to 7 or 10 volts, but that convergence is obtained with *VSCE* set to 0, 2 or 5 volts.
   (b) Using *TRANSDC* with values of *VSCE* from 5 V to 10 V, obtain a valid final d.c. solution for *FBTRIPLE* for *VSCE* = 10 V, assuming the remaining circuit and device parameters are as specified above. Note that a function such as *FSTEP* may be used as a convenient way to specify the numbers to be used for the independent parameter.
   (c) Compute the solution with *VSCE* = 10 V using *NEWTDC*.

11.4 For the circuit *LPF* in Section 11.6.1, obtain solutions with *BASECAP* over an extended frequency range in order to determine at approximately what frequency a domain error will be generated due to an out-of-range number condition in the computer.

11.5   Investigate the frequency response of the fourth-order Butterworth low-pass filter circuit shown in Fig. E11.1. Cover the frequency range from 1 Hz to 1 MHz.

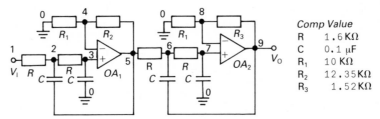

Comp	Value
R	$1.6\,\mathrm{K\Omega}$
C	$0.1\,\mu\mathrm{F}$
R_1	$10\,\mathrm{K\Omega}$
R_2	$12.35\,\mathrm{K\Omega}$
R_3	$1.52\,\mathrm{K\Omega}$

Fig. E11.1

11.6   The circuit in Fig. E11.2(a) is an emitter-coupled bipolar transistor pair, with the addition of a unity gain buffer amplifier supplying the output terminal, and

Q_1, Q_2 Parameters

$\beta_F = 10000$
$\beta_R = 1E^-6$
$I_{ES} = 1E^-14\ \mathrm{A}$
$I_{CS} = 1E^-14\ \mathrm{A}$
$C_{EB} = C_{CB} = 0\ \mathrm{F}$

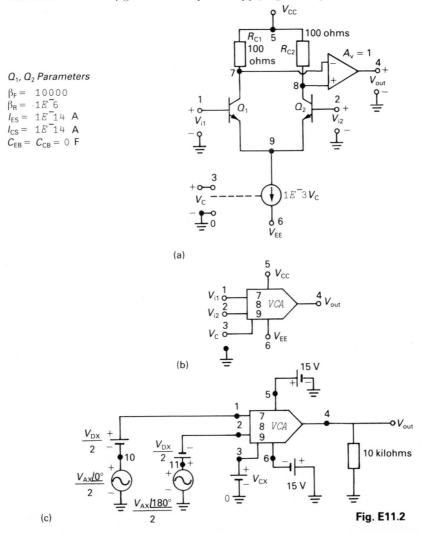

Fig. E11.2

a voltage-controlled d.c. current source at the emitters. Since the gain of a bipolar transistor is linearly dependent on d.c. bias current, this circuit arrangement may be used as the basis for a voltage-controlled amplifier as represented in Fig. E11.2(b).

Investigate the behavior of this circuit with $BASECAP$, using the convenient test arrangement shown in Fig. E11.2(c). Include the following specific results:

(a) Plot the d.c. transfer function V_{out} versus the differential input voltage V_{DX} for values of the d.c. control voltage $V_{CX} = 0.01$ V, 1 V, and 100 V respectively. A convenient voltage range for V_{DX} is $\overline{0}.2$ V to $+0.2$ V.

(b) Using a differential a.c. input voltage V_{AX} of 1 V at 1 kHz, and with $V_{DX} = 0$ V, compute the magnitude and phase angle of the a.c. voltage V_{out} for each of the following values of V_{CX} : $1 \mu V$, $0.01V$, 1.0 V, 10 V, and 100 V. Check the results for linearity.

11.7 The circuit to the right of the dotted line in Fig. E11.3 is a generalized impedance converter (GIC) configured to simulate an inductor.

(a) Determine the effective value of this inductor by computing the frequency response of the voltage at node 2 from 10 Hz to $10E8$ Hz with $R_N = 200$ kilohms and $C_N = 0$ F.

(b) Set $C_N = 1.59E^{-}9$ F and investigate the frequency response of the voltage at node 2 from 5000 Hz to $15 000$Hz.

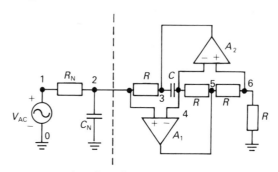

where R = 10 KΩ, C = 1.59nF
A₁ and A₂ are ideal amplifiers

Fig. E11.3

11.8 (a) Investigate whether$BASECAP$can correctly accommodate circuits containing negative resistances by computing the d.c. voltage at node 1 for the circuit in Fig. E11.4(a).

(b) Similarly compute the voltages at nodes 1 and 2 for the extended circuit in Fig. E11.4(b).

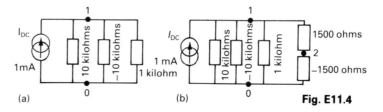

(a)   (b)   **Fig. E11.4**

11.9   The circuit in Fig. E11.5 is known as a voltage-to-current converter, where the current I_L is required to be independent of the load impedance R_{LOAD}. The op amp and the resistors R, R_1, and R_2 form a negative impedance converter whose resistance from node 2 to ground can be made equal to the negative of R_{IN}, giving

$$V_L = V_s \times R_{LOAD}/R_{IN}$$

(a)   Test the validity of this relationship by computing and plotting V_L versus R_{LOAD} for

$$1\,\Omega \le R_{LOAD} \le 1E6\,\Omega$$

(b)   Replace the d.c. source with an a.c. voltage source, set $R_{LOAD} = 1$ kilohm, and add a $0.01\mu F$ capacitor in parallel with R_{LOAD}. Plot the magnitude of the resultant a.c. load voltage from $1\,Hz$ to $1E12\,Hz$ in order to determine if the load current remains constant under these conditions.

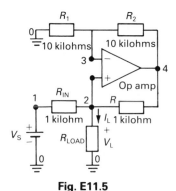

Fig. E11.5

11.10   Compute and plot the d.c. transfer function V_O versus V_I for the circuit in Fig. E11.6.

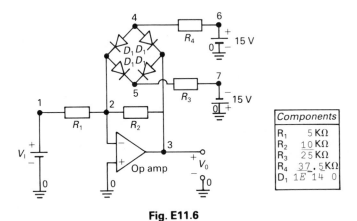

Components	
R_1	$5\,K\Omega$
R_2	$10\,K\Omega$
R_3	$25\,K\Omega$
R_4	$37.5\,K\Omega$
D_1	$1E{-}14$ 0

Fig. E11.6

11.11   Fig. E11.7 contains two ideal op amps in a Schmitt trigger arrangement. It features positive feedback (regeneration) and exhibits a hysteresis characteristic determined by R_1 and R_F. The amplitude of the output voltage is controlled by the action of the second op amp stage, which functions as a symmetrical limiter.

(a)   Compute and plot the d.c. transfer function V_0 versus V_1 for this circuit with $R_F = 10$ kilohms. Confirm that different results are obtained when
    (i) V_1 ranges from $^-20$ V to $+20$ V,
and   (ii) V_1 ranges from $+20$ V to $^-20$ V.

Note that a very large number of iterations may be required for a solution in the immediate vicinity of a high–low or low–high transition (142 iterations were required for one solution carried out by the authors).

(b)   Repeat (a) for $R_F = 50$ kilohms
(c)   Set $R_F = 5$ kilohms and note the different transfer characteristics that result when V_1 is varied as follows:

    (i) $+30$ V to $^-30$ V
    (ii) $+20$ V to $^-30$ V

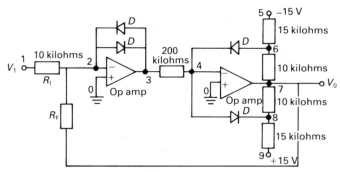

Fig. E11.7

11.12   An input stage for an integrated circuit differential amplifier is represented in Fig. E11.8. In this circuit transistors Q_3 and Q_4 function as a Widlar current

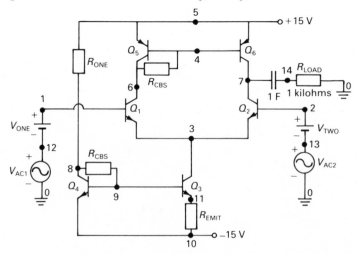

Fig. E11.8

sink for the differential pair Q_1 and Q_2. Small 'shorting' resistors $R_{CBS}=0.001$ ohms from collector to base of Q_4 and Q_5 effectively convert these transistors into diodes.

Represent each transistor using the model parameters in $QAMP$, where

	β_F	β_R	I_{ES}	I_{CS}	C_{JE}	C_{JC}
Q_AMP ← →	100	.1	$1E^-14$	$1E^-14$	$50E^-12$	$1E^-12$

(a) Compute the d.c. bias conditions for the transistors in this circuit with the d.c. sources V_{ONE} and V_{TWO} set to 1 V.

(b) Compute the small-signal a.c. mid-band voltage across R_{LOAD} if $V_{AC1} - V_{AC2} = 1|0\,V$.

11.13   The behavior of a crystal filter at or near resonance may be represented by an inductor, two capacitors, and a resistor connected as shown in Fig. E11.9. Component values for a representative 10 MHz filter are given in Fig. E11.9.

Investigate the behavior of this circuit in the vicinity of 10 MHz when supplied with a convenient a.c. current source as shown.

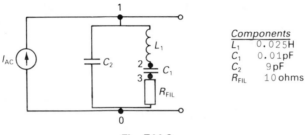

Components	
L_1	0.025H
C_1	0.01pF
C_2	9 pF
R_{FIL}	10 ohms

Fig. E11.9

11.14   Determine the small-signal a.c. frequency-dependent voltage gain between nodes 2 and 11 in the field effect transistor (FET) amplifier of Fig. E11.10.

FET Parameters		Components	
g_m	$1E^-3S$	R_{GG}	90 kilohms
r_d	40 kilohms	C_C	10 µF
C_{gs}	15 pF	R_{ss}	2.8 kilohms
C_{gd}, C	0 pF	R_D	10 kilohms

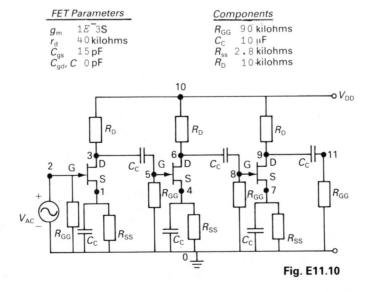

Fig. E11.10

Use the small-signal a.c. FET model given in Fig. E12.1, and the input function $ACFET$ developed in Exercise 12.4.

11.15 Compute the small-signal a.c. frequency-dependent gain of the common emitter/common base transistor pair shown in Fig. E11.11. Represent each transistor by the high-frequency hybrid-pi equivalent circuit shown in Fig. E12.2. The input function $ACBJT$ developed in Exercise 12.6 conveniently implements this device model. Appropriate parameters for this model are provided in Table E11.2.

Table E11.2

Parameter	Q_1 and Q_2
β	200
r_π	4.52 kilohms
r_0	1 megohms
r_x	200 ohms
r_c	150 ohms
r_μ	$1E12$ ohms
C_π	13.7 pF
C_μ	0.23 pF
C_{CS}	0.4 pF

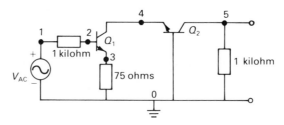

Fig. E11.11

11.16 Refer to the text by Grinich and Jackson (1975) and investigate the operation of the operational amplifier circuit found in Fig. 8.13 on page 417.

   (a)   Compute the d.c. bias conditions for each transistor in this operational amplifier circuit. Represent each transistor using the model parameters in $QMOD$, and each diode using the parameters in $DMOD$, where these parameters are given in Table E11.3.

   (b)   Specify convenient values for the a.c. current sources connected to nodes 1 and 2, and compute the small-signal difference mode gain for this circuit at 1 kHz.

   (c)   In a similar manner, compute the small-signal common-mode gain.

Table E11.3

		β_F	β_R	I_{ES}	I_{CS}	C_{JE}	C_{JC}
$QMOD$	$\leftarrow \rightarrow$	100	$.1$	$1E^-14$	$1E^-14$	$30E^-12$	$1E^-12$

		I_S	C_J				
$DMOD$	$\leftarrow \rightarrow$	$1E^-14$	$1E^-1|1$				

Solutions

11.1 Solution to *FRANKOSC* using

        RP, RF, C1
1E3 7.9E4 1.47E⁻8

    FLIN 3000 9000 50

40 60 *PLOT F,MAG ACVOUT* 2

These results show that the loop gain is slightly greater than 1.0 with $R_F = 79$ kilohms. Note that the resonance curve is very broad due to the low value of circuit Q.

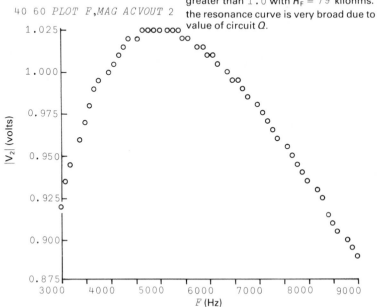

40 60 *PLOT F,ANGLE ACVOUT* 2

The following plot indicates that the phase angle changes only very gradually near resonance in this case.

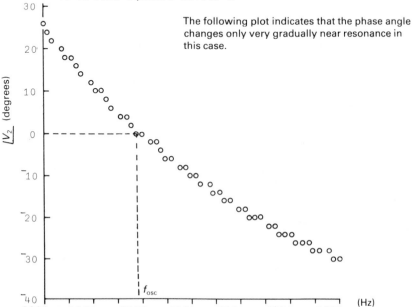

11.2   Following are approximate hand calculations of $FRANKOSC$ (see Section 11.6.3) useful for checking computer-generated results.

For oscillation (resonance) at $f_0 = 5$ kHz (assuming $L_p \approx L_s$)

$$C_1 = \frac{1}{L_p(2\pi f_0)^2} = \frac{1}{0.069 \, (2\pi 5 \times 10^3)^2} \approx 0.0146842 \,\mu\text{F}$$

$$\text{say } 0.0147 \,\mu\text{F}$$

(a)   High Q case: $R_p = 10^{20}$ ohms $\approx \infty$, Z_{in} of buffer op amp $\approx \infty$

$$\text{Loop gain} = \frac{V_2}{V_9} = \frac{V_4}{V_6} \times \frac{V_3}{V_4} \times \frac{V_6}{V_9} \times \frac{V_2}{V_3} \qquad \text{SET} = 1.0$$

$$\frac{V_4}{V_6} \approx -\frac{R_p \text{(net)}}{R_E} = \frac{(R_s^2 + X_s^2)}{R_s R_E} = -\frac{(5.5)^2 + (2\pi \times 5 \times 10^3 \times 0.069)^2}{5.5 \times 1000}$$

$$\therefore \frac{V_4}{V_6} \approx \underline{-854.4}; \text{ also } \frac{V_3}{V_4} \approx \underline{+1}$$

Now $$\frac{V_6}{V_9} \approx \frac{76\text{K}\|10\text{K}\|(\beta+1)R_E}{(76\text{K}\|10\text{K}\|(\beta+1)R_E) + 22\text{K}} = \frac{8.126\text{K}}{8.126\text{K} + 22\text{K}} \approx 0.270 \text{ for } \beta = 100$$

and $R_E = 1$ kilohm

And $$\frac{V_2}{V_3} \approx -\frac{R_F}{20\text{K}}$$

Thus, required $R_F \geq \dfrac{20000}{854.4 \times 0.27} \geq \underline{86.7 \text{ ohms}}$

Since these calculations involve simplifying approximations, choose R_F larger, say $\underline{91 \text{ ohms}}$

(b)   Low Q case: Set $R_p = 1$ kilohm

Here R_p (net) $\approx R_p = 1$ kilohm

Repeat above calculations using the same value of C_1.

$$\therefore \frac{V_4}{V_6} \approx -1 \text{ and } R_F \geq \frac{20000}{0.27} \geq \underline{74.1 \text{ kilohms}} \; (\therefore \text{ try 79 kilohms})$$

11.4   The following plot shows that no domain errors were obtained during the solution of LPF up to $5E17$ Hz. Domain errors were encountered, however, at frequencies beyond $1E18$ Hz.

```
    FLOG .1 5E17 5
    ACSOLVE
SOLVING FOR 94 FREQUENCY POINTS
CPUSECONDS=0.291
DONE


    30 50 PLOT LOGF,MAGDB ACVOUT 2
```

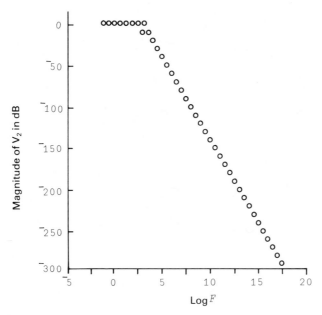

11.6   The circuit in Fig. E11.2(a) is modelled by the *BASECAP* circuit description function *VCA*. In this case, *VCA* is intended to be used within another circuit description function to specify a complete circuit. Nine circuit node numbers must be supplied as arguments to *VCA* to specify how the interconnections are made to the 'calling' circuit (in sequence from 1 to $\overline{9}$ with reference to Fig. E11.2(a).

   We also introduce a new amplifier model *OA*1 which is a modification of the op amp model *OA* to produce an ideal differential amplifier with user-specified differential voltage gain *P*[4].

```
        ∇ VCA D
[1]       QNPN P[9 1 7],QVAL←10000 1E⁻6 1E⁻14 1E⁻14 0 0
[2]       QNPN P[9 2 8],QVAL
[3]       R P[7 5],100
[4]       R P[8 5],100                High β transistors are used for Q₁ and Q₂
[5]       OA1 P[8 7 4], 1
[6]       VCIS P[6 9 3], 0 0.001
        ∇
```

```
        ∇ OA1 P
[1]       P←P,100000
[2]       VCIS P[3],0,P[1],P[2],100×P[4]
[3]       R P[3],0,0.01
        ∇
```

```
        ∇ VCT
[1]       STARTCCT
[2]       VDC 1 10 ,VDX÷2              VCT implements the complete circuit of Fig.
[3]       VDC 2 11 ,-VDX÷2            E11.2(c). Note that in this case, the node numbers
[4]       VAC 10 0 ,(VAX÷2),0         for VCA and VCT are identical
[5]       VAC 11 0 ,(VAX÷2),180
[6]       R 4 0 10000
[7]       VDC 3 0 ,VCX
[8]       VDC 6 0 ⁻15
[9]       VDC 5 0 15
```

```
[10]    VCA  1 2 3 4 5 6 7 8 9
[11]    END
        ∇
```

(a) A sample result is given below

```
        VAX ←0
        VCX ←100

        FLIN ‾.2 .2 30
        VCX
100
        30 60 PLOT TRANSDC 4
TYPE CIRCUIT NAME
VCT
TYPE NAME OF PARAMETER TO BE VARIED
VDX
TYPE PARAMETER VALUES
□:
        F
IT = 7
IT = 5
    |
    |
    |
    |
    |
    |
```

Note that *FLIN* conveniently produces a range of numbers in F which are assigned to *VDX* within *TRANSDC*

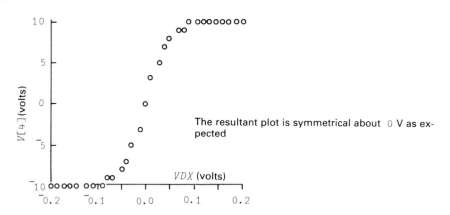

The resultant plot is symmetrical about 0 V as expected

(b) Two sample results are given

(i) For $VCX ← .001$ we get

```
        F,(MAG ACVOUT 4) ,ANGLE ACVOUT 4
    1.00E3  1.98E‾3  ‾5.00E‾15
```

(ii) For $VCX ← 1$ we get

```
        F,(MAG ACVOUT 4) ,ANGLE ACVOUT 4
    1.00E3  1.98E0  ‾5.00E‾15
```

11.7 The circuit of Fig. E11.3 is described by *GICBP* where *RNOW* and *CNOW* are used to represent R_N and C_N respectively. Each amplifier is given a gain of $1E6$ and is represented by the model *OA1* introduced in the solution to Exercise 11.6.

```
(a)      ∇GICBP
[1]      STARTCCT
[2]      VAC 1 0 1 0
[3]      R 1 2,RNOW
[4]      C 2 0,CNOW
[5]      R 2 3 10E3
[6]      C 3 4 1.59E¯9
[7]      R 4 5 10E3
[8]      R 5 6 10E3
[9]      R 6 0 10E3
[10]     OA1 2 4 5 1E6
[11]     OA1 6 4 3 1E6
[12]     END∇

         RNOW←200E3
         CNOW←0

         40 60 PLOT LOGF,MAGDB ACVOUT 2
```

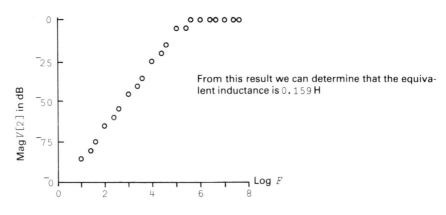

From this result we can determine that the equivalent inductance is 0.159 H

(b)

```
         CNOW←1.59E¯9
         40 60 PLOT F,MAG ACVOUT 2
```

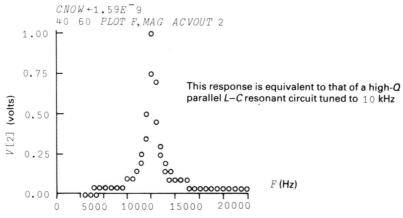

This response is equivalent to that of a high-*Q* parallel *L–C* resonant circuit tuned to 10 kHz

11.10 We use *TRANSDC*, *FLIN* and *PLOT* to obtain the transfer characteristic of this circuit with 35 solution points.

```
         ∇BRIDGE
[1]      STARTCCT
[2]      VDC 1 0 ,F
```

```
[3]      R 1 2 5000
[4]      OA   0 2 3
[5]      R 2 3 10000
[6]      D 2 5 ,D1 ← 1E‾14 0
[7]      D 4 2 ,D1
[8]      D 4 3 ,D1
[9]      D 3 5 ,D1
[10]     R 4 6 37500
[11]     VDC 6 0 15
[12]     R 5 7 25000
[13]     VDC 7 0 ‾15
[14]     END
         ∇
         FLIN ‾7 7 35
```

Note that F is produced by the execution of $FLIN$

The use of F is a simple direct way to provide solution data to $TRANSDC$

```
         30 60 PLOT TRANSDC 3
TYPE CIRCUIT NAME
BRIDGE
TYPE NAME OF PARAMETER TO BE VARIED
F
TYPE PARAMETER VALUES
□:
         F
IT = 7
IT = 4
IT = 4
```

11.12 The circuit in Fig. E11.8 is specified in the circuit description function $DCDIFAMP$ below.

```
         ∇ DCDIFAMP
[1]      STARTCCT
[2]      VDC 1 12 ,VONE
[3]      VDC 2 13 ,VTWO
[4]      QNPN 3 1 6 ,QAMP ← 100 0.1 1E‾14 1E‾14 5E‾11 1E‾12
[5]      QNPN 3 2 7 ,QAMP
[6]      QNPN 11 9 3 ,QAMP
[7]      QNPN 10 9 8 ,QAMP
[8]      R 8 9 ,RCBS ← 0.001
[9]      R 5 8 ,RONE
```

```
[10]     R 10 11 ,REMIT
[11]     VDC 10 0 ‾15
[12]     VDC 5 0 15
[13]     QPNP 5 4 6 ,QAMP
[14]     R 6 4 ,RCBS
[15]     QPNP 5 4 7 ,QAMP
[16]     R 7 5 ,RBW
[17]     VAC 12 0 2 0
[18]     VAC 13 0 1 0
[19]     C 7 14 1
[20]     R 14 0 1000
[21]     END
         ∇
```

We have included an extra resistor *RBW* in this implementation to provide a partial representation of base-width modulation effects in Q_6.

```
         ∇ HO
[1]      DCDIFAMP
[2]      DCSOLVE
[3]      BJTBIAS
[4]      ACSOLVE
[5]      MAG ACVOUT 14
         ∇
```

HO is a convenient test function. *BJTBIAS* on line 3 summarizes the transistor bias conditions required in part (a). See Section 15.5.4 for details of *BJTBIAS*. Line 5 of *HO* produces the a.c. results for part (b)

```
         REMIT ←2
         RONE ←30000
         RBW ←200000
         VONE ←VTWO ←1
         HO
CPUSECONDS=2.84
ITERATIONS=19
```

Circuit parameters

`↓↓DC NODE VOLTAGES 1 TO 14 ↓↓`   d.c. results (a)

`1 1 0.132 14.1 15 14.1 13.2 ‾14.1 ‾14.1 ‾15 ‾15 ‾4.47E‾6 ‾4.47E‾6 1.32E‾8`

TYPE	QNO	ENODE	BNODE	CNODE	VBE	VCE	IE(MA.)	IC(MA.)
NPN	1	3	1	6	.868	13.999	‾.452	.447
NPN	2	3	2	7	.868	13.112	‾.452	.447
NPN	3	11	9	3	.893	15.128	‾.913	.904
NPN	4	10	9	8	.895	.895	‾.961	.952
PNP	5	5	4	6	‾.867	‾.867	.443	‾.439
PNP	6	5	4	7	‾.867	‾1.754	.443	‾.439

```
SOLVING FOR 1 FREQUENCY POINTS
CPUSECONDS=0.17
DONE
   12.5
```
a.c. results (b)

11.15 *CECBPAIR* is used for the solution of the circuit in Fig. E11.11.

```
         ∇ CECBPAIR
[1]      STARTCCT
[2]      VAC 1 0 1 0
[3]      R 1 2 1000
[4]      R 3 0 75
[5]      R 5 0 1000
[6]      ACBJT 3 2 4 6 7 ,PARS
[7]      ACBJT 4 0 5 8 9 ,PARS
[8]      END
         ∇
```

```
        ∇ACBJT P
[1]     ⍝ PARAMETER SEQUENCE IS
[2]     ⍝ NE NB NC NBP NCP BF RPI
[3]     ⍝ RO RX RC RU [CPI CU CCS]
[4]     ⍝ []MEANS OPTIONAL IF IN SEQUENCE
[5]     ⍝
[6]       P←P, 0 0 0
[7]       R P[2 4 9]
[8]       R P[4 1 7]
[9]       R P[4 5 11]
[10]      C P[4 5 13]
[11]      R P[5 1 8]
[12]      R P[5 3 10]
[13]      C P[5],0,P[14]
[14]      C P[4 1 12]
[15]      VCIS P[1 5 4 1],P[6]÷P[7]
        ∇
        PARS
200 4.52E3 1E6 200 150 1E12 1.37E¯11 2.3E¯13 4E¯13
        CECBPAIR
        DC·SOLVE
CPUSECONDS=0.089
ITERATIONS=2

↓↓DC NODE VOLTAGES 1 TO 9↓↓

0 0 0 0 0 0 0 0 0

        FLOG 1 200E6 4
        ACSOLVE
SOLVING FOR 34 FREQUENCY POINTS
CPUSECONDS=2.68
DONE
```

30 60 *PLOT LOGF,MAG ACVOUT* 5

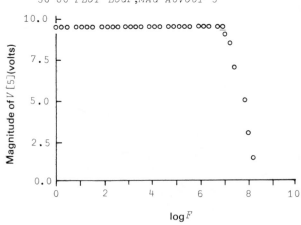

CHAPTER

12

Input Data Processing

This chapter contains a detailed study of the input portion of the $BASECAP$ package, previously introduced in Section 10.4.2 and illustrated in Fig. 10.2. The functions in the $INPUT$ group are designed to accept circuit description information, and to prepare this data for use by the d.c. and a.c. solution functions. Table 12.1 lists the names of the $INPUT$ functions and the principal global variables that will be dealt with in this chapter. For easy reference the contents of all $BASECAP$ functions are listed in Appendix 6, and a brief description of each of the principal variables used in $BASECAP$ is given in Appendix 7.

The material in Section 12.1 deals with the structure and purpose of the $INPUT$ data. An understanding of this material provides the necessary foundation for the study of the internal operation of both the $INPUT$ functions presented in Section 12.2, and the d.c. and a.c. solution functions presented in Chapters 13 and 14. The last section in this chapter is in the form of an interactive terminal session dealing with the operation and use of the various $INPUT$ functions.

Table 12.1
Composition of the $INPUT$ Group

$INPUT$	$)GRPS$					
	$)GRP\ INPUT$					
$ALPHA$	C	D	END	IAC	IDC	$INCID$
L	OA	$QCALC$	$QNPN$	$QPNP$	R	$REFNODE$
$STARTCCT$		VAC	$VCIS$	VDC		
	$)VARS$					
A	CL	G	GIN	$IACL$	$IBIN$	
IS	$ISIN$	IX	LL			
M	MIN	NH	Q	QIN	RL	$VACL$
$VCISL$						

12.1 Data Structure

The $BASECAP$ data structure is designed to accommodate the following basic information required to specify a given circuit:

(a) the number and types of elements used in the circuit
(b) parameter values for sources, devices and components
(c) node numbers, branch information, and circuit connection data
(d) reference current flow directions.

All of this data is conveniently specified by the user with the aid of the element functions R, $|L$, $|$ C, $|D$, $\backslash QNPN$, $\backslash QPNP$, $\backslash OA$, $\backslash VDC$, $|$ VAC, $|IAC$, $\backslash IDC$ and $VCIS$. (Refer to Table 11.1 for the full name of each symbol.)

As Chapter 11 showed, the various element functions are normally used within an APL holding function called a *circuit description function*. The necessary internal data arrays are set up for a particular circuit by *executing* its corresponding circuit description function. This, in turn, causes each of the individual element functions to be executed, thereby generating entries in one or more intermediate variables. After processing, the resultant data is passed on to the other groups through the global variables illustrated in Fig. 12.1.

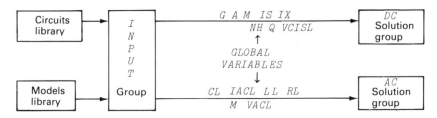

Fig. 12.1
Data variables produced by the $INPUT$ group

12.1.1 Branch Data

The basic d.c. part of this data is held in the same branch form used for many of the functions in Part 1 of this book. Thus, much of $BASECAP$'s data structure will already be familiar to the reader. These basic global variables are summarized below, and illustrated in an accompanying example using the low-pass filter circuit (LPF) from Section 11.6:

M — branch connection matrix
G — branch conductance vector
IS — pn junction reverse saturation current vector
IX — node d.c. source current vector
A — incidence matrix (branch current reference directions)

M has the shape (2,B), where B is the number of branches. The column for each branch contains its two node numbers, and the sequence indicates that current in the branch flows from the node in row 1 to the node in row 2. The columns must be in ascending branch number sequence from left to right. (See also the discussion in Section 12.1.3.)

Both the G and IS vectors must contain an entry for each branch, although entries may be zero. For example, a resistor in a branch generates a nonzero entry in G, while a diode junction generates a nonzero entry in IS. Again, an ascending branch number sequence is maintained for these vectors.

The IX vector must contain an entry for each circuit node (not counting the reference node). An entry is nonzero for a particular node if a nonzero d.c. current source exists there. Entries must be in ascending node number order.

The A matrix (from Section 1.1) contains only the numbers $0, 1$, and $\bar{1}$ and has the shape (NH, B) which is one row per node (counting the reference node), and one column per branch. The A matrix is required in the function END, and is readily derived from the M matrix using $INCID$ from Section 1.6.

The scalar quantity NH is similarly used for convenience within the $INPUT$ group functions. NH is also required subsequently by the circuit solution routines, and is readily obtained from the M matrix as follows:

$$NH \leftarrow \lceil / , M$$

NH is equal to the total number of circuit nodes, counting the reference node.

Example 12.1 Basic global data variables for LPF *(See Section 11.6)*

LPF	Execute function LPF.

M	Display M, G, IS, IX, A.

```
            M
    1  1  2  3  3
    4  2  4  4  4
            G
  1  0.001  1E¯12  0  0.001
           IS
  0  0  0  0  0
           IX
  0  0  0
            A
    1   1  0  0  0
    0  ¯1  1  0  0
    0   0  0  1  1
   ¯1   0 ¯1 ¯1 ¯1
```

Note that LPF has five branch entries, and three nodes plus ground.

12.1.2 Storage Tables

The remaining global data variables shown in Fig. 12.1 are configured as matrix tables for convenient retrieval. The matrix variables Q and $VCISL$ are used directly by the d.c. solution routines, while CL, $IACL$, LL, RL, and $VACL$ are only required by the a.c. solution routines. The detailed structure of each storage table is shown in Table 12.2.

Each of these tables holds the data for all of the elements of a particular type in the circuit, one row per element. For example, in a circuit with three capacitors, CL has the shape $(3,3)$. Similarly, for a circuit with a total of five transistors (NPN and PNP combined), Q has the shape $(5,7)$. An empty table

Table 12.2
Data storage table structures

Each table contains one row per element of its type

Table name	Description	Col. 1	Col. 2	Col. 3	Col. 4	Col. 5	Col. 6	Col. 7
CL	Capacitors	Node 1	Node 2	Capacitance (farads)				
LL	Inductors	Node 1	Node 2	Inductance (henrys)				
RL	Series resistors for inductors	Node 1	Node 2	Resistance (ohms)				
IACL	A.c. current sources	Node 1	Node 2	Magnitude (amperes)	Phase (degrees)			
VACL	A.c. voltage sources	Node 1	Node 2	Magnitude (volts)	Phase (degrees)			
VCISL	Voltage-controlled current sources	Source output-node	Source input node	Controlling node (+ve)	Controlling node (−ve)	Trans-conductance (siemens)		
Q	Transistors	Type -1 (PNP) $+1$ (NPN)	Base emitter branch no.	Emitter node	Base node	Collector node	Alpha forward	Alpha reverse

results if no element of a given type is present in the circuit. Examples of actual storage tables are found in Sections 12.2 and 12.3.

12.1.3   Intermediate Data Generation

An intermediate form of input data is normally produced by the execution of the various element description functions (exceptions are VAC and IAC as noted in Section 11.2). This intermediate $INPUT$ data is initially held in one or more of the storage tables CL, LL, RL, and $VACL$, and in five new intermediate variables, namely MIN, GIN, $ISIN$, $IBIN$, and QIN.

Subsequently in the function END, the intermediate data is processed and converted into the final form of the global variables described in Sections 12.1.1 and 12.1.2. Full details of this conversion process are given later in Section 12.2. It is sufficient to note here that each of the new intermediate data variables is a vector, and each relates to a corresponding global variable as summarized below:

Intermediate		*Final*
MIN	──converts to──▶	M
GIN	─────────────▶	G
$ISIN$	─────────────▶	IS
$IBIN$	─────────────▶	IX
QIN	─────────────▶	Q

The data generation process used in $BASECAP$ inherently allows a two-terminal circuit element to be represented internally by a compound branch containing ideal components in parallel between the specified node pair. An example is the junction capacitance in the semiconductor diode. More complex devices may be modelled (or partly modelled) in the program by two or more sets of ideal components between two or more node pairs.

As each two-terminal circuit element or device is processed, a pair of node numbers must be entered into MIN, and one corresponding entry must also be made in each of the following intermediate variables: GIN (branch conductance values), $ISIN$ (diode saturation leakage current values), and $IBIN$ (branch-formulated source current values). Zero entries may be used. For example, a zero entry in $ISIN$ signifies that no diode is present. The processing for some elements, such as capacitors, inductors, and a.c. voltage sources, additionally generates entries in one or more of the storage tables listed in Table 12.2.

Consider the example of a 5-V d.c. voltage source between nodes 2 and 6. This source is represented in $BASECAP$ as a d.c. current source of 5 A in parallel with a 1.0 S conductance, requiring the following data generation operations:

```
MIN  ←MIN,  2 6
ISIN ←ISIN, 0
GIN  ←GIN,  1
IBIN ←IBIN, 5
```

Several additional examples appear in Sections 12.2 and 12.3.

Transistors are represented in *BASECAP* by a simplified Ebers–Moll model (Ebers and Moll 1954) which includes two junction diodes and two dependent current generators. Accordingly, each transistor causes two diodes to be added to the circuit, resulting in two sets of sequential entries in *MIN*, *ISIN*, *IBIN*, and *GIN*. The junction capacitance terms for diodes and transistors are assumed to be constant, and are stored in the regular capacitance storage table CL. Finally, the processing of each transistor also generates a total of seven entries in the intermediate data vector *QIN*. These entries are as follows: type number (-1 PNP, or $+1$ NPN), base–emitter branch number, three temporary node numbers, alpha forward, and alpha reverse. Consult Section 12.2.5 for examples of *QIN*. Further information on the theoretical basis of the transistor model is given in Section 13.4.2.

Some device models are programmed entirely in terms of other more basic elements, and thus do not require a separate data storage table. An example is the operational amplifier which contains three resistors and two voltage-controlled current sources.

The voltage-controlled current source has a single branch between the nodes containing the controlled source. *VCIS* generates an additional row (containing five columns as per Table 12.2) in the matrix *VCISL*. One set of entries is also generated in each of the intermediate data variables *GIN*, *ISIN*, *IBIN*, and *MIN*. See Section 12.2.6 for examples.

12.2 Program Structure

This section contains an overview of the structural arrangement of functions within the *INPUT* group, followed by a detailed discussion of the purpose and operation of each function.

12.2.1   Overview

The main functions are listed below in the overall sequence in which they must normally be used:

(1) *STARTCCT*	Initialize variables
(2) *IDC,R,VDC C*	Element description functions called in
L,IAC,VAC,D	any order
QNPN, QPNP, VCIS, OA	
(3) *END*	Data conversion and final processing

This sequence ensures that the data variables will be cleared by *STARTCCT* prior to entering data for a new circuit. Final formatting takes place in *END* after all circuit elements have been entered. There is no hierarchy among the element description functions. They can be used in any order to describe a given circuit. Each function can also be tested separately, if desired, to reveal its internal operation.

The simple open program structure, combined with the fact that most of the functions in this group are invoked by the user, results in the very simple tree structures shown in Fig. 12.2. These tree structures document the calls to other functions. For example, the functions R and $VCIS$ are used by OA.

With the exception of $INCID$, which was previously discussed in Section 1.6, each $INPUT$ function is discussed in detail and illustrated in examples in the following subsections. The order of presentation follows the general sequence listed at the beginning of this section. Many of the examples in this chapter make use of a utility function called $DISPLAY$, which is also included with $BASECAP$. The function $DISPLAY$ conveniently generates a tabular listing of the contents of one or more variables (scalars, vectors, or simple matrices) whose names are given as arguments.

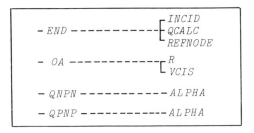

Fig. 12.2
Tree Structures in the $INPUT$ Group

12.2.2   The Function $STARTCCT$

Table 12.3
The function $STARTCCT$

$\nabla STARTCCT$

The purpose of $STARTCCT$ is to initialize each of the circuit *input* data variables to an empty state

[1] $MIN \leftarrow IBIN \leftarrow ISIN \leftarrow GIN \leftarrow QIN \leftarrow \iota 0$

The intermediate data variables are emptied

[2] $LL \leftarrow RL \leftarrow CL \leftarrow 0\ 3\ \rho 0$

Empty tables are defined for LL, RL, and CL Each has three columns. This structure is used to allow entries to be added easily to a given matrix table one row at a time by the process of lamination

[3] $IACL \leftarrow VACL \leftarrow 0\ 4\ \rho 0$

Empty four-column tables are structured for $IACL$ and $VACL$

[4] $VCISL \leftarrow 0\ 5\ \rho 0\ \nabla$

An empty five-column table is created for $VCISL$

STARTCCT initializes the input data variables for a new circuit as explained in Section 12.2.1. Table 12.3 describes in detail the internal operation of this function. The results produced by *STARTCCT* are then illustrated in an example.

Example 12.2 *STARTCCT* (Table 12.3)

STARTCCT| Execute *STARTCCT*

DISPLAY'MIN IBIN ISIN GIN QIN LL RL CL IACL VACL VCISL'

Item		Contents
MIN	↔	
IBIN	↔	
ISIN	↔	
GIN	↔	
QIN	↔	
LL	↔	
RL	↔	
CL	↔	
IACL	↔	
VACL	↔	
VCISL	↔	

This display shows that after the excution of *STARTCCT* all intermediate data variables and storage tables are empty.

12.2.3   The functions *IDC*, *R*, *VDC*

The functions *IDC*, *R* and *VDC* specify an independent d.c. current source, a resistor, and an independent d.c. voltage source. They provide the capability for describing a simple d.c. circuit in *BASECAP* . The detailed design and oper-

Table 12.4
The function *IDC*

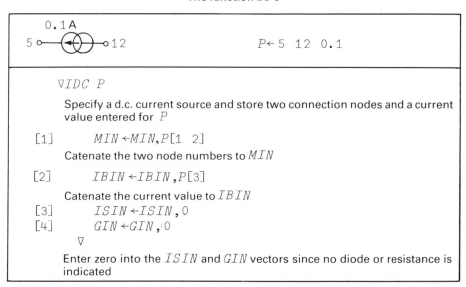

```
    ∇IDC P
        Specify a d.c. current source and store two connection nodes and a current
        value entered for P
[1]        MIN ←MIN,P[1 2]
        Catenate the two node numbers to MIN
[2]        IBIN ←IBIN,P[3]
        Catenate the current value to IBIN
[3]        ISIN ←ISIN,0
[4]        GIN ←GIN,0
        ∇
        Enter zero into the ISIN and GIN vectors since no diode or resistance is
        indicated
```

ation of these functions are described in the accompanying tables. The data variables produced by each function are illustrated in an example.

Example 12.3 *IDC*   (Table 12.4)

STARTCCT	Empty input data variables
IDC 5 12 .1	Specify a d.c. current source
DISPLAY'MIN IBIN ISIN GIN'	Display pertinent variables

Item		*Contents*	
MIN	↔	5 12	The two node numbers
IBIN	↔	0.1	The d.c. current value
ISIN	↔	0	These are 0 since no diode or con-
GIN	↔	0	ductance is used, and the branch numbering sequence must be preserved

Table 12.5
The function R

1.0 kohm 7 o—▭—o 10	$P \leftarrow$ 7 10 1000

∇ *R P*

Specify a resistor, and store two connection nodes and a resistance value entered for P

[1]   *MIN ← MIN, P[1 2]*

Catenate the two node numbers to *MIN*

[2]   *IBIN ← IBIN, 0*
[3]   *ISIN ← ISIN, 0*

Catenate a zero onto the *IBIN* and *ISIN* vectors since no diode or d.c. current source is included in this branch

[4]   *GIN ← GIN, ÷ P[3]*
    ∇

Compute the conductance and catenate it to *GIN*

Example 12.4   (Table 12.5)

STARTCCT	Empty input data variables
R 7 10 1*E*3	Specify a resistor (execute *R*)
DISPLAY'MIN IBIN ISIN GIN'	Display resultant variables

Item		Contents	
MIN	↔	7 10	The two node numbers appear
IBIN	↔	0	The 0 entries preserve the branch
ISIN	↔	0	numbering sequence . This con-
GIN	↔	0.001	ductance corresponds to the 1000 ohm resistor value

Table 12.6
The function *VDC*

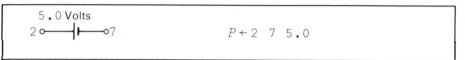

5.0 Volts
2 o———|⊢———o7 $P \leftarrow 2\ 7\ 5.0$

∇*VDC P*

    Specify a d.c. voltage source and store two connection nodes and a voltage value entered for *P*

[1]   *MIN ←MIN ,P*[1 2]

    Catenate the two node numbers to *MIN*

[2]   *IBIN ←IBIN ,P*[3]÷1

    Convert the voltage to a current source by dividing by a one ohm resistor, and catenate it to *IBIN*

[3]   *ISIN ←ISIN ,*0
[4]   *GIN ←GIN ,*÷1
       ∇

    Catenate zero to *ISIN*;catenate the conductance associated with *VDC* to *GIN*

Example 12.5 *VDC*   (Table 12.6)

    STARTCCT   Empty input data variables
    VDC 2 7 5   Execute the function *VDC*
    DISPLAY'MIN IBIN ISIN GIN'   Display affected variables

Item		Contents	
MIN	↔	2 7	The node numbers appear
IBIN	↔	5	A 5 A current source in parallel with
ISIN	↔	0	1 S results
GIN	↔	1	

12.2.4   The functions C, L, IAC, VAC

These are the basic a.c. element and source description functions that respectively specify a capacitor, an inductor, an independent a.c. current source, and an independent a.c. voltage source. Each function is described in detail in a table, and the data produced by each function is illustrated in an example.

Table 12.7
The function C

$1E^-6F$ 4 o———||———o 19	$P \leftarrow 4\ \ 19\ \ 1E^-6$

$\nabla\ C\ P$

    Specify a capacitor, and store two connection nodes and a capacitance value entered for P

[1] $MIN \leftarrow MIN, P[1\ 2]$

    Catenate the two node numbers to MIN

[2] $IBIN \leftarrow IBIN, 0$
[3] $ISIN \leftarrow ISIN, 0$

    $IBIN$ and $ISIN$ each get a zero

[4] $GIN \leftarrow GIN, \div 1000000000000$

    A 10^{12} ohm resistor is automatically placed in parallel with each capacitor to maintain the branch structure for d.c. solutions. The conductance is catenated to GIN

[5] $CL \leftarrow CL, [1]\ \ P$
        ∇

    The vector consisting of two node numbers and the capacitance value is laminated as a row on the bottom of the CL matrix

Example 12.6 C (Table 12.7)

$STARTCCT$	Empty input data variables
$C\ 4\ \ 19\ \ 1E^-6$	Input a capacitor
$DISPLAY'MIN\ IBIN\ ISIN\ GIN\ CL'$	
	Display relevant variables

Item		*Contents*

MIN	$\leftrightarrow$	4 19	Node numbers 4 and 19
IBIN	$\leftrightarrow$	0	
ISIN	$\leftrightarrow$	0	
GIN	$\leftrightarrow$	$1E^-12$	The large parallel resistance
CL	$\leftrightarrow$	$4.00E0$ $1.90E1$ $1.00E^-6$	Node numbers and capacitance value in the *CL* table

Table 12.8
The function L

$1E^-3$H

6 $\circ$——⌇⌇⌇——$\circ$ 8 $P \leftarrow 6\ \ 8\ \ 1E^-3$

$\nabla\ \ L\ \ P$

Specify an inductor, and store two connection nodes and an inductance value entered as parameters

[1] $MIN \leftarrow MIN, P[1\ \ 2]$
[2] $IBIN \leftarrow IBIN, 0$
[3] $ISIN \leftarrow ISIN, 0$

Catenate the two nodes to MIN, and a zero to $IBIN$ and $ISIN$

[4] $GIN \leftarrow GIN, \div 0.00001$

A $1E^-5$ ohm resistor is converted to a conductance, and entered into the GIN vector to preserve the branch structure of the circuit for d.c. solutions. See also line 6 below

[5] $LL \leftarrow LL, [1]\ P$

Catenate the two nodes and inductor value onto the LL matrix as an additional row

[6] $RL \leftarrow RL, [1]\ P[1\ \ 2], 0.00001\ \nabla$

Extend the RL matrix to have the same shape as LL. The purpose of RL is to keep track of the $1E^-5$ ohm resistors used in inductor positions, so they can be removed after the d.c. solution

Example 12.7   (Table 12.8)

STARTCCT	Execute *STARTCCT*
$L\ \ 6\ \ 8\ \ 1E^-3$	Input an inductor
DISPLAY'MIN IBIN ISIN GIN LL RL '	Display relevant variables

Item		Contents			
MIN	↔	6 8			Node numbers 6 and 8
IBIN	↔	0			
ISIN	↔	0			
GIN	↔	1*E*5			The small resistor used for d.c. only
LL	↔	6	8	0.001	Node numbers and *L* appear in *LL*
RL	↔	6.00*E*0	8.00*E*0	1.00*E*¯5	Node numbers and the small place holding resistor appear in *RL*

Table 12.9
The function *IAC*

$$P \leftarrow 4 \quad 0 \quad 1 \quad 10$$

∇ *IAC* P

   Specify an a.c. current source and store two connection nodes plus a current magnitude and phase angle entered in the parameter list *P*.

[1]   *MIN←MIN,P*[1 2]
[2]   *IBIN←IBIN,*0
[3]   *ISIN←ISIN,*0
[4]   *GIN←GIN,*0

   The two node numbers are catenated to the *MIN* vector. *IBIN*, *ISIN*, and *GIN* each gets a zero

[5]   *IACL←IACL,*[1] P
     ∇
   The four entries in the parameter list are laminated as an additional row onto the bottom of the *IACL* matrix

Example 12.8 *IAC*   (Table 12.9)

STARTCCT	Execute *STARTCCT*
IAC 4 0 1 10	Input an a.c. current source
DISPLAY'MIN IBIN ISIN GIN IACL'	Display relevant variables

Item		Contents	
MIN	↔	4 0	Node numbers 4 and 0
IBIN	↔	0	These 0 entries preserve the
ISIN	↔	0	d.c. branch numbering
GIN	↔	0	

$$IACL \quad \leftrightarrow \quad 4 \quad 0 \quad 1 \quad 10$$

The two node numbers and the a.c. current source magnitude and angle appear in $IACL$

Table 12.10
The function VAC

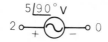

$$5\underline{|90°}\,\text{V}$$
$$2 \circ \text{---}(\sim)\text{---}\circ 0 \qquad\qquad P \leftarrow 2 \quad 0 \quad 5 \quad 90$$

$\nabla \quad VAC \; P$

Specify an a.c. voltage source and store two connection nodes plus a voltage magnitude and phase angle entered in the parameter list P

[1] $MIN \leftarrow MIN, P[1 \; 2]$
[2] $IBIN \leftarrow IBIN, 0$
[3] $ISIN \leftarrow ISIN, 0$

This processing is also found in the function IAC

[4] $GIN \leftarrow GIN, 1$

An a.c. current source will later be defined from the a.c. voltage source and parallel 1 S conductance. Thus 1 is catenated to GIN

[5] $VACL \leftarrow VACL, [1] \; P$
         ∇

The four entries are laminated directly onto the bottom of the $VCISL$ matrix as an additional row

Example 12.9 VAC (Table 12.10)

$STARTCCT$ Empty input data variables
$VAC \; 2 \; 0 \; 5 \; 90$ Input an a.c. voltage source
$DISPLAY\,'MIN \; IBIN \; ISIN \; GIN \; VACL'$

$Item$		$Contents$	
MIN	$\leftrightarrow$	2 0	Nodes 2 and 0 as entered
$IBIN$	$\leftrightarrow$	0	
$ISIN$	$\leftrightarrow$	0	
GIN	$\leftrightarrow$	1	1 S
$VACL$	$\leftrightarrow$	2 0 5 90	The a.c. voltage source data

12.2.5   The functions $ALPHA$, D, $QNPN$, $QPNP$

The various device-related functions are described in detail in the tables in this subsection. $ALPHA$ is used as a supporting function for the NPN and PNP bipolar transistor functions $QNPN$ and $QPNP$. The data produced by each function is also illustrated in an example.

Table 12.11
The function $ALPHA$

  ∇ $Z \leftarrow ALPHA\ B$

    Compute the common-base short-circuit current gain α for a transistor whose current gain β is entered as the argument.

[1] $Z \leftarrow B \div 1 + B$

  ∇

    The explicit result is $\alpha = \beta /(1 + \beta)$

Example 12.10 $ALPHA$ (Table 12.11)

             $ALPHA$ 50 100 200 300   The input transistor
0.98 0.99 0.995 0.997 β values are con-
                                     verted to α values

Table 12.12
The function D

   7 ⟶▷⟶ 10 $P \leftarrow 7\ \ 10,IS,CJ$

  ∇ $D\ P$

    Specify a diode junction, and store two node numbers, a saturation leakage current and a junction capacitance entered as parameters

[1] $MIN \leftarrow MIN,P[1\ 2]$
[2] $IBIN \leftarrow IBIN,0$
[3] $ISIN \leftarrow ISIN,P[3]$

    Catenate the two node numbers to the MIN vector; catenate zero to the $IBIN$ vector, and the diode saturation leakage current to the $ISIN$ vector

[4] $GIN \leftarrow GIN,0.000000000001$

    A 10^{-12} S conductance is catenated to GIN. This corresponds to a very large resistance (10^{12}) ohms) placed in parallel with the diode as an artificial but practical means of speeding up the convergence during d.c. solution

[5] $P \leftarrow P,0$
[6] $CL \leftarrow CL,[1]\ \ P[1\ \ 2\ \ 4]$

  ∇

    The vector consisting of the two nodes and the capacitance value is laminated as a row on the bottom of the CL matrix

---

Example 12.11 *D*   (Table 12.12)

STARTCCT	Empty input data variables
D 7 10 1E¯10 2.5E¯12	Input a diode
DISPLAY'MIN IBIN ISIN GIN CL'	

Item		*Contents*		

MIN	↔	7 10			Nodes 7 and 10 are stored in *MIN*
IBIN	↔	0			
ISIN	↔	1E¯10			The saturation leakage current
GIN	↔	1E¯12			The large parallel resistor; see *D*[4]
CL	↔	7.00E0	1.00E1	2.50E¯12	

The junction capacitance entries appear in *CL*

Table 12.13
The function *QNPN*

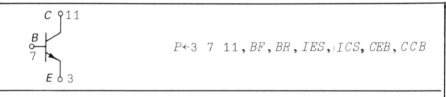

$P \leftarrow 3 \ 7 \ 11 , BF , BR , IES , ICS, CEB , CCB$

∇ *QNPN P*

Specify an NPN transistor and input the following parameters in order: *E, B, C, BF, BR, IES, ICS, CEB,* and *CCB*, as defined in Chapter 11

[1]   *MIN←MIN,P*[2 1 2 3]

Store in *MIN* two sets of node numbers in the correct sequence for an NPN transistor; *B, E, B, C*

[2]   *IBIN←IBIN,* 0,0
[3]   *ISIN←ISIN,P*[6 7]

Store two zeros in *IBIN*, one for each branch assigned to the transistor, and store the saturation leakage currents in *ISIN*

[4]   *GIN←GIN*,0.000000000001, 0.000000000001

Store in *GIN* the conductance corresponding to a large 10^{12} ohm resistance across each junction. As in the case of the diode, these speed up convergence during a d.c. solution

[5]   *QIN←QIN*,1,((ρ*ISIN*)−1),2,1,2, *ALPHA P*[4 5]

Store seven entries in the *QIN* vector as follows:
(1) the type number 1 to signify NPN
(2) the base-emitter branch number

(3) the temporary dummy node numbers $2,1,2$; note that MIN contains the actual node numbers, and that the dummy node numbers will be adjusted in the function END
(4) alpha forward and alpha reverse

[6] $P \leftarrow P, 0 \ 0$

Catenate two zeros onto the list of incoming parameters P. In the event that the user does not wish to specify CEB and CCB, then $P[8,9]$ would take on the default values of zero for the two junction capacitances. Otherwise, a zero value is stored for $P[10]$ and for $P[11]$, and these are never used

[7] $CL \leftarrow CL, [1] \ P[1 \ 2 \ 8]$

Store *E, B* and CEB as an additional row in the CL matrix

[8] $CL \leftarrow CL, [1] \ P[3 \ 2 \ 9]$
         ∇

Store *C, B* and CCB as a second additional row in the CL matrix

Example 12.12 $QNPN$ (Table 12.13)

$STARTCCT$ Execute $STARTCCT$
$QNPN \ 3 \ 7 \ 11 \ 100 \ 1.5 \ 1E^-10 \ 1E^-11 \ 25E^-12 \ 3E^-12$

$DISPLAY \ 'MIN \ IBIN \ ISIN \ GIN \ QIN \ CL'$ The transistor data

Item		*Contents*	
MIN	$\leftrightarrow$	7 3 7 11	Node numbers for each diode
$IBIN$	$\leftrightarrow$	0 0	No d.c. source currents
$ISIN$	$\leftrightarrow$	$1E^-10 \ 1E^-11$	Two saturation leakage currents
GIN	$\leftrightarrow$	$1E^-12 \ 1E^-12$	Two large parallel resistors; see $QNPN[4]$
QIN	$\leftrightarrow$	1 1 2 1 2 0.99 0.6	Intermediate form of transistor data
CL	$\leftrightarrow$	$3.00E0 \qquad 7.00E0 \qquad 2.50E^-11$ $1.10E1 \qquad 7.00E0 \qquad 3.00E^-12$	Two junction capacitances

Table 12.14
The function $QPNP$

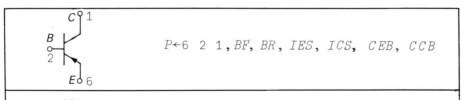

$P \leftarrow 6 \ 2 \ 1, BF, \ BR, \ IES, \ ICS, \ CEB, \ CCB$

$\nabla \ QPNP \ P$

Specify a PNP transistor and input the following parameters in order as defined in Chapter 11: *E, B, C,* BF, BR, IES, ICS, CEB, and CCB

```
[1]    MIN←MIN,P[1 2 3 2]
[2]    IBIN←IBIN,0,0
[3]    ISIN←ISIN,P[6 7]
[4]    GIN←GIN,0.000000000001, 0.000000000001
[5]    QIN←QIN,¯1,((ρISIN)-1),1,2,1, ALPHA P[4 5]
[6]    P←P, 0 0
[7]    CL←CL,[1] P[1 2 8]
[8]    CL←CL,[1] P[3 2 9]
       ∇
```

QPNP is seen to be identical to QNPN except for three minor differences:
(1) the node sequence in each pair in line 1 is reversed to correspond with the different diode directions in the PNP transistor Ebers–Moll model
(2) in line 5 the type number is −1 for this PNP transistor, and
(3) also in line 5, a different sequence of temporary dummy node numbers is used for processing convenience

Example 12.13 *QPNP*   (Table 12.14)

```
QPNP  6 2 1 100 1.5 1E¯10 1E¯11 25E¯12 3E¯12
```

STARTCCT is *not* executed in this example

```
DISPLAY'MIN IBIN ISIN GIN QIN CL'
```

Item   *Contents*

```
MIN   ↔   7 3 7 11 6 2 1 2
IBIN  ↔   0 0 0 0
ISIN  ↔   1E¯10 1E¯11 1E¯10 1E¯11
GIN   ↔   1E¯12 1E¯12 1E¯12 1E¯12
QIN   ↔   1 1 2 1 2  0.99 0.6 ¯1 3 1 2 1 0.99 0.6
CL    ↔      3.00E0      7.00E0      2.50E¯11
             1.10E1      7.00E0      3.00E¯12
             6.00E0      2.00E0      2.50E¯11
             1.00E0      2.00E0      3.00E¯12
```

These are the cumulative results for the NPN transistor from the previous example as well as the PNP transistor described above. The intermediate variables and storage tables are correspondingly extended

12.2.6   The functions *VCIS*, *OA*

The operational amplifier function *OA* contains only resistor functions *R* and voltage-controlled current source functions *VCIS*. The function *R* was described in Section 12.2.3. The functions *VCIS* and *OA* are described in detail in the accompanying tables. An example illustrates the data produced by each function.

Table 12.15
The function $VCIS$

```
    ——o 10              o 5
    +
                 ┌───────┐
    V      GM×V  ( ↑ ) GM=1E⁻3 S    P←5  4  10  11  1E⁻3
                 └───────┘
    —         
    ——o 11              o 4
```

∇ $VCIS$ P

Specify a voltage-controlled current source, and store the following parameters entered in the parameter list P as defined in Chapter 11:

NTO, NTI, NFP, NFN, GM

[1] $MIN←MIN,P[1\ 2]$

One branch is assigned to this element between the nodes NTO and NTI. Hence these nodes are stored in MIN

[2] $IBIN←IBIN,0$
[3] $ISIN←ISIN,0$
[4] $GIN←GIN,0$

Store zero values in $IBIN, ISIN$ and GIN. This is not a simple branch, and must be processed separately in a manner discussed in Chapter 13

[5] $VCISL←VCISL,[1]\ P$
    ∇

Store all five of these parameters in an additional row of the $VCISL$ matrix

---

Example 12.14 $VCIS$ (Table 12.15)

$STARTCCT$	Clear input data variables
$VCIS$ 5 4 10 11 1E⁻3	Describe a voltage-controlled
$DISPLAY$'MIN IBIN ISIN GIN VCISL'	current source and display
	pertinent variables

Item		*Contents*				
MIN	↔	5 4				
$IBIN$	↔	0				
$ISIN$	↔	0				
GIN	↔	0				
$VCISL$	↔	5	4	10	11	0.001

Nodes at the current source location

All four nodes and the trans-conductance value appear in the $VCISL$ table

Table 12.16
The function *OA*

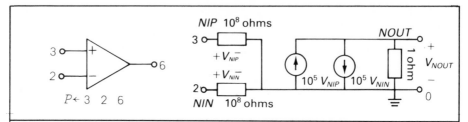

∇ *OA P*

Specify an operational amplifier and the three node numbers for the following interconnections: *NIP, NIN, NOUT*. The reference node, number 0, is
also used in the model implementation (see Section 11.2)
This function calls only the element description functions R and $VCIS$
which handle the detailed data storage as described previously

[1] R $P[1]$, 0 100000000
[2] R $P[2]$, 0 100000000

Specify a 10^8 ohm resistor between nodes *NIP* and 0, and also between
NIN and 0

[3] R $P[3]$, 0 , 1

Specify a 1 ohm resistor in parallel with the dependent current source
between nodes *NOUT* and 0, to convert the output to a voltage

[4] $VCIS$ 0, $P[3]$, $P[2]$, 0, 100000
[5] $VCIS$ $P[3]$, 0, $P[1]$, 0, 100000
    ∇

Specify two voltage-controlled current sources in parallel opposing
between nodes *NOUT* and 0, each with *GM* equal to 10^5 . One of these is
controlled by the voltage at *NIP* and the other at *NIN*

Example 12.15 *OA*   (Table 12.16)

STARTCCT	Clear input data variables
OA 3 2 6	Describe an op amp
DISPLAY'*MIN IBIN ISIN GIN VCISL*'	Display pertinent variables

Item	Contents											
MIN $\leftrightarrow$	3 0 2 0 6 0 0 6 6 0										Note the five node pairs
IBIN $\leftrightarrow$	0 0 0 0 0										Five 0 entries required
ISIN $\leftrightarrow$	0 0 0 0 0										Five 0 entries required
GIN $\leftrightarrow$	$1E^-8$ $1E^-8$ 1 0 0										Three resistors here
VCISL $\leftrightarrow$	0	6	2	0 100000							Two voltage-controlled
	6	0	3	0 100000							current sources appear in *VCISL*

12.2.7   The functions *REFNODE*, *QCALC*, *END*

REFNODE and *QCALC* are supporting functions used in the final circuit data formatting function *END* (see Section 12.2.1). *REFNODE* adjusts the circuit reference node number entries in *MIN*, while *QCALC* adjusts and reformats transistor data. The accompanying tables contain detailed descriptions of these three functions, and examples illustrate the results they produce.

Table 12.17
The function *REFNODE*

∇ *MNEW ←REFNODE MZ*; *GNODEI*

> *REFNODE* takes as its argument a vector of node numbers and produces an explicit result which is an adjusted node number vector. It also uses the global variable *NH*. See *END*. The specific purpose of *REFNODE* is to detect all zeros in the input node list and replace each zero with the number in *NH*. This is required to enable the solution routines to produce the correctly dimensioned nodal conductance matrix

[1]   *MNEW ←MZ*

Assign the input node vector to the result *MNEW*

[2]   *GNODEI←(MZ = 0)/ι ρMZ*

Compute the index number of each zero entry in the vector of node numbers

[3]   *MNEW[GNODEI] ← NH*
∇

Assign the number in *NH* to each of these zero locations in the result

Example 12.16 *REFNODE*   (Table 12.17)

$NH←9$
MIN
3 0 2 0 6 0 0 6 6 0
$REFNODE\ MIN$
3 9 2 9 6 9 9 6 6 9

Typical input data for the function *REFNODE*

The value in *NH* now replaces all ground node (0) entries in the result

Table 12.18
The function $QCALC$

	Inputs			Outputs	
QIN	M	NQ	$QCALC$		Q

∇ $Q \leftarrow QCALC\ QIN$; QNO; EBN; CN; $BEBRI$

> $QCALC$ takes as its argument the vector of transistor data produced by $QNPN$ and $QPNP$, and produces a matrix result having seven entries per row, and one row per transistor in the circuit. $QCALC$ also requires the global variables M and NQ. See END
> In addition to reformatting the data, $QCALC$ replaces the dummy temporary emitter, base, and collector node numbers in the input list with correct values obtained from M

[1] $QNO \leftarrow 1$

> Initialize the QNO looping index to 1 to start with the first transistor

[2] $Q \leftarrow (NQ,7) \rho QIN$

> Format the transistor data into NQ rows and 7 columns, and assign to the result matrix Q

[3] $BEBRI \leftarrow Q[;2]$

> Extract the base-emitter branch number for each transistor and assign to $BEBRI$ as an index vector

[4] $\rightarrow ((1 \uparrow \rho QIN) = 0)/0$

> Drop out of the function if there are no transistors in the circuit

[5] $L: EBN \leftarrow M[Q[QNO;\ 3\ 4]; BEBRI[QNO]]$

> For the first transistor, extract from M the correct final emitter and base node numbers; assign to the vector EBN

[6] $CN \leftarrow M[Q[QNO;5]; BEBRI[QNO]+1]$

> Similarly extract the correct final collector node number and assign to CN

[7] $Q[QNO;\ 3\ 4\ 5] \leftarrow EBN,CN$

> For the first transistor, replace the dummy E, B and C node numbers in Q by the correct ones

[8] $\rightarrow L \times \iota (QNO \leftarrow QNO+1) \leq NQ$
     ∇

> Increment the transistor index number QNO, and loop back to line 5 to carry out the node number processing for the next transistor if one is present. Repeat for all remaining transistors

Table 12.19
The function END

Inputs				Outputs		
GIN	MIN	$IBIN$		G	M	IX
QIN	$ISIN$	$VCISL$	END	Q	IS	$VCISL$
CL	LL	RL		CL	LL	RL
				A	NH	

$\quad \nabla\ END ; NQ ; MINO$

The purpose of END is to compute final forms for various global variables containing circuit data. This consists essentially of reformatting the inter-mediate data variables, and also performing some housekeeping opera-tions on node numbers with the help of the functions $REFNODE$, $QCALC$. and $INCID$

For inputs, END requires the following global variables: MIN, GIN, $IBIN$, QIN, $ISIN$, $VCISL$, CL, LL, RL

As outputs, END produces the following global variables: G, A, M, IS, IX, NH, Q, $VCISL$, CL, LL, RL

[1] $NH \leftarrow (\lceil /, MIN) + 1$

Compute the total number of nodes in the circuit. This equals 1 + the highest user-assigned node number in MIN

[2] $MINO \leftarrow REFNODE\ MIN$

Eliminate all zero entries (reference nodes) in MIN and replace them by the value in NH; assign the resultant vector to $MINO$

[3] $M \leftarrow Q((\rho\ GIN), 2) \rho\ MINO$

Convert $MINO$ to a two-column matrix, then transpose it into the desired two-row format for the branch connection matrix M

[4] $A \leftarrow\ NH\ INCID\ M$

Call the function $INCID$ with arguments NH and M, to produce the incidence matrix A

[5] $IX \leftarrow \bar{1} \downarrow A + . \times IBIN$

Compute the reduced node source current vector IX from the branch current formulation in the intermediate variable $IBIN$

[6] $NQ \leftarrow (\rho QIN) \div 7$

Determine the total number of bipolar transistors, NQ

[7] $Q \leftarrow QCALC\ QIN$

Produce the desired transistor data storage table Q by calling the function $QCALC$ with the intermediate variable QIN as its argument

[8] $G \leftarrow GIN$
[9] $IS \leftarrow ISIN$

The branch vectors for conductance and saturation leakage current are identical to their respective intermediate forms

[10] $VCISL[;\iota 4] \leftarrow ((1\uparrow\rho VCISL),4)\rho REFNODE, VCISL[;\iota 4]$

Apply the function $REFNODE$ to the vector of four node numbers in each row of the $VCISL$ matrix in order to replace all entries of 0 by the value in NH. This result is reshaped into its original matrix form and used to replace the node number entries in the original $VCISL$ matrix

[11] $CL[; 1\ 2] \leftarrow ((1\uparrow\rho CL),2)\rho REFNODE, CL[; 1\ 2]$

[12] $LL[; 1\ 2] \leftarrow ((1\uparrow\rho LL),2)\rho REFNODE, LL[; 1\ 2]$

Perform the same operation on the node numbers in the CL and LL matrices as described in line 10, except that only two node numbers are involved for capacitors and inductors

[13] $RL[; 1\ 2] \leftarrow LL[; 1\ 2]$
         ∇

Obtain the same node number adjustment in the RL table by taking a copy from LL

Example 12.17 END and $QCALC$ (Tables 12.18, 12.19)

```
STARTCCT
Q←IS←M←G←ι0

R 5 6 1E6
QPNP 7 8 9 100 1 1E⁻9 2E⁻9 1E⁻11 2E⁻11

DISPLAY  'G IS M Q'
```

Clear input data variables and final data variables

Input a resistor and a transistor

Item		*Contents*
G	↔	
IS	↔	
M	↔	
Q	↔	

Confirm that the final data variables are still empty

```
END
DISPLAY'GIN G ISIN IS MIN M QIN Q CL'
```

Carry out final formatting in END

Item		*Contents*
GIN	↔	1E⁻6 1E⁻12 1E⁻12
G	↔	1E⁻6 1E⁻12 1E⁻12
ISIN	↔	0 1E⁻9 2E⁻9
IS	↔	0 1E⁻9 2E⁻9
MIN	↔	5 6 7 8 9 8
M	↔	5 7 9
		6 8 8
QIN	↔	⁻1 2 1 2 1 0.99 0.5
Q	↔	⁻1 2 7 8 9 0.99 0.5
CL	↔	7.00E0 8.00E0 1.00E⁻11
		9.00E0 8.00E0 2.00E⁻11

Compare intermediate data variables generated by R and $QNPN$ with the corresponding final data produced by END (and $QCALC$)

Comments

(1) $G, IS, M,$ and Q are produced in *END* from the intermediate variables *GIN, ISIN, MIN* and *QIN* respectively.

(2) Q is directly generated by *QCALC*. The Q matrix has the correct base–emitter branch number (2) and the correct emitter, base and collector node numbers (7, 8, 9) obtained from M.

12.3 Interactive Terminal Session

This section is devoted to examples which deal with the content and structure of the *INPUT* data variables, as well as the internal operation of the various *INPUT* functions. Complete but simple circuits are featured. The examples are in the form of annotated records of actual terminal sessions at an interactive computer. Readers are encouraged to similarly work these and other examples, and to interactively explore additional features of this material, while actually using the *BASECAP* package on the computer.

12.3.1   Circuit Input Data – Using *INPUT* Functions

In this session we focus on the data generated to describe a simple one-transistor d.c. circuit (Fig. 12.3) using the *BASECAP INPUT* functions. The function *MH133IF*, listed below, describes this circuit.

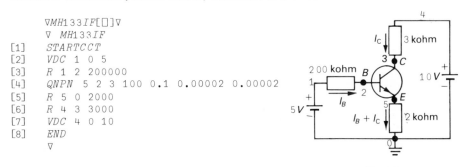

```
        ∇MH133IF[□]∇
        ∇ MH133IF
  [1]    STARTCCT
  [2]    VDC  1  0  5
  [3]    R  1  2  200000
  [4]    QNPN  5  2  3  100  0.1  0.00002  0.00002
  [5]    R  5  0  2000
  [6]    R  4  3  3000
  [7]    VDC  4  0  10
  [8]    END
        ∇
```

Fig. 12.3

Next we execute the circuit description function to generate the input data, and then display the results.

```
        MH133IF
        DISPLAY'G IS IX M Q'
```

Item		*Contents*
G	↔	1 $5E^-6$ $1E^-12$ $1E^-12$ 0.0005 0.000333 1
IS	↔	0 0 0.00002 0.00002 0 0 0
IX	↔	5 0 0 10 0
M	↔	1 1 2 2 5 4 4
		6 2 5 3 6 3 6·
Q	↔	1 3 5 2 3 0.99 0.0909

It is instructive to relate the above data to the circuit as well as the circuit description function. Note that the sequence of the branch entries in G, IS, M etc. is determined by the order in which each element or device is entered in the function $MH1331F$.

The circuit d.c. solution is now easily obtained, whereupon the results may be checked separately by hand if desired, to confirm the validity of both the input data and the solution process.

```
      DCSOLVE
CPUSECONDS = 0.288
ITERATIONS = 21


↓↓DC  NODE  VOLTAGES 1 TO 5 ↓↓


5 4.0|6 4.1 10 3.94
```

```
        VBN                          Branch d.c. voltages
5 0.942 0.116 ‾0.0424 3.94 5.9 10
```

```
     M1,M2,SMIN,TOL,VBMAX,VT,ETA ,TEMP    Solution parameters
2 10 1E‾12 0.1 1 0.026 1 20              (see PARMSET )
```

12.3.2   Circuit Input Data — No $INPUT$ Functions

This session illustrates an alternate approach to the solution of the circuit previously solved in Section 12.3.1. In this case, however, the $BASECAP$ $INPUT$ functions are not used, in order to demonstrate more clearly what processing they perform for the user.

In preparation, it is first necessary to redraw the circuit as shown below with all voltage sources replaced by their Norton equivalents. The nodes must be numbered from 1 and the reference node must be assigned the highest node number in the circuit (6). The user must also assign branch numbers in sequential order as indicated. Equivalent current source values and conductances are also needed, as are alpha forward and alpha reverse for the transistor.

The data required by the $BASECAP$ solution routines must then be generated directly using APL statements as illustrated in the function $MH133$.

```
     ∇MH133[□] ∇
     ∇ MH133
[1]  G← 1 5E‾6 0 0 0.0005 0.000333 1
[2]  IS← 0 0 0.00002 0.00002 0 0 0
[3]  IX← 5 0 0 10 0
[4]  M←Q 7 2 ρ 1 6 1 2 2 5 2 3 5 6 4 3 4 6
[5]  Q← 1 7 ρ 1 3 5 2 3 0.99 0.001
     ∇
```

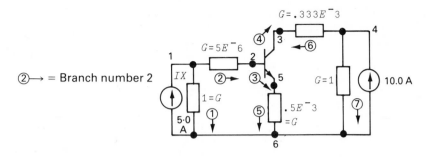

Fig. 12.4

See, for example, $END[3,5]$ and $QCALC[2]$ for corresponding processing steps in the $INPUT$ functions.

   We now execute $MH133$ and display the resultant data generated for this circuit.

$$MH133$$
$$DISPLAY'G\ IS\ IX\ M\ Q'$$

Item		*Contents*
G	$\leftrightarrow$	1 5E^-6 0 0 0.0005 0.000333 1
IS	$\leftrightarrow$	0 0 0.00002 0.00002 0 0 0
IX	$\leftrightarrow$	5 0 0 10 0
M	$\leftrightarrow$	1 1 2 2 5 4 4
		6 2 5 3 6 3 6
Q	$\leftrightarrow$	1 3 5 2 3 0.99 0.001

   A comparison with the data generated in Section 12.3.1 confirms that these numbers agree, allowing for the small approximations made above in specifying the input data. Also note that no $1E12$ ohm resistors appear across the transistor junctions in this case.

   This session serves to confirm how useful the various $INPUT$ functions really are in a package such as $BASECAP$. They relieve the user of a significant amount of tedious and error-prone preparation work, and make circuit solutions much faster, easier and more convenient for the user.

12.3.3   Circuit Example — Intermediate Data

This session deals with the circuit shown in the accompanying figure and described in the function $FRANK$ listed below. A 'stop' is placed at line 12 of $FRANK$ so execution will be terminated prior to invoking END in order to display only the initial data in the storage tables CL, $VCISL$, etc. before final formatting takes place.

$\nabla FRANK[\Box]\nabla$
$\nabla FRANK$
[1] $STARTCCT$
[2] $R\ 6\ 0\ 76000$
[3] $R\ 6\ 7\ 10000$
[4] $R\ 2\ 7\ 1000$
[5] $R\ 5\ 0\ 5.5$
[6] $L\ 4\ 5\ 0.069$
[7] $C\ 6\ 1\ 1E^-6$
[8] $VDC\ 0\ 7\ \ 15$
[9] $VAC\ 1\ 0\ \ 1\ 0$
[10] $OA\ 4\ 3\ 3$
[11] $QNPN\ 2\ 6\ 4\ \ ,Q1$
[12] END
      ∇

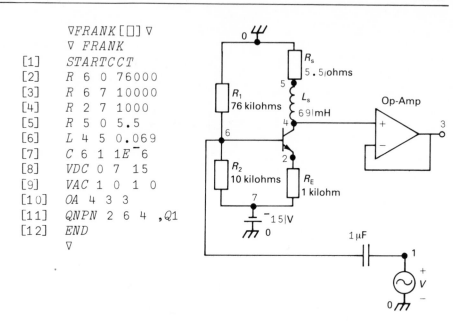

Fig. 12.5

$Q1\leftarrow100\ .1\ 1E^-16\ \ 1E^-15\ \ 1E^-11\ \ 1E^-13$
                                   Transistor data
$S\triangle FRANK\leftarrow12$ Place a 'stop' at line 12

$FRANK$ Execute circuit description function
$FRANK[12]$ APL System message confirming 'stop'

$\Box PW\leftarrow68$ Set print width

$DISPLAY'GIN\ MIN\ IBIN\ QIN\ ISIN\ VCISL\ CL\ LL\ RL\ VACL'$

Item		Contents
GIN	$\leftrightarrow$	$0.0000132\ 0.0001\ 0.001\ 0.182\ 1E5\ 1E^-12\ 1\ 1\ 1E^-8\ 1E^-8\ 1\ 0\ 0$
		$1E^-12\ 1E^-12$
MIN	$\leftrightarrow$	$6\ 0\ 6\ 7\ 2\ 7\ 5\ 0\ 4\ 5\ 6\ 1\ 0\ 7\ 1\ 0\ 4\ 0\ 3\ 0\ 3\ 0\ 0\ 3\ 3\ 0\ 6\ 2\ 6\ 4$
$IBIN$	$\leftrightarrow$	$0\ 0\ 0\ 0\ 0\ 0\ 15\ 0\ 0\ 0\ 0\ 0\ 0\ 0\ 0$
QIN	$\leftrightarrow$	$1\ 14\ 2\ 1\ 2\ 0.99\ 0.0909$
$ISIN$	$\leftrightarrow$	$0\ 0\ 0\ 0\ 0\ 0\ 0\ 0\ 0\ 0\ 0\ 0\ 0\ 1E^-16\ 1E^-15$
$VCISL$	$\leftrightarrow$	$0\quad\quad\ 3\quad\quad 3\quad\quad 0\ 100000$
		$3\quad\quad\ 0\quad\quad 4\quad\quad 0\ 100000$
CL	$\leftrightarrow$	$6.00E0\quad 1.00E0\quad 1.00E^-6$
		$2.00E0\quad 6.00E0\quad 1.00E^-11$
		$4.00E0\quad 6.00E0\quad 1.00E^-13$
LL	$\leftrightarrow$	$4\quad\quad 5\quad\quad 0.069$
RL	$\leftrightarrow$	$4.00E0\quad 5.00E0\quad 1.00E^-5$
$VACL$	$\leftrightarrow$	$1\ 0\ 1\ 0$

$S\triangle|FRANK\leftarrow\iota0$ Remove the 'stop'

Comments

(1) The above display shows all the intermediate data variables, and the initial contents of the storage tables generated for the circuit $FRANK$. All except $VACL$ are subsequently used by END.

(2) GIN, MIN, $IBIN$, QIN, and $ISIN$ are in vector form.

(3) $VCISL$, CL, LL, RL and $VACL$ are in matrix form.

(4) Note that the node number 0 appears in several places in MIN and $VCISL$.

(5) See Exercise 12.1 for a further investigation of this circuit.

Exercises

12.1   This exercise uses the circuit description function *FRANK* and the corresponding circuit from Section 12.3.3. Compute and display the following final data variables and storage tables for this circuit: *G*, *M*, *IX*, *Q*, *IS*, *VCISL*, *CL*, *LL*, *RL*, *VACL*, *A*, and *NH*. Compare these results with the intermediate data previously generated for this circuit in Section 12.3.3. Comment on the main similarities and differences between final and corresponding intermediate data values.

12.2   The functions *VAC*, *VDC*, *L*, *C*, *D*, *QNPN* and *QPNP* produce data which may under certain circumstances affect the accuracy of a circuit solution, and it is important to be aware of these limitations. Identify the specific causes involved, and briefly note the circumstances under which a significant error might be caused.

12.3   An essential requirement of the *BASECAP* solution routines is that the circuit nodes be numbered consecutively with no missing numbers. Gaps in the numbering sequence can easily be made by the user, especially in large circuits. Write and test an APL function, for use in *END*, that will test for this node-numbering error and provide a warning message to the user. The message should indicate specifically all missing node numbers.

12.4   Figure E12.1 contains a small-signal linear a.c. circuit model useful for both MOS field-effect transistors (MOSFETS) and junction field-effect transistors (JFETS). This model, which contains elements already defined in *BASECAP*, can be readily accommodated in an input function in the same manner as the operational amplifier. Write and test a new APL function called *ACFET* which performs the input data processing for these transistors in accordance with this model using the element description functions *C*, *R* and *VCIS*.

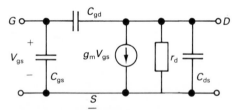

Fig. E12.1
Small-signal linear a.c. FET model

12.5   The existing Op-Amp model has no built-in frequency compensation, and is restricted to a predetermined value of voltage gain. Design and implement a new Op-Amp model called *OA1*, which overcomes these limitations. Use the existing function *OA* as a guide, but note that an additional node and one or more elements will be required to implement frequency compensation. Both the gain and the break frequency should be user-specified parameters in the new model. Suitable default values should be supplied for these parameters within *OA1*, so their specification by the user becomes optional.

12.6   Design and test an APL function called *ACBJT* for use with *BASECAP* to implement the small-signal a.c. bipolar transistor model shown in Fig. E12.2. This is the conventional hybrid-π model with the addition of collector resistance r_c, and collector–substrate capacitance C_{cs}. $ACBJT$ should accept node and parameter information as follows:

$$E\ B\ C\ B'\ C'\ \beta\ r_\pi\ r_0\ r_x\ r_c\ r_\mu\ C_\pi\ C_\mu\ C_{CS}$$

Provide suitable default values for the last seven parameters beginning with r_0, so their specification by the user becomes optional. See Exercise 11.17 for representative parameter values for this model.

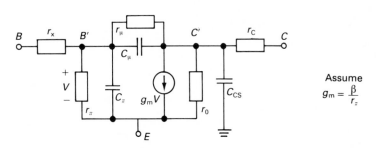

Fig. E12.2
Small-signal a.c. BJT circuit model for $ACBJT$

Solutions

12.1   See Section 12.3.3 for the function $FRANK$. Execute the circuit description function $FRANK$ and display selected final data variables and storage tables as follows:

```
      Q1
100 0.1 1E^-16 1E^-15 1E^-11 1E^-13
      FRANK

      DISPLAY'G M IX Q IS VCISL CL LL RL VACL A NH'
```

Item		*Contents*

G $\leftrightarrow$ 0.0000132 0.0001 0.001 0.182 1E5 1E^-12 1 1 1E^-8 1E^-8 1 0 0
1E^-12 1E^-12

M $\leftrightarrow$ 6 6 2 5 4 6 8 1 4 3 3 8 3 6 6
8 7 7 8 5 1 7 8 8 8 3 8 2 4

IX $\leftrightarrow$ 0 0 0 0 0 0 $^-$15

Q $\leftrightarrow$ 1 14 2 6 4 0.99 0.0909

IS $\leftrightarrow$ 0 0 0 0 0 0 0 0 0 0 0 0 0 0 1E^-16 1E^-15

$VCISL$ $\leftrightarrow$ 8.00E0 3.00E0 3.00E0 8.00E0 1.00E5
3.00E0 8.00E0 4.00E0 8.00E0 1.00E5

CL $\leftrightarrow$ 6.00E0 1.00E0 1.00E^-6
2.00E0 6.00E0 1.00E^-11
4.00E0 6.00E0 1.00E^-13

LL $\leftrightarrow$ 4 5 0.069

RL $\leftrightarrow$ 4.00E0 5.00E0 1.00E^-5

$VACL$ $\leftrightarrow$ 1 0 1 0

A $\leftrightarrow$
```
 0  0  0  0  0 ^-1  0  1  0  0  0  0  0  0  0
 0  0  1  0  0  0  0  0  0  0  0  0  0  1  0
 0  0  0  0  0  0  0  0  0  1  1 ^-1  1  0  0
 0  0  0  0 ^-1  0  0  0  1  0  0  0  0  0 ^-1
 0  0  0  1 ^-1  0  0  0  0  0  0  0  0  0  0
 1  1  0  0  0  1  0  0  0  0  0  0  0  1  1
 0 ^-1 ^-1  0  0  0 ^-1  0  0  0  0  0  0  0  0
^-1  0  0 ^-1  0  0  1 ^-1 ^-1 ^-1 ^-1  1 ^-1  0  0
```

NH $\leftrightarrow$ 8

Comments
(1) G and IS are identical to GIN and $ISIN$ respectively.
(2) IX is the node source equivalent of $IBIN$.
(3) M is the matrix form of the data in MIN with ground nodes (0) replaced by NH (8).
(4)   The matrix variables CL, LL, and RL are identical in both intermediate and final forms in this particular circuit because no ground nodes are involved.
(5) Q, derived from QIN, now contains valid node numbers (2, 6, 4) for the NPN transistor. The program has assigned 14 as the base—emitter branch number.
(6)   The ground nodes (0) in $VCISL$ are replaced by the contents of NH (8), which is one more than the highest user-assigned node number (7).
(7) $VACL$ is not affected by END. This is a special case where the second node is always required to be ground.
(8) A contains 8 rows (see NH,) and 15 branches (equal to the number of entries in GIN or G).

12.2   In VAC and VDC, the corresponding voltage source is represented by a current source and a parallel conductance of 1 S. The resultant source resistance may affect node voltages in low-impedance circuits (100 ohms or less). Both cases apply under d.c. and a.c. conditions, so this effect may be frequency dependent.
   The function L causes a $1E^-$5 ohm resistor to be used in place of an inductor, under d.c. conditions. The drop across this resistor could be significant in some extremely low-resistance circuit solutions. This resistor is removed for the a.c. solution.
   The functions C, D, $QNPN$ and $QPNP$ all introduce extra $1E12$ ohm resistors into the circuit solution that may affect both d.c. and a.c. solution accuracies. In the capacitor case, this may

be of concern in very high-resistance d.c. circuits and very high-frequency a.c. solutions. The diode and transistor implementations are less accurate in reverse-biased junction situations where the current in the $1E12$ ohm resistor could be significant, either for d.c. or high-frequency a.c. solutions.

CHAPTER

13

D.C. Circuit Analysis

In this chapter we consider the detailed design and operation of the functions comprising $BASECAP$'s DC group. This group was briefly introduced in Section 10.4.3, and its relationship to the rest of the $BASECAP$ package can be seen in Fig. 10.2. Both linear and nonlinear d.c. solutions are supported by these functions, which provide an essential capability for a general-purpose circuit analysis package such as $BASECAP$.

Because the d.c. solutions produced by these functions contain vital information for the electronic circuit designer, it is important to gain a thorough understanding of the operation of this part of the package. Consider for example, that d.c. operating point information provided by a d.c. solution is required in order to confirm the correct d.c. biasing of all nonlinear semiconductor devices in a circuit. Similarly, certain device parameters, which are essential for carrying out a small-signal a.c. solution of a circuit, must first be extracted from a d.c. solution.

In this group, two main d.c. solution functions provide for two different solution formats as follows:

$DCSOLVE$ produces a single nonlinear d.c. solution for all node voltages in the circuit

$TRANSDC$ produces a nonlinear d.c. transfer-function type of solution composed of any number of sets of individual node voltage solutions. One or more nodes may be designated as output nodes. One set of such results is produced for each value of an independent (circuit or device) parameter. These results are automatically formatted in a form suitable for plotting

In all, there are 24 different APL functions in the DC group (four of these are used only as alternatives to other functions). The names of all of these functions are listed in Table 13.1, along with the relevant global variable names. See Appendix 7 for a brief description of the principal variables in this group. Complete listings of all the DC group functions are included in Appendix 6.

In contrast with the structure of the $INPUT$ group, most of the d.c. functions are organized into two highly structured arrangements controlled by $DCSOLVE$ and $TRANSDC$. These structures are discussed in Section 13.1. The design and operation of $DCSOLVE$ and $TRANSDC$, together with their various

Table 13.1
Composition of the DC **Group**

```
          )GRPS
DC
          )FNS
ALTBASE1      DCSOLVE DIA      DIFDIODE      DIFNETL DIODE
GTRANS  IEQUIV  INCID  INITIALIZE  ITRANS  NEWTALT  NEWTALT1
NEWTDC   NEWTDC1   PARMSET   PARMTELL   TRANS   TRANSDC UNIT
VBCHKCAN         VBSTART            VBSTART2            VCFORM

          )VARS
A       ALTREF    BEBRI   CCT    DBI     DPI    ETA    G
GMAT    GT        GTI     ICOM   IDNEW   IQ     IQI    IS
IT      IX        JF      M      M1      M2     NH     PAR
PVAL    Q         SLOPE   SMIN   S1      TEMP   TOL    TYPE
VALL    VANS      VBMAX   VBN    VCISL   VT     ΔVB
```

supporting functions, are described in Sections 13.2 and 13.3. Following this in Section 13.4, are the methods used to implement the various devices in $BASECAP$, including semiconductor diodes, bipolar transistors, and voltage-controlled current sources. Finally, in Section 13.5, we introduce some alternative d.c. solution functions, and then present a sample interactive terminal session featuring several functions from this group.

13.1 DC Group Structure

This section contains an overview of the DC group, beginning with a description of the global input and output data variables used.

13.1.1 Data Structure

The various functions in this group have been implemented on the assumption that the following global data variables exist in the workspace for the circuit whose d.c. solution is desired: G, A, M, IS, IX, NH, Q, $VCISL$. These were shown as outputs from the $INPUT$ group in Fig. 12.1, and comprise the essential input data for the DC group as shown in Fig. 13.1.

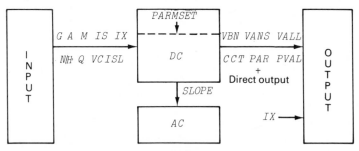

Fig. 13.1
DC **group I/O variables**

As indicated by the list of variables in Table 13.1, a large number of global variables remain in the workspace after a d.c. solution. Most of these are of interest to the user only for diagnostic, checking or instructional purposes. The major d.c. output variables are shown in Fig. 13.1, and are briefly described here. *DC SOLVE* produces

direct output   all node voltages are displayed on the terminal.

VANS   all node voltages are stored in the vector *VANS*.

VBN   a vector of all branch voltages updated at the last iteration step.

Executing *TRANSDC* produces

direct output   the selected node voltage results, together with the values of the independent parameters, are formatted and provided as explicit output in a form suitable for use by *PRINT* or *PLOT*

VAL L   a matrix consisting of one row of all circuit node voltages for each value of the independent parameter

CCT   a character variable containing the name of the circuit solved by *TRANSDC*

PAR   a character variable containing the name of the (circuit or device) parameter varied by *TRANSDC*

PVAL   a vector of parameter values used for *PAR*

An additional very important result of the d.c. solution process, and generated by both *DC SOLVE* and *TRANSDC*, is the linearized nodal conductance matrix (*SLOPE*) for the entire circuit. *SLOPE* is updated at each iteration step to accommodate the linearized representations for each of the nonlinear devices. The final composition of *SLOPE* corresponds to the last iteration step (i.e. at d.c. convergence). The small-signal a.c. solution functions require *SLOPE* since it is the real part of the nodal admittance matrix, and is automatically adjusted for the d.c. operating points of the nonlinear devices in the circuit.

Other global variables containing solution control parameters are also used by the *DC* group. These were presented previously in Section 11.4, and are documented on-line in the function *PARMTELL*. Any of these may be changed at any time by the user, or set to their default values by executing the function *PARMSET*.

Of these parameters, *ALTREF*, *ETA*, *TEMP*, *SMIN*, *VBMAX*, and *TOL* are required by both *DC SOLVE* and *TRANSDC*. The remaining parameters, *M1*, *M2* and *VT*, are required only by the function *VBCHKCAN* which is an alternative to the function $ALTBASE1$. The effects of many of these parameters are illustrated in the interactive terminal session and in the exercises in this chapter.

13.1.2   Program Structure

Tree structures are given in Fig. 13.2 for the functions $DCSOLVE$ and $TRANSDC$. These tree structures show the hierarchy of calls to other functions. For example, it can be seen that $NEWTALT$ and $VBSTART$ are utilized directly by $DCSOLVE$, and also that $INITIALIZE$, $ALTBASE1$, $IEQUIV$, and $DIFNETL$ are directly used by $NEWTALT$. Both trees in Fig. 13.2 are very similar, and are in fact identical on levels 3–5. The pair $NEWTALT$ and $NEWTALT1$ and the pair $VBSTART$ and $VBSTART2$ have the same respective overall purposes, and have only minor internal differences. Major differences do exist,

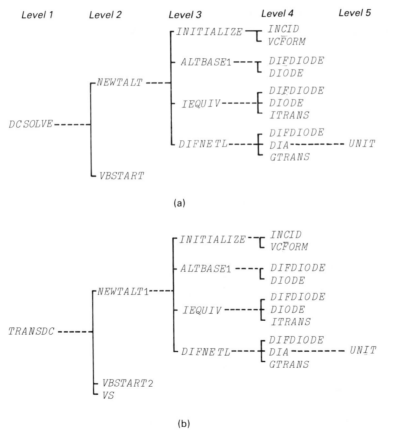

Fig. 13.2

Tree structures for (a) the function $DCSOLVE$ **(b) the function** $TRANSDC$

however, in the internal structure and operation of *DC SOLVE* and *TRANSDC*, as will be seen in Section 13.2. *VS* on level 2 of Fig. 13.2 is a formatting utility function considered to belong to the *OUT* group, and is discussed in Chapter 15.

The reader will recognize many function names that are common to Part 1 of this book. These comprise many of the functions found on levels 4 and 5 of the tree structures in Fig. 13.2, and include: *UNIT*, *DIA* (see *COND*), *DIODE*, *DIFNODE*, *DIFNETL*, and *INCID* . These routines serve the same overall purposes as their counterparts in Part 1, and are either identical to, or extended versions of those same routines.

13.2 User-executed *DC* Functions

In this section we examine in detail the two main functions *DC SOLVE* and *TRANSDC* which appear on level 1 of the tree structures for the *DC* group in Fig. 13.2. Subsequent sections deal with the functions on successively lower levels in the tree structures using a 'top down' approach which involves the reader in progressively greater detail as each lower level of functions is discussed.

13.2.1 *DC SOLVE*

The niladic function *DC SOLVE* is normally executed directly by the *BASECAP* user to obtain the d.c. solution for a circuit. Examples of the use of *DC SOLVE* are found in Section 11.6. The detailed internal operation of *DC SOLVE* is presented in Table 13.2. Clearly this is a very simple high-level function in which most of the actual processing is performed by the function *NEWTALT*.

Table 13.2
The function *DCSOLVE*

∇	*DC SOLVE*
	DC SOLVE is called directly by the user to compute and display one set of d.c. node voltages for a given circuit.
[1]	*VANS ← IX NEWTALT VBSTART 2*
	Compute the node voltages in *NEWTALT* and store them in the vector *VANS*. *IX* is the node source current vector, and *VBSTART 2* produces the initial branch voltages to initiate the solution
[2]	' '
[3]	' '
[4]	'↓↓ *DC NODE VOLTAGES* 1 *TO* '; (*NH*-1) ;'↓↓'
[5]	' '
	Display header information. Lines 2, 3, and 5 merely produce blank lines in the resultant display for emphasis
[6]	*VANS*
∇	
	Display the node voltages in sequence

13.2.2 *TRANSDC*

The function *TRANSDC* is designed to produce and store many sets of d.c. node voltage solutions for a given circuit, and is considerably more complex than *DCSOLVE*. Each set of node voltages corresponds to a different value of an independent parameter (device or circuit), whose effect on circuit behavior is being investigated.

In order to use *TRANSDC*, it is required that the name of the independent parameter be specified as a variable in the related circuit description function. The variable *R*1 in the circuit *FBTRIPLE* in Section 11.6.2 is an example of this requirement.

TRANSDC causes the entire circuit description function to be executed repeatedly, once for each value of the independent parameter. In this sense, *TRANSDC* essentially 'reaches back' into the *INPUT* group and controls the circuit set-up process performed by that group.

A consequence of this solution approach is that it is not mandatory to set up the data variables by executing the circuit description function before using *TRANSDC*, as it is for *DCSOLVE*. It is generally advisable to do so, however, for checking purposes. It is also good practice to complete a nominal d.c. solution for a new circuit using *DCSOLVE* first, before using *TRANSDC*. The multiple solutions carried out by *TRANSDC* can use considerable amounts of CPU time, and therefore can be costly.

The flow chart in Fig. 13.3 shows the overall sequence of operations within the function *TRANSDC*, and the accompanying material (Table 13.3) presents its internal operation in detail. Note that *VBN* is used to provide the starting voltages for successive d.c. solutions (*TRANSDC* line 1 5) for optimum efficiency. See Exercises 13.1 and 13.2 for further information. An example of the use of *TRANSDC* is given in Section 11.6.2.

13.3 Intermediate Level Functions

The functions on levels 2 and 3 of the tree structures in Fig. 13.2 are discussed in detail in this section. The primary solution function in the entire group is *NEWTALT*, or its equivalent *NEWTALT*1. If desired, either function may be executed directly by the user. *VBSTART* and *VBSTART*2 compute initial branch starting voltages for the solution routines. The remaining functions discussed in this section are all invoked by *NEWTALT* and *NEWTALT*1, and include *INITIALIZE*, *ALTBASE*1, *IEQUIV*, and *DIFNETL*.

13.3.1 *NEWTALT* **and** *NEWTALT*1

These two functions are identical except that lines 1 and 1 0 of *NEWTALT*, pertaining to CPU time calculation and display, are not included in *NEWTALT*1. Thus, the use of *NEWTALT*1 in *TRANSDC* produces a neater display and faster operation.

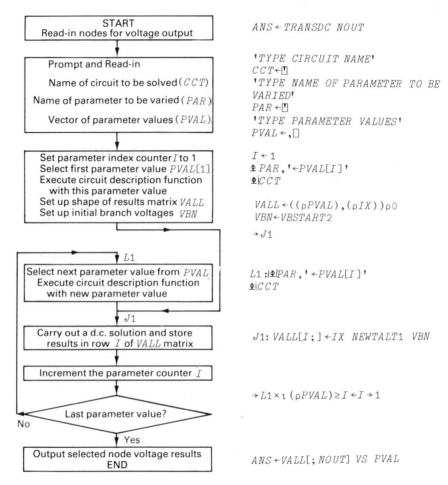

Fig. 13.3
Flow chart for the function $TRANSDC$

Table 13.3
The function $TRANSDC$

∇　$ANS \leftarrow TRANSDC\ NOUT$

$TRANSDC$ is designed to produce a series of d.c. solutions for a specified circuit as a function of a range of values for a chosen independent circuit or device parameter. A sequence of node voltages is produced and output for one or more nodes specified in the input argument list. The user is prompted to enter the following:

(1) the name of the circuit
(2) the name of the parameter to be varied
(3) the parameter values

[1]   *'TYPE CIRCUIT NAME'*
[2]   *CCT←*▯

Prompt the user and store the name of the circuit into the character variable CCT

[3]   *'TYPE NAME OF PARAMETER TO BE VARIED'*
[4]   *PAR ←*▯

Prompt the user and store the name of the independent parameter in PAR

[5]   *'TYPE PARAMETER VALUES'*
[6]   *PVAL ←,*▯

Prompt the user and store the vector of values to be used for this parameter in $PVAL$

[7]   *I←1*

Initialize the counter I to 1. This is used to index through each of the parameter values in turn

[8] $\pounds PAR$*,'←PVAL[I]'*

Execute the assignment statement to cause the independent parameter in PAR to take on the first parameter value in $PVAL$

[9] $\pounds CCT$

Execute the circuit description function, whose name is held in CCT, with the first value of the independent parameter

[10]   *VALL←((ρPVAL),(ρIX))ρ0*

Set up the empty results matrix $VALL$ to have one row per value of the independent parameter, and one column for each node aside from the reference node

[11]   *VBN←VBSTART2*

Execute $VBSTART2$ to specify starting branch voltages in VBN

[12]   *→J1*

Jump to the line labelled $J1$

[13]   *L1:*$\pounds PAR$*,'←PVAL[I]'*

Label this line $L1$ and assign the next value in $PVAL$ to the independent parameter in PAR

[14] $\pounds CCT$

Execute the circuit description function with this value of the independent parameter

[15]   *J1:VALL[I;]←IX NEWTALT1 VBN*

Label this line $J1$, solve for the node voltages using $NEWTALT1$, and store the results in row number I

[16] $\rightarrow L1 \times \iota\ (\rho PVAL) \ge I \leftarrow I+1$

Increment the counter and go to the line labled $L1$ if an unused parameter value remains, otherwise execute the following line

[17] $ANS \leftarrow VALL[;NOUT]\ \ VS\ PVAL$
    ∇

When all solutions have been carried out, select the columns corresponding to the specified output node(s), and format the results ready for plotting

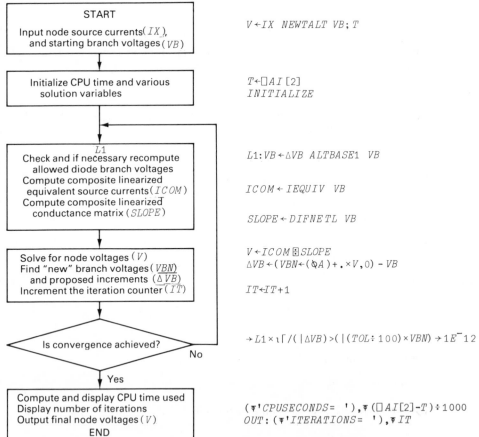

Fig. 13.4
Flow chart for the function $NEWTALT$

The remainder of our discussion in this subsection focusses only on $NEWTALT$ which implements a modified Newton–Raphson algorithm for the solution of a nonlinear d.c. circuit. The essential details of this method were presented in Chapters 2 and 7. $NEWTALT$ incorporates a modification which limits each diode junction branch voltage excursion to a 'safe' value from one iteration step to the next. The method used for this is known as the method of alternating bases, described later in the function $ALTBASE1$.

Figure 13.4 contains a flow chart of $NEWTALT$, and a detailed line-by-line description is included in Table 13.14. See Section 13.5 for examples involving $NEWTALT$.

Table 13.4
The function $NEWTALT$

$\quad \nabla \ \ V \leftarrow IX \ NEWTALT \ VB;T$

    This function carries out the nonlinear d.c. solution of the circuit, and returns explicit node voltage results. Input arguments required in the header are node source currents (IX) and starting branch voltages (VB). VB may be a scalar which is applied to all branches. The other data items required are shown in Fig. 13.1. T is a local variable.

[1] $T \leftarrow \Box AI[2]$

    Record the time on the CPU system clock before commencing the main solution.

[2] $INITIALIZE$

    Compute initial values for a number of variables required to carry out the d.c. solution. See $INITIALIZE$.

[3] $L1:VB \leftarrow \triangle VB \ ALTBASE1 \ VB$

    Assign the label $L1$ to permit looping to this line. Check and, if necessary, limit the diode function branch voltage excursions to help ensure convergence by calling $ALTBASE1$.

[4] $ICOM \leftarrow IEQUIV \ VB$

    Compute and store in $ICOM$ the circuit's composite linearized equivalent source current vector (including contributions from the nonlinear devices), by means of the function $IEQUIV$.

[5] $SLOPE \leftarrow DIFNETL \ VB$

    Compute and store in $SLOPE$, the circuit's linearized conductance matrix (including contributions from the nonlinear devices), by means of the function $DIFNETL$.

[6] $V \leftarrow ICOM \boxplus SLOPE$

    Carry out a matrix solution for the node voltages and store in V.

[7] $\triangle VB \leftarrow (VBN \leftarrow (\lozenge A) + . \times V, 0) - VB$

    Convert from node to branch voltages (VBN) and also compute the size of the branch voltage increments to be used for the next iteration

---

[8] $IT \leftarrow IT + 1$

Increment the iteration counter

[9]
$\rightarrow L1 \times \iota \lceil /(|\Delta VB) > (|(TOL \div 100) \times VBN)| + 0.000000000001$

Check if convergence has been met within the tolerance criterion for all branches. If so, go to the next line. If not, loop back to the line labelled $L1$, and carry out another solution using the most recent results as input

[10] $(\Phi'CPUSECONDS= \ '), \Phi(\square AI[2]-T) \div 1000$

Compute and display the CPU time in seconds required to carry out the solution

[11] $OUT:(\Phi'ITERATIONS= \ '), \Phi IT$
     ∇

Display the number of iterations required to produce convergence

13.3.2 $VBSTART$ **and** $VBSTART2$

These functions are essentially alternatives. Either may be used with $NEWTALT$ or $NEWTALT1$ to generate initial branch voltages with which to begin the Newton–Raphson iteration solution for a circuit. Detailed descriptions are contained in Tables 13.5 and 13.6.

Table 13.5
The function $VBSTART$

∇ $VB \leftarrow VBSTART$

This function produces an initial set of branch voltages and passes them on as an explicit result

[1] $VB \leftarrow (\rho G)\rho 0.4$
     ∇

The arbitrary value of 0.4 V is replicated to give a vector of the correct number of entries for VB

Table 13.6
The function $VBSTART2$

∇ $VB \leftarrow VBSTART2$

This function is identical in purpose to $VBSTART$

[1] $VB \leftarrow (\rho G)\rho 0.4$

This line is identical to line 1 of $VBSTART$

[2] $VB[Q[;2]+1] \leftarrow ^-1$
   ▽

The base–collector branch of each transistor is initially reverse biased at $^-1$ V. This is sometimes a more efficient starting condition for transistor amplifier circuits

Example 13.1   *VBSTART* **and** *VBSTART2* **(Tables 13.5, 13.6)**

The example compares the initial branch voltages produced by *VBSTART* and *VBSTART2* for *FBTRIPLE* (see Section 11.6.2).

```
        FBTRIPLE                        Execute FBTRIPLE
        VBSTART
0.4 0.4 0.4 0.4 0.4 0.4 0.4 0.4 0.4 0.4 0.4 0.4
        0.4 0.4 0.4 0.4 0.4 0.4 0.4 0.4
        VBSTART2
0.4 0.4 ¯1 0.4 0.4 0.4 ¯1 0.4 0.4 0.4 ¯1 0.4 0.4
        0.4 0.4 0.4 0.4 0.4 0.4 0.4
```

Note the differences in the collector–base branch voltage of each transistor.

13.3.3 *INITIALIZE*

The function *INITIALIZE* is used at an early point in *NEWTALT* and *NEWTALT1* to prepare various data variables. It is very easy to check the global variable results produced by *INITIALIZE* for a given circuit as the following example illustrates.

Example 13.2 *INITIALIZE*

The circuit *FBTRIPLE* is again used as a vehicle to provide test data for *INITIALIZE*. The sample result displayed from *INITIALIZE* is *DBI*, which is a vector containing the branch numbers assigned to all diode junctions in the circuit. By relating these results to the circuit description function for *FBTRIPLE* in Section 11.6.2, the reader can confirm, for example, that diode *D2* is in branch number 12.

```
        FBTRIPLE          Execute a circuit description function
        VB ← VBSTART      Compute initial branch voltages
        INITIALIZE        Execute INITIALIZE
        DBI               Display DBI (diode branch number
2 3 6 7 8 10 11 12 13     index)
```

Within *INITIALIZE* there is some duplication involved in computing *A* and *NH*, since these are produced also in *END* in the *INPUT* group. The reason for this is to permit the d.c. solution functions to be executed for instructional purposes using simple input test data obtained without the use of the normal *INPUT* functions. For an example of such testing see Section 12.3.2. *INITIALIZE* is described in detail in Table 13.7.

It is instructive for the reader to explore interactively on the computer some of the internal processing steps specified in the various function description tables. The following simple example contains such an exploration pertaining to *INITIALIZE*.

Example 13.3   Exploring the calculation of *DBI*

This example uses the same circuit (*FBTRIPLE*) as the previous example. The processing steps used to obtain *DBI* are explored with reference to line 3 of *INITIALIZE*,

```
        FBTRIPLE          Execute FBTRIPLE
        IS                Display diode saturation currents
0 1E‾9 1E‾9 0 0  1E‾9 1E‾9 1E‾9 0  0 1E‾9 1E‾9 1E‾9  1E‾9 0 0 0 0 0 0 0
        ρIS               Compute the number of branches
20

        ιρIS              Obtain a vector for indexing
1 2 3 4 5 6 7 8 9 10 11 12 13 14 15 16 17 18 19 20
        (IS≠0)            Identify diode junction locations

0 1 1 0 0 1 1 1 0 1 1 1 1 0 0 0 0 0 0 0
        (IS≠0)/ιρIS       Compute DBI by logical compression
2 3 6 7 8 10 11 12 13
```

Table 13.7
The function *INITIALIZE*

   ∇   *INITIALIZE*

The purpose of *INITIALIZE* is to prepare for the solution by computing initial values for a number of vectors and matrices. These are generated in the form of global variables

[1] $NH \leftarrow (\lceil /, M)$.

Compute the total number of circuit nodes including the reference node

[2] $A \leftarrow NH \ INCID \ M$

Compute the incidence matrix A

[3] $DBI \leftarrow (IS \neq 0)/ι\rho IS$

Compute an index vector of branch numbers containing diode or transistor junctions

[4] $\Delta VB \leftarrow (\rho IS)\rho IT \leftarrow 0$

Set the iteration counter IT to zero, and initialize the incremental branch voltage vector to all zeros

[5] $VB \leftarrow (\rho \Delta VB)\rho VB$

If VB is a scalar, replicate it for each branch. If VB is already a vector, this statement has no effect

[6] $TYPE \leftarrow Q[;1]$

Extract the transistor type numbers from the Q matrix and assign to $TYPE$

[7] $IQI \leftarrow (1+\rho IX)\rho 0$

The initial transistor node source current components are set to zero

[8] $GTI \leftarrow ((1+\rho IX),(1+\rho IX))\rho 0$

Initialize to zero, the conductance matrix used for the transistor conductance gain components

[9] $GMAT \leftarrow (NH, NH)\rho 0$

An initial conductance matrix, having NH rows and columns, is set to zero

[10] $BEBRI \leftarrow Q[;2]$

Create a base–emitter junction branch number index vector

[11] $JF \leftarrow 11594.2 \div ETA \times TEMP + 273$

Compute the junction factor JF using existing values for ETA and $TEMP$

[12] $VCFORM\ VCISL$
     ∇

Compute the nodal conductance matrix for all voltage-controlled current sources, and update the global variable $GMAT$ using the function $VCFORM$

13.3.4 $ALTBASE$ 1

$ALTBASE$ 1 plays an important moderating role in the Newton–Raphson iteration process for a nonlinear circuit containing diodes and transistors, and helps to speed convergence to a valid d.c. solution. The branch voltage excursions between iteration steps are effectively controlled by the action of $ALTBASE$ 1. An important consequence of this action is that during the critical initial stages of the iteration process, the diode junction branches cannot sporadically become excessively forward biassed. It is very important to avoid this condition since movement towards subsequent convergence at lower current levels is extremely slow, as will be demonstrated later in Section 13.5.

$ALTBASE$ 1 is an implementation of the method of alternating bases presented in Chapter 7. This method involves an additional processing step in which the inverse of the exponential diode relation is used to compute revised junction voltages. Figure 13.5 shows the structure of $ALTBASE$ 1, and a detailed description is contained in Table 13.8. Examples illustrating the internal operation of $ALTBASE$ 1 are included in Section 13.5.

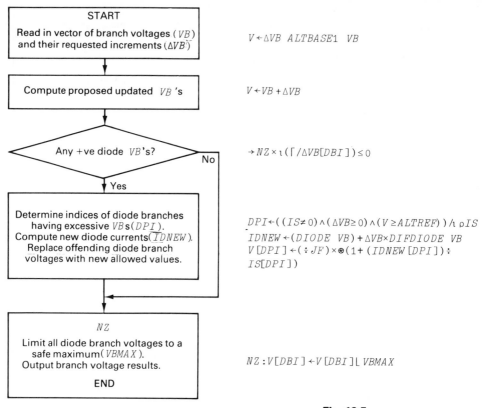

$$V \leftarrow \Delta VB \ ALTBASE1 \ VB$$

$$V \leftarrow VB + \Delta VB$$

$$\rightarrow NZ \times \iota (\lceil / \Delta VB[DBI]) \leq 0$$

$$DPI \leftarrow ((IS \neq 0) \wedge (\Delta VB \geq 0) \wedge (V \geq ALTREF)) \wedge \rho IS$$
$$IDNEW \leftarrow (DIODE \ VB) + \Delta VB \times DIFDIODE \ VB$$
$$V[DPI] \leftarrow (\div JF) \times \circledast (1 + (IDNEW[DPI]) \div$$
$$IS[DPI])$$

$$NZ:V[DBI] \leftarrow V[DBI] \lfloor VBMAX$$

Fig. 13.5
Flow chart for the function $ALTBASE1$

Table 13.8
The function $ALTBASE1$

$\quad \nabla \quad V \leftarrow \Delta VB \ ALTBASE1 \ VB$

The purpose of this function is to check the size and polarity of the proposed branch voltage increment VB. If this is positive and excessive, only a portion of the requested increment is allowed. Input arguments are the previous branch voltage vector VB, and the proposed branch voltage increments obtained from the previous iteration. $ALTBASE1$ returns the explicit result V

[1] $\quad V \leftarrow VB + \Delta VB$

Sum the branch voltage increments and the previous branch voltages to obtain the proposed new branch voltages, V

[2] $\quad \rightarrow NZ \times \iota (\lceil / \Delta VB[DBI]) \leq 0$

Jump to the last line labelled NZ if no diode junctions have positive VB's

[3] $\quad DPI \leftarrow ((IS \neq 0) \wedge (\Delta VB \geq 0) \wedge (V \geq ALTREF))/\iota \rho IS$

Compute DPI, the index vector of diode branches having VBs which are both positive and greater than the boundary reference value in $ALTREF$. The default value for $ALTREF$ is 0.26 V

[4] $IDNEW \leftarrow (DIODE\ VB) + \Delta VB \times DIFDIODE\ VB$

Based on a knowledge of the diode equation, compute revised diode *currents* for the given conditions of ΔVB and VB

[5] $V[DPI] \leftarrow (\div JF) \times \circledast (1 + (IDNEW[DPI]) \div IS\ [DPI])$

Compute revised junction voltages and update V for those branches identified in the vector DPI which would otherwise have excessive voltage increments

[6] $NZ: V[DBI] \leftarrow V[DBI] \lfloor VBMAX$
      ∇

Ensure that all diode branch voltages are below a safe maximum $VBMAX$. The default value for $VBMAX$ is 1 V

13.3.5 $IEQUIV$

As discussed in Section 2.3, the companion-model approach represents each nonlinear component or device by an appropriate linearized equivalent conductance and parallel current source. The linearized equivalent current sources used to represent diodes and transistors in $BASECAP$ are computed in $IEQUIV$ and combined with the actual node source currents (IX). This process must be carried out once during each iteration step, in order to update the composite node source currents before each new circuit solution. A detailed description of $IEQUIV$ is contained in Table 13.9.

Table 13.9
The function $IEQUIV$

∇ $ICOM \leftarrow IEQUIV\ VB; IEQ; QNO$

$IEQUIV$ returns as an explicit result in $ICOM$, the composite linearized equivalent node source current vector for the complete nonlinear circuit, valid for a particular set of d.c. operating conditions specified in the branch voltage vector VB. IEQ and QNO are used as local variables in this function

[1] $IEQ \leftarrow (DIODE\ VB) - VB \times DIFDIODE\ VB$

Compute the linearized equivalent current source for each diode junction and assign to IEQ

[2] $ICOM \leftarrow IX - {}^{-}1 \downarrow A + . \times IEQ$

Convert the vector IEQ from branch form to node form, add with appropriate sign to the node source current vector IX, and assign the composite vector to $ICOM$

[3] $QNO \leftarrow 1$

Set QNO to 1 to select the first transistor

[4] $L1: ICOM \leftarrow ICOM + IQ \leftarrow ITRANS$

Compute the linearized equivalent gain-dependent current source components for the first transistor and add to the result matrix $ICOM$

[5] $\rightarrow L1 \times \iota (QNO \leftarrow QNO + 1) \leq 1 \uparrow \rho Q$

▽

Increment the transistor counter index QNO and repeat the calculations in line 4 above for each additional transistor

---

Example 13.4   Diode companion model components

In this example we compute representative numerical values for the linearized equivalent current source IEQ, the diode current ID, and the linearized diode conductance GD shown in the diode circuit representation below. We assume a value for $IS = IE^-11$ amperes and a forward bias of 0.5 V. Nominal values are used for ETA and temperature. The processing for IEQ is defined in $IEQUIV$ line 1.

The various labelled components in the circuit representation are defined in the test function $DTEST$, which is useful for the computation of the required numerical results. Note that $DTEST$ line 4 and $IEQUIV$ line 1 are equivalent.

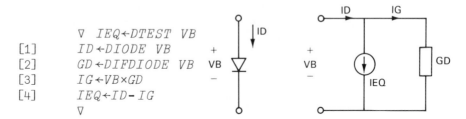

```
        ▽ IEQ←DTEST VB
[1]     ID ←DIODE  VB
[2]     GD ←DIFDIODE VB
[3]     IG ←VB×GD
[4]     IEQ←ID- IG
        ▽
```

Fig. 13.6

We now specify input test data (IS, DBI, VB), and compute IEQ. IS and VB are in amperes and volts respectively, and are specified in vector form as required by $DIODE$ and $DIFDIODE$ (see Section 13.4.1).

```
        IS ←,1E^-11
        DBI ←1
        □←IEQ←DTEST ,.5
  ‾0.0110056
```

The remaining component values used in the calculation of IEQ are readily displayed as follows:

$$IG \qquad\qquad\qquad\qquad ID$$
$$0.0116535 \qquad\qquad\qquad 0.000647898$$
$$GD \qquad\qquad\qquad\qquad IEQ + .5 \times GD$$
$$0.023307 \qquad\qquad\qquad 0.000647898$$

Since IEQ is negative, its actual direction is upwards in the figure above.

13.3.6 $DIFNETL$

This is an extended version of the function $DIFNET$ introduced in Section 7.2. It has a similar overall purpose, but is designed to produce the linearized equivalent conductance matrix corresponding to a given set of d.c. branch voltages (VB) for circuits containing resistors, diodes, bipolar transistors, and voltage-controlled current sources. $DIFNETL$ must be called once within each iteration step in functions such as $NEWTALT$ and $NEWTALT1$ to accommodate the changing voltage conditions as the solution converges.

The material in Table 13.10 constitutes a detailed description of $DIFNETL$. Recall that $GMAT$ on line 2 is a reduced partial nodal conductance matrix produced by $VCFORM$ on line 12 of $INITIALIZE$ (Section 13.3.3). $GMAT$ contains only the conductance contributions from the voltage-controlled current sources in the circuit. The internal operation of $VCFORM$ is covered later in Section 13.4.4.

Table 13.10
The function $DIFNETL$

$\nabla \quad S \leftarrow DIFNETL\ VB;\ QNO$

This function returns, as an explicit result S, the linearized conductance matrix for the entire circuit corresponding to the nonlinear device operating points contained in the branch voltage vector VB. QNO is a local variable in $DIFNETL$

[1] $S1 \leftarrow DIFDIODE\ VB$

Compute the linearized slope conductances for the diodes and make entries in the appropriate positions in the vector $S1$. Non-junction branches have zero-valued entries in this vector

[2] $S \leftarrow GMAT + \ ^-1\ ^-1\ \downarrow A + . \times (DIA\ S1 + G) + . \times \lozenge A$

Produce the partial nodal conductance result matrix containing components due to resistors (G), diode junctions ($S1$), and voltage-controlled current sources. This is a reduced matrix of shape(NR, NR), produced via the diagonal matrix in the function DIA, and the incidence matrix A

[3] $QNO \leftarrow 1$

Set QNO to 1 to select the first transistor

[4] $L1:\ S \leftarrow S + GT \leftarrow GTRANS$

Add the gain-dependent conductance contributions into S for the first transistor

[5] $\rightarrow L1 \times \iota (QNO \leftarrow QNO + 1) \le 1 \uparrow \rho Q$

∇

Increment the transistor counter index QNO and repeat the calculations in line 4 above for each additional transistor

13.4 Device Functions

The remaining functions from levels 4 and 5 of the tree structures in Fig. 13.2 are considered in detail in this section. Most of these low-level functions are directly concerned with one or more of the devices implemented in $BASECAP$. These devices include the semiconductor diode, the bipolar transistor and the voltage-controlled current source.

The d.c. behavior of the diode is computed with the functions $DIODE$ and $DIFDIODE$ discussed in Section 13.4.1. The theoretical basis for the bipolar transistor model is presented in Section 13.4.2, while Section 13.4.3 describes $GTRANS$ and $ITRANS$, which are used in the implementation of the transistor model. The model for the voltage-controlled current source and the function $VCFORM$ are the final topics in this section.

The tree structures in Fig. 13.2 contain three additional basic-level functions: $INCID$, DIA (see $COND$ in Table 1.1), and $UNIT$. These are functionally identical to corresponding routines presented in Section 1.6. They are included in the listings in Appendix 5, but are not discussed here.

13.4.1 $DIODE$ and $DIFDIODE$

The functions described here are similar to those developed in Exercise 2.2. Specifically, $DIODE$ and $DIFDIODE$ represent the nonlinear semiconductor diode by an equivalent constant current source and linearized slope conductance during the Newton–Raphson d.c. solution process. Differences from the previous functions pertain mainly to the use of indexing (DBI). Indexing improves efficiency since all but the diode junction branches are excluded from the exponential I-V calculations. Recall that the diode branch index (DBI) is computed in $INITIALIZE$ (see section 13.3.3).

Table 13.11
The function $DIODE$

∇ $I \leftarrow DIODE \ V$

This function computes the exponential current–voltage relationship for any number of diodes. The input argument V is the vector of circuit branch voltages. The result argument I holds the diode currents. I has the same shape as the vector V, with non-diode branches having zero entries. $DIODE$ also requires the following global variables: IS, DBI, JF and $VBMAX$

[1] $I \leftarrow IS \times 0$

Initialize the result vector to the correct shape containing all zeros

[2] $I[DBI] \leftarrow IS[DBI] \times {}^- 1 + \ast JF \times V[DBI] \lceil (-VBMAX)$

   ∇

For each diode branch $(DBI)_{|}$, compute the current according to the diode equation and enter these currents in the appropriate positions in the result vector. Reverse bias voltages are limited to $-|VBMAX$

Table 13.12
The function $DIFDIODE$

    ∇ $S \leftarrow DIFDIODE \; VB$

This function is very similar in overall design to the $DIODE$ function, and uses the same arguments with the addition of the global variable $SMIN$. The purpose of $DIFDIODE$ is to compute a vector of diode slope conductances corresponding to the d.c. operating voltages VB in the input argument list

[1] $S \leftarrow (\rho IS) \rho 0$

The result vector is initialized

[2] $S[DBI] \leftarrow SMIN \lceil IS[DBI] \times JF \times \ast JF \times VB[DBI] \lceil (-VBMAX)$

   ∇

Using the derivative of the diode equation, compute the diode slope conductances; reverse bias voltages are limited to $-VBMAX'$; all diode slope conductances are made to be equal to or greater than the parameter limit $SMIN$. The slope conductances are entered into the appropriate diode positions in the result vector S

Example 13.5 $DIODE$

The use of indexing in the function $DIODE$ is illustrated for the simple one-transistor circuit shown in Fig. 13.7 with its circuit description function $EXGTRANS$.

```
      ∇EXGTRANS[□] ∇
      ∇ EXGTRANS
[1]     STARTCCT
[2]     R 3 0 10000
[3]     R 4 3 70000
[4]     R 1 0 2000
[5]     R 4 5 10000
[6]     QNPN 1 3 5 ,QMOD13
[7]     R 4 2 10
[8]     VDC 2 0 20
[9]     END
      ∇
      QMOD13
100 0.1 1E‾13 1E‾13 1E‾10 1E‾12
```

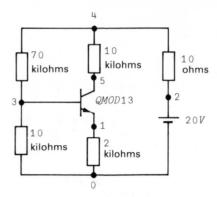

Fig. 13.7

We now set a trace on $DIODE$ line 2 and carry out the d.c. solution as follows:

$$ETA \leftarrow 1.2$$
$$TEMP$$

20

$$T \triangle DIODE \leftarrow 2$$

$$EXGTRANS$$
$$DCSOLVE$$
$DIODE[2]$ $5.35E^-8$ $5.35E^-8$
$DIODE[2]$ $5.35E^-8$ $5.35E^-8$
$DIODE[2]$ $4.16E^-6$ $^-1E^-13$
$DIODE[2]$ $4.16E^-6$ $^-1E^-13$
$DIODE[2]$ 0.000213 $^-1E^-13$
$DIODE[2]$ 0.000213 $^-1E^-13$
$DIODE[2]$ 0.000842 $^-1E^-13$
$DIODE[2]$ 0.000842 $^-1E^-13$
$DIODE[2]$ 0.000865 $^-1E^-13$
$CPUSECONDS = 3.46$
$ITERATIONS = 5$

$\downarrow\downarrow DC\ NODE\ VOLTAGES\ 1\ TO\ 5 \downarrow\downarrow$

$1.73\ 20\ 2.42\ 20\ 11.4$

$$DBI$$

5 6

$$\rho VBN$$

8

Nine sets of results are produced by $DIODE$ during the five iterations. Only the diode junction branches 5 and 6 are selected. Without indexing, four times as many computations would be required since there are eight

branches in the entire circuit. Similar conclusions apply for the function
DIFDIODE.

13.4.2 Bipolar Transistor d.c. Model

We now establish the model and derive the theoretical relationships used to
compute transistor behavior in *BASECAP*. These concepts are embodied
directly in the functions *GTRANS* and *ITRANS* presented in Section 13.4.3.
For simplicity we will work with one type of transistor (PNP), and later relate
these results to the NPN transistor.

The following analysis is based on the node and branch arrangement
shown for the PNP transistor in Fig. 13.8, and is compatible with the approach

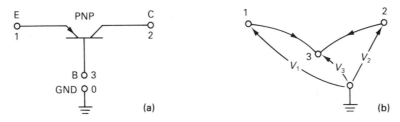

Fig. 13.8
PNP transistor node and branch designations

used in Chapter 3. The emitter (E) collector (C), and base (B) nodes are
numbered 1, 2 and 3 respectively. For generality, none of the transistor nodes is
connected to ground.

We represent the transistor by the simplified Ebers–Moll model shown in
Fig. 13.9. It consists of two diodes, two current-controlled current sources, and

α_F = Forward common-base
    short circuit current gain
α_R = Reverse common-base
    short circuit current gain

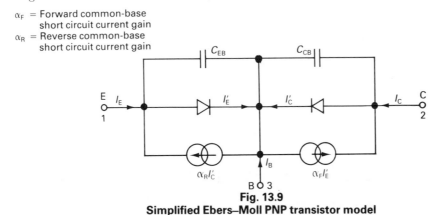

Fig. 13.9
Simplified Ebers–Moll PNP transistor model

two capacitances to account for charge storage effects. In accordance with the
requirements for a PNP transistor, the two diodes in the model have their
cathodes at the base, and this is compatible with the positive branch current
directions assigned in Fig. 13.8(b).

This model clearly neglects many important second-order effects in the
interests of simplicity, such as (1) contact resistances associated with the

emitter, base, and collector, (2) breakdown effects, (3) base-width modulation, (4) the dependence of α_F or α_R on current and temperature, and (5) the non-linear nature of the charge storage capacitances C_{EB} and C_{CB}. However, this model does account for essential transistor behavior such as gain, and the automatic control of junction currents for any combination of terminal voltages. Furthermore, as will be shown in the next chapter, it inherently supports a.c. transistor behavior as well as d.c. Hence, this is a very useful arrangement with which to begin the transistor modelling process.

In this chapter we are concerned only with d.c. behavior so we will not consider the capacitive components C_{EB} and C_{CB} further, except to note the following. C_{EB} and C_{CB} are assumed to be fixed capacitances, and are readily accounted for by conventional entries in the CL table. They will, of course, have no impact on the d.c. equations.

Recalling that the diode is represented during the Newton–Raphson non-linear solution process by an equivalent linearized conductance and current source in parallel, we apply the same approach to the diodes in the transistor model in Fig. 13.9. This results in the linearized d.c. equivalent representation of Fig. 13.10, where G_E and I_{EEQ} represent the emitter–base diode, and G_C and I_{CEQ} represent the collector–base diode. All four of these terms may change value at each iteration step during the Newton–Raphson solution.

We now use the configuration in Fig. 13.10 to obtain a convenient mathematical representation of the transistor with a set of d.c. node voltage equations. With no external current sources as shown, the algebraic sum of the currents at node 1 (the emitter) must be zero. Thus,

$$I_E = 0 = (V_1 - V_3) G_E + I_{EEQ} - \alpha_R I_C' \tag{13.1}$$

Similarly at the collector, node 2,

$$I_C = 0 = (V_2 - V_3) G_C + I_{CEQ} - \alpha_F I_E' \tag{13.2}$$

and also at the base, node 3,

$$\begin{aligned} I_B &= (V_3 - V_1) G_E + \alpha_R I_C' - I_{EEQ} \\ &+ (V_3 - V_2) G_C + \alpha_F I_E' - I_{CEQ} = 0 \end{aligned} \tag{13.3}$$

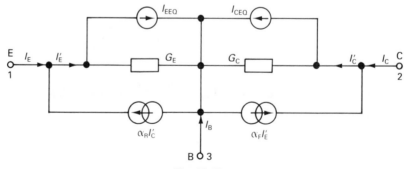

Fig. 13.10
Linearized equivalent d.c. transistor model

Rearranging and collecting terms in equations (13.1)–(13.3) gives

$$G_E V_1 + 0\ V_2 + (-G_E)\ V_3 \qquad\qquad = \alpha_R I_C' - I_{EEQ} \qquad\qquad (13.4)$$

$$0\ V_1 + G_C V_2 + (-G_C)\ V_3 \qquad\qquad = \alpha_F I_E' - I_{CEQ} \qquad\qquad (13.5)$$

$$(-G_E) V_1 + (-G_C) V_2 + (G_E + G_C) V_3$$
$$= -\alpha_R I_C' + I_{EEQ} - \alpha_F I_E' + I_{CEQ} \qquad (13.6)$$

As expected, the use of ground for our reference has produced considerable redundancy in these equations.

Equations (13.4)–(13.6) can be made more useful by eliminating the two diode current terms I_E' and I_C', where from Fig. 13.10 we note that

$$I_E' = I_{EEQ} + G_E V_1 - G_E V_3 \qquad\qquad (13.7)$$

and

$$I_C' = I_{CEQ} + G_C V_2 - G_C V_3 \qquad\qquad (13.8)$$

Substituting equations (13.7) and (13.8) into equations (13.4)–(13.6), and rearranging in matrix form, we get the following result which we can then analyze:

$$\begin{bmatrix} G_E & 0 - \alpha_R G_C & -G_E + \alpha_R G_C \\ 0 - \alpha_F G_E & G_C & -G_C + \alpha_F G_E \\ -G_E + \alpha_F G_E & -G_C + \alpha_R G_C & G_E + G_C - \alpha_F G_E - \alpha_R G_C \end{bmatrix} \begin{bmatrix} V_1 \\ V_2 \\ V_3 \end{bmatrix}$$
$$= \begin{bmatrix} -I_{EEQ} + \alpha_R I_{CEQ} \\ -I_{CEQ} + \alpha_F I_{EEQ} \\ I_{EEQ} + I_{CEQ} - \alpha_F I_{EEQ} - \alpha_R I_{CEQ} \end{bmatrix} \qquad (13.9)$$

It is interesting to note that equation (13.9) now contains only the simplest possible terms. All of these terms contain the direct effects of one or both of the diodes in the model, through either an equivalent conductance or a current source. The transistor gain-dependent behavior is represented by the presence of the terms α_F and α_R. These are unique to the transistor, and distinguish our model and mathematical representation from the much simpler (and not very useful) behavior of two diodes connected back-to-back.

The significance of this arrangement is now readily confirmed by temporarily setting $\alpha_F = \alpha_R = 0$. All that remain are the simple conductance and current source terms, corresponding only to the diode components in the model. These are given in equation (13.10):

$$\begin{bmatrix} G_E & 0 & -G_E \\ 0 & G_C & -G_C \\ -G_E & -G_C & G_E + G_C \end{bmatrix} \begin{bmatrix} V_1 \\ V_2 \\ V_3 \end{bmatrix} = \begin{bmatrix} -I_{EEQ} \\ -I_{CEQ} \\ I_{EEQ} + I_{CEQ} \end{bmatrix} \qquad (13.10)$$

All of these are automatically computed for such a diode-only connection by means of the functions *DIODE* and *DIFDIODE*.

In addition to these components, all we need to complete the transistor's d.c. representation are the conductance and source current terms shown in equation (13.11):

$$\begin{bmatrix} 0 & -\alpha_R G_C & \alpha_R G_C \\ -\alpha_F G_E & 0 & \alpha_F G_E \\ \alpha_F G_E & \alpha_R G_C & -\alpha_F G_E - \alpha_R G_C \end{bmatrix} \begin{bmatrix} V_1 \\ V_2 \\ V_3 \end{bmatrix} = \begin{bmatrix} \alpha_R I_{CEQ} \\ \alpha_F I_{EEQ} \\ -\alpha_F I_{EEQ} - \alpha_R I_{CEQ} \end{bmatrix} \quad (13.11)$$

These we compute in the functions $GTRANS$ and $ITRANS$ in the manner to be explained in detail in Section 13.4.3.

In summary, the transistor representation used here takes advantage of the regular diode model to compute the terms in equation (13.10). These individual terms are stored as global variables in the functions $DIFNETL$ (see $S1$) and $IEQUIV$ (see IEQ). Each component is then additionally multiplied by the appropriate alpha terms (see $GTRANS$ lines 4 and 5, and $ITRANS$ lines 4 and 5) in accordance with equation (13.11). Finally, the terms from both equations (13.10) and (13.11) are added into the appropriate positions in the overall circuit conductance matrix ($DIFNETL$,line 4), and the node source current vector ($IEQUIV$ line 4).

In practice, these particular positions are determined by the actual node numbers assigned to emitter, collector and base. In $BASECAP$, these may be any valid node numbers, and need not be sequentially assigned. The corresponding results for an NPN transistor are identical except for the signs of the current source terms on the right-hand side of equation (13.11), which are all reversed.

13.4.3 $GTRANS$ and $ITRANS$

Tables 13.13 and 13.14 contain detailed descriptions of the functions $GTRANS$ and $ITRANS$ which are specifically used to implement the transistor's gain-dependent terms derived previously in equation (13.11). Both functions appear on level 4 of the tree structures in Fig. 13.2, and are similar in their internal implementation.

Table 13.13
The function $GTRANS$

∇ $GT \leftarrow GTRANS; R; ECB; F$

$GTRANS$ computes the gain-dependent conductance terms for a single transistor and makes the appropriate entries into an initially zeroed result matrix. The result argument GT has the same shape as $SLOPE$. The following global variables are used by $GTRANS$: Q, GTI, $BEBRI$, and $S1$

[1] $GT \leftarrow {}^-1 \ {}^-1 \ \downarrow GTI$

Initialize the result argument to all zeros with the shape (NR, NR)

[2] $\rightarrow ((1 \uparrow \rho Q) = 0)/0$

Exit from the function if there are no transistors in the circuit, and return the above result argument with all zeros

[3] $ECB \leftarrow Q[QNO;\ 3\ 5\ 4]$

For a given transistor selected by the index QNO, extract from Q, the emitter, collector and base node numbers and assign as a three-element vector to ECB

[4] $F \leftarrow Q[QNO;\ 6] \times S1[BEBRI[QNO]]$

Extract α forward from Q, and GE from $S1$; compute their product and assign to F

[5] $R \leftarrow Q[QNO;\ 7] \times S1[BEBRI[QNO]\ +1]$

Assign to R the similar product of α_R and GC

[6] $GT \leftarrow GTI$

The result vector GT is re-initialized to all zeros with the shape (NH, NH)

[7] $GT[ECB;\ ECB] \leftarrow\ 3\ 3\ \rho 0,(-R),R,(-F),0,F,F,R,-R+F$

Formulate the 3×3 matrix shown in equation (13.11) with the R and F terms computed above. Place these entries in the result matrix according to the E, C and B node numbers

[8] $GT \leftarrow\ {}^{-}1\ {}^{-}1\ \downarrow GT$
   ∇

Drop the last row and column from the result matrix to produce the shape $NR \times NR$; return the result matrix

---

Table 13.14
The function $ITRANS$

$\nabla\ IQ \leftarrow ITRANS;\ ECB;\ FE;\ RC$

$ITRANS$ computes the gain-dependent equivalent source current terms for a single transistor, and makes the appropriate entries into the result vector which otherwise contains zeros. The shape of the result argument corresponds to the shape of IX. $ITRANS$ uses the following global variables: $IQI, Q, BEBRI$ and IEQ

[1] $IQ \leftarrow {}^{-}1 \downarrow IQI$

Initialize the result vector to all zeros with the shape NR

[2] $\rightarrow ((1 \uparrow \rho\ Q) = 0)/0$

Exit from the function if there are no transistors in the circuit, and return the above result vector with all zeros

[3] $ECB \leftarrow Q[QNO;\ 3\ 5\ 4]$

For a given transistor selected by the index QNO, extract from Q the emitter, collector, and base node numbers, and assign these as a three-element vector to ECB

[4] $FE \leftarrow Q[QNO;\ 6] \times IEQ[BEBRI[QNO]]$

Extract alpha forward from Q, and I_{EEQ} from IEQ; compute their product and assign to FE

[5] $RC \leftarrow Q[QNO; 7] \times IEQ[BEBRI[QNO]+1]$

Assign to RC the similar product of alpha reverse and I_{CEQ}

[6] $IQ \leftarrow IQI$

Reinitialize the result vector IQ to all zeros with shape NH

[7] $IQ[ECB] \leftarrow (-RC),(-FE),RC+FE$

Make the appropriate entries in the result vector as shown in equation (13.11) according to the E, C and B node numbers

[8] $IQ \leftarrow {}^-1 \downarrow IQ \times TYPE[QNO]$

    ∇

Choose the correct multiplier from the vector $TYPE$ ($^-1$ for PNP, $+1$ for NPN), and multiply the various entries in IQ; drop the last entry in IQ and return the result vector having the shape NR

Example 13.6 $GTRANS$ (Table 13.13)

The purpose of this example is to illustrate the internal operation of $GTRANS$ using the circuit $EXGTRANS$ from the example in Section 13.4.1.

We first obtain the final branch voltage vector VBN for use in a subsequent step.

```
        EXGTRANS
        DCSOLVE
    CPUSECONDS = 3.34
    ITERATIONS = 5

   ↓↓ DC NODE VOLTAGES 1 TO 5 ↓↓

   1.73 20  2.42 20  11.4
            VBN
   2.42 17.6 1.73 8.56 0.694  ¯9  ¯0.0111 20
```

We now set a trace on lines 3 to 8 of $GTRANS$ and use $NEWTALT$ to repeat the d.c. solution of $EXGTRANS$. The use of VBN with $NEWTALT$ conveniently ensures that this final solution is obtained with only one iteration.

```
        TΔGTRANS←3 4 5 6 7 8
        IX NEWTALT VBN
   GTRANS[3]   1 5 3
   GTRANS[4]   0.0282
   GTRANS[5]   9.09E¯14
   GTRANS[6]
```

```
      0  0  0  0  0  0
      0  0  0  0  0  0
      0  0  0  0  0  0
      0  0  0  0  0  0
      0  0  0  0  0  0
      0  0  0  0  0  0
GTRANS[7]
   0.00E0       ⁻9.09E⁻14      9.09E⁻14
  ⁻2.82E⁻2      0.00E0         2.82E⁻2
   2.82E⁻2      9.09E⁻14      ⁻2.82E⁻2
GTRANS[8]
   0.00E0       0.00E0      9.09E⁻14     0.00E0    ⁻9.09E⁻14
   0.00E0       0.00E0      0.00E0       0.00E0     0.00E0
   2.82E⁻2      0.00E0     ⁻2.82E⁻2      0.00E0     9.09E⁻14
   0.00E0       0.00E0      0.00E0       0.00E0     0.00E0
  ⁻2.82E⁻2      0.00E0      2.82E⁻2      0.00E0     0.00E0
CPUSECONDS =1
ITERATIONS =1
  1.73  20  2.42  20  11.4
```

It is interesting to observe that the 3×3 matrix of transistor conductance terms produced by line 7 is inserted into appropriate locations in the final GT matrix. Note that columns of the matrix from $GTRANS$ line 7 are in $\overline{ECB}$ node number order $(1\ \ 5\ \ 3)$, and that columns 2 and 3 relate to columns 5 and 3 respectively in GT.

13.4.4   Voltage-controlled Current Source and $VCFORM$

We now discuss the detailed implementation used in $BASECAP$ to realize a voltage-controlled current source (VCCS). This processing is carried out in the function $VCFORM$ which is the last of the low-level functions from Fig. 13.2 to be discussed. $\overline{VCFORM}$ makes direct use of the data in the storage table $VCISL$, created by executing the element description function $VCIS$ once for each voltage-controlled current source in the circuit.

The theoretical basis for the $VCCS$ was presented in Section 3.5. This is summarized in Fig. 13.11, using the $BASECAP$ model and notation from Section 11.2. The nodal conductance matrix for a circuit must have two positive and two negative GM entries added to it for each $VCCS$. Corresponding rows in the matrix are determined by the 'to' nodes while the 'from' or controlling nodes determine the columns in accordance with the assigned node numbers.

A single loop is used in the function $VCFORM$ to process one row (i.e. one $VCCS$) at a time as shown in the flow chart in Fig. 13.12. The various GM entries are easily added into the matrix in this function by the use of APL's simple and very powerful indexing capabilities. A detailed explanation of $VCFORM$ is given in Table 13.15.

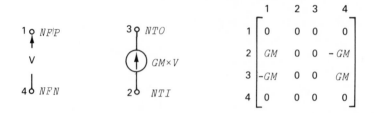

Fig. 13.11 $BASECAP$
VCCS nodal conductance matrix

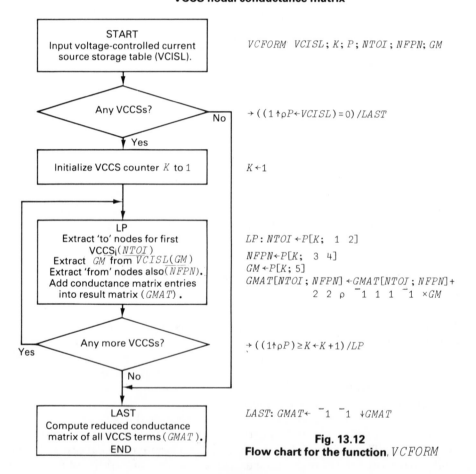

Flowchart box	APL code
START Input voltage-controlled current source storage table (VCISL).	$VCFORM\ VCISL; K; P; NTOI; NFPN; GM$
Any VCCSs? — No	$\rightarrow((1\uparrow\rho P\leftarrow VCISL)=0)/LAST$
Initialize VCCS counter K to 1	$K\leftarrow 1$
LP Extract 'to' nodes for first VCCS$_i$ ($NTOI$) Extract GM from $VCISL$ (GM) Extract 'from' nodes also ($NFPN$). Add conductance matrix entries into result matrix ($GMAT$).	$LP: NTOI\leftarrow P[K;\ 1\ 2]$ $NFPN\leftarrow P[K;\ 3\ 4]$ $GM\leftarrow P[K;5]$ $GMAT[NTOI;NFPN]\leftarrow GMAT[NTOI;NFPN]+$ $\quad 2\ 2\ \rho\ ^-1\ 1\ 1\ ^-1\ \times GM$
Any more VCCSs? — Yes	$\rightarrow((1\uparrow\rho P)\geq K\leftarrow K+1)/LP$
LAST Compute reduced conductance matrix of all VCCS terms ($GMAT$). END	$LAST: GMAT\leftarrow\ ^-1\ ^-1\ \downarrow GMAT$

Fig. 13.12
Flow chart for the function $VCFORM$

Table 13.15
The function $VCFORM$

$\nabla\quad VCFORM\quad VCISL; K; P; NTOI; NFPN; GM$

$VCFORM$ takes the data in the $VCISL$ storage table and produces appropriate entries in the global result matrix $GMAT$ having the shape $(NR\times NR)$. All other values in $GMAT$ are zero

[1] $\rightarrow ((1 \uparrow \rho P \leftarrow VCISL) = 0) / LAST$

Determine how many VCCSs are in the circuit (number of rows in $VCISL$), and assign to P; go to the line labelled $LAST$ if there are none

[2] $K \leftarrow 1$

Set K to 1 to select the first device

[3] $LP : NTOI \leftarrow P[K; \ 1 \ 2]$

Label this line LP: extract 'to' nodes for this device

[4] $NFPN \leftarrow P[K; \ 3 \ 4]$

Extract 'from' nodes for this device

[5] $GM \leftarrow P[K; 5]$

Extract GM for this device

[6] $GMAT[NTOI; NFPN] \leftarrow GMAT[NTOI; NFPN] + 2 \ 2 \ \rho \ ^-1 \ 1 \ 1 \ ^-1 \times GM$

First generate a 2×2 matrix of GM entries with appropriate signs: then add individual entries into appropriate locations of $GMAT$

[7] $\rightarrow ((1 \uparrow \rho \ P) \geq K \leftarrow K + 1) / LP$

Test for last device; loop to line 3 and repeat as necessary

[8] $LAST : GMAT \leftarrow \ ^-1 \ ^-1 \ \downarrow GMAT$

    ∇

Compute reduced form of result matrix $GMAT$ and exit

13.5 Interactive Terminal Session

This section is devoted mainly to an interactive terminal session illustrating the operation and use of several of the d.c. functions. Also briefly described are four additional functions which are alternatives to functions described previously. These are not normally required for regular solution work with $BASECAP$, and are presented for instructional purposes.

13.5.1 Alternative Solution Functions: $VBCHKCAN$, $NEWTDC$, $NEWTDC1$, $TRANS$

The four alternative solution functions introduced here all involve the use of a different method of controlling the diode junction voltage increments during the Newton–Raphson iteration process. The fundamental function in this set is $VBCHKCAN$ which is an alternative to the function $ALTBASE1$. The remaining three functions directly or indirectly call $VBCHKCAN$, and each is otherwise identical to an equivalent function presented in Sections 13.2 and 13.3.

The contents of these four functions are listed in Appendix 6. It can be seen that *NEWTDC* corresponds to the function *NEWTALT*, except for line 3 which calls *VBCHKCAN* instead of *ALTBASE1*. *NEWTDC1* corresponds to *NEWTALT1* but also calls *VBCHKCAN* on line 2. *TRANS* is the alternative d.c. transfer-function routine and is similar to *TRANSDC* except that it uses *NEWTDC1* in place of *NEWTALT1* on line 15.

The function *VBCHKCAN* incorporates a 'brute-force' type of diode-branch voltage-limiting algorithm. It is similar to the approach illustrated in Chapter 2 in Fig. 2.8, and is based on the method used in an early CACD package called *CANCER* from the University of California at Berkeley (Nagel and Roher 1971).

In this section, only the input–output behavior of *VBCHKCAN* is emphasized. *VBCHKCAN* monitors the diode branch voltages, and may restrict their changes from one iteration to the next. Two multiplier coefficients *M1* (nominally = 2) and *M2* (nominally = 10), together with a threshold voltage *VT* (nominally = 0.026 V), are used for this purpose. A diode branch voltage above the value $M2 \times VT$ is restricted to a positive increment of not more than $M1 \times VT$ at any given iteration step. Just as with the function *ALTBASE1*, the overall objective is to speed convergence by preventing all diode junctions from becoming excessively forward-biassed, particularly during the early stages of the Newton–Raphson iteration process.

VBCHKCAN is implemented in parallel form using logic vectors and Boolean arithmetic to test for the possible conditions applying to each diode branch. This approach takes advantage of APL's parallel processing capability, and also the relatively fast logical operations in APL as compared to the use of multiple loops. This algorithm is the subject of more detailed study in a later exercise for the student.

13.5.2   Comparing *NEWTALT* and *NEWTDC*

Nonlinear d.c. solution tests are carried out below with the circuit *CACDCCT* (Fig. 13.13) comparing the operation of *NEWTALT* (with *ALTBASE1*) and *NEWTDC* (with *VBCHKCAN*) during the iteration process. These are two alter-

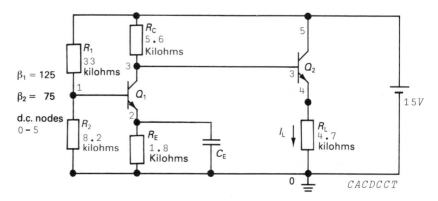

Fig. 13.13

nate methods used in *BASECAP* to limit the junction voltage excursions during the iteration solutions.

Both methods are seen to work, but starting conditions significantly affect the iteration efficiency in each case. It is instructive to investigate the operation of these routines further using a variety of other starting conditions.

```
        ∇CACDCCT[□] ∇
        ∇ CACDCCT          Circuit description function for the
[1]     STARTCCT           above circuit. Note that C_E is not
[2]     R 1 0 8200         included here
[3]     R 1 5 33000
[4]     R 2 0 1800
[5]     R 5 3 5600
[6]     QNPN 2 1 3 125 0.1 4E⁻9 4E⁻9
[7]     R 4 0 4700
[8]     QNPN 4 3-5 75 0.1 4E⁻9 4E⁻9
[9]     VDC 5 0 15
[10]    END
        ∇
```

```
M1 ,M2 ,VT,SMIN,VBMAX ,ETA ,TEMP,ALTREF,TOL  Solution parameters
 2 10 0.026 1E⁻12 1 2 20 0.26 0.1       (see PARMTELL)
```

```
        CACDCCT            Set up the circuit
```

```
        IX NEWTDC VBSTART
CPUSECONDS = 0.199
ITERATIONS = 8
2.92 2.28 7.85 7.2 15
```
VBSTART automatically sets each transistor B–E junction and each B–C junction to $0.4\,V$ forward bias for the first iteration step

```
        IX NEWTDC .4
CPUSECONDS = 0.195
ITERATIONS = 8
2.93 2.28 7.85 7.2 15
```

```
        IX NEWTALT VBSTART2
CPUSECONDS = 0.095
ITERATIONS = 4
2.92 2.28 7.85 7.2 15
```
VBSTART2 automatically sets each transistor B–E junction to $0.4\,V$ forward bias and each B–C junction to $1.0\,V$ reverse bias to start the iteration

```
        IX NEWTALT .4
CPUSECONDS = 0.071
ITERATIONS = 3
 2.92 2.28 7.85 7.2 15
```
NEWTALT (and *ALTBASE1*) with the arbitrary starting condition of all VB 's $= + 0.4\,V$ produces the fastest solution time for this circuit. This is

```
        IX NEWTALT VBSTART
CPUSECONDS = 0.07
ITERATIONS = 3
2.92 2.28 7.85 7.2 15
```
interesting since *final* solutions have the transistor B–C junctions reverse biassed (negative)

13.5.3   Tracing the Operation of *ALTBASE*1

By setting a trace on the lines of *ALTBASE*1 before carrying out a solution for *CACDCCT*, it is possible to observe how large potential forward-biassing voltages are restricted to safer more practical levels by *ALTBASE*1.

```
        VBMAX ←.8
        ETA ←2
        CACDCCT
        INITIALIZE
        DBI
5  6  8  9                                                              (1)
        T∆ ALTBASE1 ←ι7                                                 (2)

        IX NEWTALT .4
ALTBASE1[1]  0.4  0.4  0.4  0.4  0.4  0.4  0.4  0.4  0.4  0.4  0.4
ALTBASE1[2]  →6
ALTBASE1[6]  0.4  0.4  0.4  0.4                                         (3)
ALTBASE1[1]  7.32  ‾7.68  2.04  4.63  5.28  ‾3.05  5.28  5.09  ‾4.63  15
ALTBASE1[2]  →3                      VB[5 6]              VB[8 9]        (4)
ALTBASE1[3]  5  8                                                       (5)
ALTBASE1[4]  6.92E‾12  ‾8.08E‾12  1.64E‾12  4.23E‾12  0.00107  ‾0.000735
        4.88E‾12  0.00103  ‾0.00108  1.46E‾11                           (6)
ALTBASE1[5]  0.631  0.629                        (7)
ALTBASE1[6]  0.631  ‾3.05  0.629  ‾4.63
ALTBASE1[1]  2.92  ‾12.1  2.28  7.14  0.641  ‾4.93  7.2  0.654  ‾7.14  15
ALTBASE1[2]  →3
ALTBASE1[3]  5  8
ALTBASE1[4]  ‾4.4E‾12  ‾4.4E‾12  2.37E‾13  2.51E‾12  0.00127  ‾4E‾9
        1.92E‾12  0.00153  ‾4E‾9  ‾3.25E‾18
ALTBASE1[5]  0.64  0.65
ALTBASE1[6]  0.64  ‾4.93  0.65  ‾7.14                                   (8)
CPUSECONDS = 0.114
ITERATIONS = 3
2.92  2.28  7.85  7.2  15
```

Comments (refer to numbers on the right above)

(1)   The diode junction branches are 5 6 8 and 9,
(2)   A trace is set on all lines of *ALTBASE*1.
(3)   The initial diode junction branch voltages are 0.4 V as expected.
(4)   Notice the extremely large forward-biassing voltages that are requested for the diode junctions in branches 5 and 8 at the beginning of the next iteration step (5.28 V and 5.09 V respectively!)   .
(5)   Branches 5 and 8 are detected by *ALTBASE*1 as having positive branch voltage increments during this iteration step.
(6)   Revised currents in *IDNEW*.
(7)   Recalculated values for $\overline{VB[5\ 8]}$. These new 'allowed' junction voltages are greater than 0.4 V, but are no longer excessively large.
(8)   The final junction voltages that result when convergence is achieved.

13.5.4   Convergence with *VBCHKCAN*

In this interactive session we illustrate the internal operation of *VBCHKCAN* and compare the effects of using different values of the solution control parameter $M1$. Recall that $M1 \times VT$ represents the largest 'allowed' positive ΔVB for a forward biassed junction whose voltage at the previous step is at or above the value $M2 \times VT$.

To illustrate what is happening, we put a trace on line 15 of *VBCHKCAN* which displays resultant values for the diode junction voltages $VB[DBI]$ at each step throughout the solution.

```
          ETA ←1.8
          VBMAX ←1
          CACDCCT                   Set up the circuit CACDCCT
          INITIALIZE
          JF
22
          DBI
  5  6  8  9                        Diode junction branches


          TΔ VBCHKCAN←15            Trace line 15 of VBCHKCAN


          M1,M2,VT
  2 10 0.026                        Solution control parameters


          IX NEWTDC 0              Solve using NEWTDC
VBCHKCAN[15]  0  0  0  0
VBCHKCAN[15]  0.26 ⁻12 0.26 ⁻0.00746    Note the 0.26V
VBCHKCAN[15]  0.312 ⁻11.6 0.312 ⁻0.417  at iteration 2
VBCHKCAN[15]  0.364 ⁻10.9 0.364 ⁻1.16   Subsequently ΔVB
VBCHKCAN[15]  0.416 ⁻9.32 0.416 ⁻2.72   for branches 5 and
VBCHKCAN[15]  0.468 ⁻7.2 0.468 ⁻4.86    8 do not exceed
VBCHKCAN[15]  0.52 ⁻5.6 0.52 ⁻6.47      2×0.026=0.052
VBCHKCAN[15]  0.572 ⁻4.91 0.572 ⁻7.17   volts at each step
VBCHKCAN[15]  0.578 ⁻4.75 0.586 ⁻7.33
VBCHKCAN[15]  0.577 ⁻4.75 0.584 ⁻7.33
CPUSECONDS = 0.277
ITERATIONS = 10
2.92 2.34 7.66 7.08 15




          M2 ←10


          M1 ←10
          M1,M2,VT
  10 10 0.026                       M1 is increased to 10
```

```
      IX NEWTDC 0
VBCHKCAN[15]  0  0  0  0
VBCHKCAN[15]  0.26 ¯12 0.26 ¯0.00746
VBCHKCAN[15]  0.52 ¯11.6 0.52 ¯0.417
VBCHKCAN[15]  0.631 ¯4.91 0.663 ¯7.17
VBCHKCAN[15]  0.6 ¯4.81 0.625 ¯7.27
VBCHKCAN[15]  0.582 ¯4.76 0.598 ¯7.32
VBCHKCAN[15]  0.578 ¯4.75 0.586 ¯7.33
VBCHKCAN[15]  0.577 ¯4.75 0.584 ¯7.33
CPUSECONDS = 0.218
ITERATIONS = 8
2.92 2.34 7.66 7.08 15
```

The maximum ΔVB is now $+0.26$ volts $(10×0.026)$:

Note the slight overshoot in VB[5 8]

Fewer iterations are required

```
      M1 ←50
      M1,M2,VT
50 10 0.026
```

$M1$ is increased to 50

```
      IX NEWTDC 0
VBCHKCAN[15]  0  0  0  0
VBCHKCAN[15]  1 ¯12 1 ¯0.00746
VBCHKCAN[15]  0.955 ¯5.86 0.955 ¯6.21
VBCHKCAN[15]  0.909 ¯5.72 0.909 ¯6.35
VBCHKCAN[15]  0.864 ¯5.59 0.864 ¯6.48
VBCHKCAN[15]  0.818 ¯5.46 0.818 ¯6.62
VBCHKCAN[15]  0.773 ¯5.32 0.773 ¯6.75
VBCHKCAN[15]  0.728 ¯5.19 0.728 ¯6.89
VBCHKCAN[15]  0.684 ¯5.06 0.685 ¯7.02
VBCHKCAN[15]  0.643 ¯4.94 0.644 ¯7.14
VBCHKCAN[15]  0.608 ¯4.84 0.611 ¯7.24
VBCHKCAN[15]  0.586 ¯4.77 0.591 ¯7.31
VBCHKCAN[15]  0.578 ¯4.75 0.585 ¯7.33
VBCHKCAN[15]  0.577 ¯4.75 0.584 ¯7.33
CPUSECONDS = 0.372
ITERATIONS = 14
2.92 2.34 7.66 7.08 15
```

$VBMAX$ causes the limit on VB at iteration 2 Considerable overshoot occurs in both VB[5 8] values

Efficiency is notably worse here

Exercises

13.1 Reference to the listing for the function *TRANSDC* in Section 13.2.2 indicates that the circuit description function is executed on both lines 9 and 14. Explain why this duplication appears.

13.2 Temporarily modify *NEWTALT*1 to display resultant values for the base–emitter voltage and the change in this voltage specifically for transistor $Q3$ in the circuit *FBTRIPLE*, along with the number of iterations required to obtain d.c. convergence. Using a d.c. transfer function solution (*TRANSDC* and *NEWTALT*1) for *FBTRIPLE* over the range of $R1$ values contained in the vector $R1\,VAL$ (see Section 11.6.2), determine if the results displayed by the modified *NEWTALT*1 can be used to help explain the pattern observed in the number of iterations. Use the following additional circuit and device parameters:

$ETA \leftarrow 1$

$TEMP \leftarrow 20$

$ALTREF \leftarrow 0.26$

$VBMAX \leftarrow 0.9$

$SMIN \leftarrow 1E^-12$

$Q1 \leftarrow 50 \ 0.1 \ 1E^-13 \ 1E^-13 \ 8E^-11 \ 3E^-12$

$QN \leftarrow 50 \ 0.1 \ 1E^-13 \ 1E^-13 \ 8E^-11 \ 3E^-12$

$D1 \leftarrow 1E^-13 \ 3E^-12$

$R2 \leftarrow 8.2E3$

$RI \leftarrow 1E3$

$C1 \leftarrow .0001$

$C2 \leftarrow 1E^-12$

13.3 The reference parameter *ALTREF* used for the method of alternating bases in the function *ALTBASE*1 has a nominal value of 0.26 volts. This value is not necessarily optimum for all types of circuits and starting conditions, and it can be expected that the value of *ALTREF* will significantly affect the number of iterations required to obtain a d.c. solution.

    Determine the optimum value of *ALTREF* for the circuit *FBTRIPLE* for the specific case where solutions are started at zero volts ($IX\ NEWTALT$ 0). Use values of *ALTREF* over the range from 0–1 0V and tabulate the resultant number of iterations required. The following device and circuit parameters should also be used for this investigation:

$$ETA \leftarrow 1.0$$
$$TEMP \leftarrow 20$$
$$Q1 \quad \leftarrow QN \leftarrow 50 \ .1 \ 1E^-9 \ 1E^-9 \ 1E^-12 \ 1E^-12$$
$$R1 \quad \leftarrow 2E4$$
$$R2 \quad \leftarrow 8.2E3$$
$$RI \quad \leftarrow 1000$$
$$C1 \quad \leftarrow C2 \leftarrow 1E^-6$$

13.4 Draw and label a flow chart of the internal operation of the function $VBCHKCAN$.

13.5 Investigate and explain why d.c. convergence is very slow for a nonlinear circuit such as $CACDCCT$ when the Newton–Raphson solution is started with relatively large initial branch voltages of say 1.5 V.

Use $NEWTDC$ for this investigation, placing a trace on line 7 to show ΔVB at each iteration step. Compare solutions obtained with different values of ETA, say $1.0, 1.4$ and 1.8. How do these results differ from those given in Section 13.5.4 which are started at 0 V?

13.6 In Section 2.4.1 the method of linearizing the semiconductor diode exponential relationship beyond a critical voltage was described as one means that may be employed to help prevent d.c. convergence problems.

(a) Develop suitable algorithms incorporating this method for use in place of $DIODE$ and $DIFDIODE$ in the $BASECAP$ package. Use indexing for the junction branches where appropriate in the new functions.

(b) Compare d.c. solution efficiency obtained by this method with that now obtained by $NEWTALT$. Use a variety of starting conditions with several test circuits (including $CACDCCT$ and $FBTRIPLE$) for this investigation.

13.7 Develop and test an algorithm to be used with the method of linearization (Exercise 13.6) at the completion of the d.c. solution to automatically check if all junction voltages are below the critical voltage. This algorithm should give a suitable warning message to the user if the specified condition is violated.

Solutions

13.1   The circuit description function is executed on line 9 for initialization purposes prior to entering the main loop on line 13. Specifically, the variables IX and G for the given circuit are required on lines 10 and 11 to compute the correct shape of the result matrix $VALL$ and the initial branch starting voltages in VBN.

The circuit description function must also be executed within the loop (line 14) immediately following the selection of each new value of the independent parameter in order to set up the new data arrays prior to the solution step (line 15).

13.2   A modified function $NEWTALT1$ appears below:

```
      ∇ V←IX NEWTALT1 VB;T
[1]      VBTEMP←VB
[2]      INITIALIZE
[3]   L1: VB←ΔVB ALTBASE1 VB
[4]      ICOM←IEQUIV VB
[5]      SLOPE←DIFNETL VB
[6]      V←ICOM⊟SLOPE
[7]      ΔVB←(VBN←(⍉A)+.×V,0)-VB
[8]      IT←IT+1
[9]      →L1×ι⌈/(|ΔVB)>(|(TOL÷100)×VBN)+0.000000000001
[10]     'IT= ';IT;' ';VBN[10];' ';(VBN-VBTEMP)[10]
      ∇
```

Following are the results obtained for $IT, VBN[10]$, and $VB[10]$ for $FBTRIPLE$ using the circuit and parameter data shown (note that branch number 10 is the base—emitter junction of $Q3$):

	∇FBTRIPLE[☐]∇		ETA
	∇ FBTRIPLE	1	
[1]	STARTCCT		TEMP
[2]	R 1 0 ,R2	20	
[3]	QNPN 0 1 3 ,Q1		ALTREF
[4]	R 9 3 14600	0.26	
[5]	R 9 5 14600		VBMAX
[6]	QNPN 4 3 5 ,QN	0.9	
[7]	D 4 0 ,D1		SMIN
[8]	R 9 8 14600	1E⁻12	
[9]	QNPN 7 5 8 ,QN		Q1
[10]	D 7 6 ,D1	50 0.1 1E⁻13 1E⁻13	
[11]	D 6 0 ,D1		QN
[12]	VDC 9 0 20	50 0.1 1E⁻13 1E⁻13	
[13]	R 8 2 ,R1		D1
[14]	R 2 1 ,R1	1E⁻13	
[15]	R 10 11 ,RI		R1
[16]	C 11 1 ,C1	2E4	
[17]	C 2 0 ,C2		R2
[18]	VAC 10 0 1 0	8.2E3	
[19]	END		RI
	∇	1E3	
			C1
		0.0001	
			C2
		1E⁻12	

```
      ANS←TRANSDC 8
TYPE CIRCUIT NAME
FBTRIPLE
TYPE NAME OF PARAMETER TO BE VARIED
```

```
R1
TYPE PARAMETER VALUES
□:

    □←R1VAL
2E5 1.5E5 1E5 7.5E4 5E4 4E4 3E4 2.5E4 2E4 1.8E4 1.6E4 1.4E4 1.2E4 1E4
    8E3 6E3 4E3 2E3 1E3 500
IT= 9   0.208  ¯0.192
IT= 6   0.208  5.52E¯6
IT= 5   0.212  0.00438
IT= 22  0.546  0.333
IT= 3   0.568  0.0222
IT= 3   0.573  0.00532
IT= 3   0.578  0.00439
IT= 2   0.58   0.00194
IT= 2   0.582  0.0018
IT= 2   0.582  0.000686
IT= 2   0.583  0.000668
IT= 2   0.584  0.00065
IT= 2   0.584  0.000634
IT= 2   0.585  0.000619
IT= 2   0.585  0.000604
IT= 3   0.586  0.00059
IT= 11  0.587  0.000589
IT= 17  0.605  0.0184
IT= 4   0.603  ¯0.0016
IT= 4   0.6    ¯0.0034
```

Two main iteration peaks are observed for solutions corresponding to $R1VAL[4\ 18]$, which required 22 and 17 iterations respectively. These are noted below.

	IT	$VBN[10]$	$\Delta VB[10]$	Comments
1)	22	0.546	0.333	$\Delta VB[10]$ is very large, also $VBN[10]$ is now large enough to cause significant transistor current flow.
2)	17	0.605	0.0184	$\Delta VB[10]$ is larger than its neighbors, though much smaller than in the case above. However, $VBN[10]$ is large, and even small changes in the base–emitter voltage have significant effects.

13.3 Results are as shown in Table E13.1.

The optimum value of $ALTREF$ for the circuit and conditions tested is approximately 0.5.

Table E13.1

$ALTREF$	Iterations	$ALTREF$	Iterations
0	192	.45	31
.01	192	.5	22
.1	53	.55	22
.2	51	.6	30
.26	36	.65	30
.3	36	.7	30
.35	90	.75	43
.375	97	.8	53
.385	97	.85	$DE\star$
.39	97	.9-9	$\overline{DE\star}$
.393	51	10	53
.40	31		

$\star DE$ – Domain error occurred, solution not obtained

13.4   A flowchart for $VBCHKCAN$ is shown in Fig. E13.1.

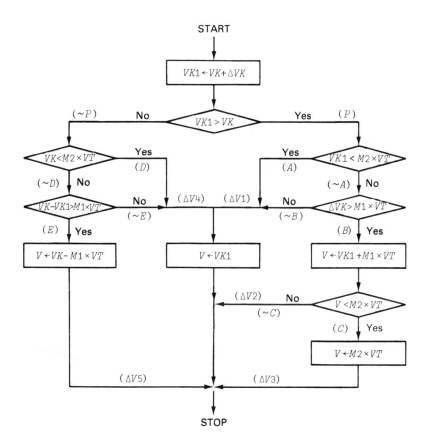

Fig. E13.1

(1) $VK1$, VK and ΔVK are junction branch voltage vectors having one entry for each entry in DBI.

(2) VK represents the previous voltages; ΔVK represents the 'requested' voltage increments.

(3) V represents "allowed" final branch voltages.

(4) A, B, C, D, E, and P are logic vectors used in the parallel implementation of the algorithm.

A.C. Circuit Analysis

In this chapter we consider in detail the design and implementation of the functions in the AC group. As noted in Section 10.4.4, these functions are designed to produce small-signal node-voltage solutions for a circuit in the frequency domain. Important fundamental material, which is essential for an understanding of the design and operation of these functions, is contained in Chapter 4. Significant interaction also exists with the $INPUT$ functions in Chapter 12 and the DC functions in Chapter 13, on which the AC functions depend for essential device and circuit data.

The principal function in the AC group is $ACSOLVE$, which performs the following processing:

$ACSOLVE$ produces any number of sets of small-signal a.c. node-voltage solutions corresponding to the entries contained in the global frequency vector F. The node voltage results are stored in real and imaginary form in the result array $VREIM$.

The composition of the AC group is given in Table 14.1, which lists 12 different functions. For convenient reference, complete listings of the AC group functions are included in Appendix 6, and a brief description of each of the principal variables used in this group is given in Appendix 7.

Table 14.1
Composition of the AC Group

)$GRPS$							
AC							
)FNS							
$ACPREP$	$ACSOL$	$ACSOLVE$	BB	$FLIN$	$FLOG$	$FORM$	$FSTEP$
$IFORM$	$IREIM$	$REALSOLV$		$YNET$			
)$VARS$							
BC	BGL	BL	CL	F	$IACL$	$ICOMPLEX$	
IIM	IRE	LL	M	NH	RL	$SLOPE$	$VACL$
$VREIM$	W	$YCOMP$	$YIMAG$	YRE	$YREACT$		

As the previous paragraph suggests, most of the functions in this group are controlled by *AC SOLVE* in a strongly structured arrangement. The data and program structures used for this group are presented in Section 14.1. The design and operation of the *AC* solution functions are discussed in Sections 14.2 and 14.3, and supplementary functions used to select the solution frequencies are presented in Section 14.4.

14.1 AC Group Structure

The overall design and organization of the global input and output data variables are considered first in this section, followed by a description of the structural arrangement used for the functions in this group.

14.1.1   Data Structure

The principal data items associated with the *AC* group are shown in Fig. 14.1.

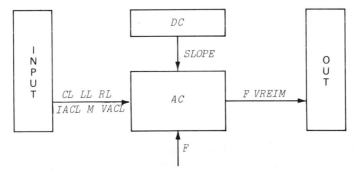

Fig. 14.1
AC **group data I/O variables**

The items required from the *INPUT* group (Chapter 12) are the various a.c.-related component and source storage tables. These include *CL*, *LL*, and *RL* for capacitors and inductors, as well as *IACL* and *VACL* for the a.c. current and voltage sources. The branch connection matrix *M* is required in the function *FORM*, although *NH* could be used directly instead for the same purpose if desired.

The nodal conductance matrix for the overall circuit is also required during the a.c. solution process. This is available following the d.c. solution for the given circuit in the global variable *SLOPE* (see *NEWTALT*, Section 13.3). This matrix is extremely significant for the a.c. solution process and is discussed further in subsequent paragraphs.

The last remaining requirement prior to using *AC SOLVE* is for the user to specify in *F* one or more frequency points at which solutions are desired.

Only the variables *VREIM* and *F* are normally required by the *OUT* group for subsequent displays of a.c. circuit performance. As specified in Section 11.5, *VREIM* holds the set of node voltages obtained at each frequency in a separate column. *VREIM* has the shape (2,*NR*,ρ*F*), where the real parts are stored in plane 1 and the imaginary parts are in plane 2.

A similar approach has been adopted as standard throughout *BASECAP* for all complex a.c. data variables, both local and global. Thus, complex a.c. quantities such as *ICOMPLEX* in *ACSOL*, and *YCOMP* in the function *YNET* are defined as arrays having two planes containing the real and imaginary parts respectively. Each plane in such an array must have the same shape, i.e., the same number of rows and columns.

The matrix *SLOPE* is fundamentally important to the a.c. solution process as explained in the remainder of this subsection. Specifically, the a.c. analysis method employed in *BASECAP* produces a linear small-signal (i.e. incremental) a.c. solution. However the transistors and diodes in the circuit are inherently nonlinear, and the values of the small-signal parameters used to represent the transistor under a.c. conditions are highly operating-point dependent. Thus, it is mandatory to carry out the d.c. nonlinear solution first in order to establish the correct a.c. solution parameters for these devices.

In *BASECAP*, the necessary small-signal a.c. conductance parameters for the nonlinear devices are contained in *SLOPE* as an automatic byproduct of the linearized models used in the d.c. solution process. It is assumed that the convergence criteria for the d.c. solution have been met, in which case the linearized conductance components representing the nonlinear devices in *SLOPE* will be valid at the final d.c. operating conditions in the circuit.

The other remaining consideration for the transistor and diode a.c. models pertains to the device capacitance components. The values of these capacitances must be adjusted (manually in the case of *BASECAP*) to reflect the resultant d.c. operating conditions, since they are also operating-point dependent. On the other hand, no special considerations are required to use the voltage-controlled current source under a.c. solution conditions. It is linear by definition, and all the necessary entries for this device are contained in *SLOPE* so they are passed on to the a.c. solution programs automatically.

14.1.2 Program Structure

The majority of the *AC* group functions are directly or indirectly called by *ACSOLVE* as the tree structure indicates in Fig. 14.2. *ACPREP* and *ACSOL* on

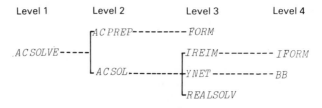

Supplementary functions: *FLIN*; *FSTEP*; *FLOG*

Fig. 14.2
AC **group tree structure**

level 2 are called directly by *AC SOLVE* to respectively prepare the data for solution, and manage the a.c. solution process.

The functions *ACPREP* and *FORM* deal only with *R*, *L* and *C* components, and produce frequency-independent matrices. The functions *YNET* and *BB* later compute the required frequency-dependent admittance matrices and update them at each frequency point. *IREIM* and *IFORM* process all of the a.c. sources for the circuit, and *REALSOLV* implements the actual nodal voltage solution.

ACPREP is also responsible for making inductance-related adjustments to the nodal conductance matrix *SLOPE|*. Specifically, a nodal conductance matrix *BGL* is formed from the small place-holding resistors substituted for the inductors in the d.c. solution. This matrix is then subtracted from *SLOPE* on an element-by-element basis to eliminate the effects of these resistors at each of the various inductor locations.

In a similar manner, the effects of the large resistors substituted in place of capacitors for the d.c. solution could also be removed in *ACPREP*. In the interests of simplicity, however, this is not currently done in *BASECAP*. Their values are sufficiently large that they will normally have negligible effect on a.c. circuit behavior.

14.2 High-level *AC* functions

In this section we consider the functions *ACSOLVE*, *ACPREP*, and *ACSOL* shown on levels 1 and 2 of the tree structure in Fig. 14.2. As indicated previously, it is assumed that a valid d.c. solution has been obtained for the circuit in order to produce the nodal conductance matrix *SLOPE* prior to executing *ACSOLVE*. It is also essential that one or more values be specified for the frequency variable *F*.

14.2.1 *AC SOLVE*

AC SOLVE is executed directly by the user as explained and illustrated earlier in Sections 11.5 and 11.6. The entire a.c. solution process is controlled by *ACSOLVE* through *ACPREP* and *ACSOL*. Any number of a.c. node voltage solutions can be produced and automatically stored in real and imaginary form in the result array *VREIM*. The CPU time required for the solutions is displayed for information, and the user is notified when the processing is completed. Further internal details of *AC SOLVE* are given in Table 14.2.

Table 14.2
The function *AC SOLVE*

∇   *AC SOLVE*

The purpose of *AC SOLVE* is to access and prepare all the required circuit data and carry out an a.c. solution for each entry in the frequency vector *F*, and store the real and imaginary parts of the node voltages in two planes in the result array $VREIM$

[1] $ACPREP$

Call $ACPREP$ which forms the capacitive and reciprocal inductive matrices, and prepares the net real part of the reduced admittance matrix

[2] $(\Psi'SOLVING\ FOR\ \ ')\,,(\Psi\rho,F)\,,\Psi'FREQUENCY\ POINTS'$

Print a confirming message

[3] $AC\overline{SO}L$

Compute the necessary a.c. solutions and store the results in $VREIM$

[4] $'DONE'$
   ∇

Inform the user when the task is complete

14.2.2 $ACPREP$

This function is called once prior to the actual solution ($ACSOLVE$ line 1) to process the capacitive and inductive data, and to correct $SLOPE$ by removing the effects of the small resistors used in place of inductors for the d.c. solution. Refer to Table 14.3 and to Example 14.1 for further details on the internal operation of $ACPREP$.

Example 14.1 $ACPREP$ *(Table 14.3)*

The circuit RLC (Fig. 14.3) is used in this example to illustrate the nature of the following data variables produced during the operation of $ACPREP$: BC, BL, BGL and YRE.

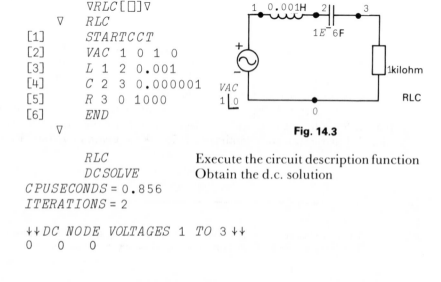

```
        ∇RLC[□]∇
    ∇   RLC
[1]     STARTCCT
[2]     VAC 1 0 1 0
[3]     L 1 2 0.001
[4]     C 2 3 0.000001
[5]     R 3 0 1000
[6]     END
    ∇
```

Fig. 14.3

   RLC Execute the circuit description function
   $DCSOLVE$ Obtain the d.c. solution
$CPUSECONDS = 0.856$
$ITERATIONS = 2$

$\downarrow\downarrow DC\ NODE\ VOLTAGES\ 1\ TO\ 3\ \downarrow\downarrow$
   0   0   0

$$G$$
1 1E5 1E^-12 0.001   Conductance vector (1E5 and 1E^-12
                                  terms relate to L and C respectively)

$$SLOPE$$

1E5	$^-$1E5	0E0
$^-$1E5	1E5	$^-$2E^-12
0E0	$^-$2E^-12	1E^-3

The nodal conductance matrix

$F \leftarrow 1000$ Set the frequency to 1000 Hz

$ACPREP$ Execute $ACPREP$

$$BC$$ Capacitance matrix (unreduced)

0	0	0	0
0	0.000001	$^-$0.000001	0
0	$^-$0.000001	0.000001	0
0	0	0	0

$$BL$$ Reciprocal inductance matrix

1E3	$^-$1E3	0E0	0E0
$^-$1E3	1E3	0E0	0E0
0E0	0E0	0E0	0E0
0E0	0E0	0E0	0E0

(unreduced)

Conductance matrix containing entries
from the place-holding resistor at the
inductor location (unreduced)

$$BGL$$

100000	$^-$100000	0	0
$^-$100000	100000	0	0
0	0	0	0
0	0	0	0

$$YRE$$

1E0	0E0	0E0
0E0	0E0	$^-$2E^-12
0E0	$^-$2E^-12	1E^-3

YRE is identical to $SLOPE$ minus the
effects of the place-holding resistor
at the inductor location

Table 14.3
The function $ACPREP$

∇ $ACPREP$

The purpose of $ACPREP$ is to access and prepare all the required circuit data prior
to the actual solution phase. Key results are left in the global matrices BC, BL,
and YRE. Necessary input global variables are: CL, LL, RL, and $SLOPE$, along
with NH

[1] $BC \leftarrow FORM$ CL

Form the circuit capacitive matrix and store in BC

[2] $BL \leftarrow FORM\ LL[;\ 1\ 2], \div LL[;3]$

Form the circuit reciprocal inductive matrix and store in BL

[3] $BGL \leftarrow FORM\ RL[;\ 1\ 2], \div RL[;3]$

Convert to conductances, the resistor values in the RL,storage table and form the corresponding nodal conductance matrix and store in BGL. These entries account for the small place-holding resistors used in place of inductors for the d.c. solution

[4] $YRE \leftarrow SLOPE-\ ^-1\ ^-1\ \downarrow BGL$
      $\triangledown$

Subtract their effect from $SLOPE$ and store in YRE the net real part of the composite reduced admittance matrix

14.2.3 $ACSOL$

The complex a.c. node voltage solutions are actually produced in $ACSOL$. A single loop is implemented in this function to index through the frequency variable F, and repeat the solution once for each frequency in turn as required. A flowchart for $ACSOL$ is shown in Fig. 14.3, and a line-by-line description follows in the box below. Several internal results produced by $ACSOL$ are illustrated in Example 14.2.

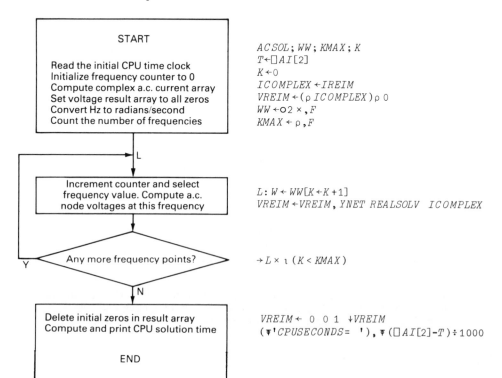

START

Read the initial CPU time clock
Initialize frequency counter to 0
Compute complex a.c. current array
Set voltage result array to all zeros
Convert Hz to radians/second
Count the number of frequencies

$ACSOL;WW;KMAX;K$
$T \leftarrow \square AI[2]$
$K \leftarrow 0$
$ICOMPLEX \leftarrow IREIM$
$VREIM \leftarrow (\rho\ ICOMPLEX)\rho\ 0$
$WW \leftarrow \circ 2 \times , F$
$KMAX \leftarrow \rho, F$

L

Increment counter and select
frequency value. Compute a.c.
node voltages at this frequency

$L: W \leftarrow WW[K \leftarrow K+1]$
$VREIM \leftarrow VREIM, YNET\ REALSOLV\ ICOMPLEX$

Any more frequency points?

$\rightarrow L \times \iota\ (K < KMAX)$

N

Delete initial zeros in result array
Compute and print CPU solution time

END

$VREIM \leftarrow\ 0\ 0\ 1\ \downarrow VREIM$
$(\triangledown 'CPUSECONDS=\ '), \triangledown (\square AI[2]-T) \div 1000$

Fig. 14.4 Flowchart for the function $ACSOL$

Example 14.2 *AC SOL (Table 14.4)*

This example continues the computations of Example 14.1 to illustrate selected results produced by the execution of *AC SOL* for the test circuit *RLC*.

RLC	Establish the starting point for this
DC SOLVE	example

$CPUSECONDS = 0.837$

$ITERATIONS = 2$

$\downarrow\downarrow DC\ NODE\ VOLTAGES\ 1\ TO\ 3\downarrow\downarrow$

0   0   0

       $F \leftarrow 1000$

       AC PREP

AC SOL	Execute *AC SOL*

$CPUSECONDS = 0.293$

VREIM	The result array contains real and
0.999	imaginary parts. There is only one
1	column since only one frequency point
0.976	was specified

$^-0.000149$

$^-0.00628$

0.149

W	The frequency in radians/second
$6.28E3$	

YNET   The complete admittance array

$1.00E0$	$0.00E0$	$0.00E0$	
$0.00E0$	$0.00E0$	$^-2.00E^-12$	The real part is just
$0.00E0$	$^-2.00E^-12$	$1.00E^-3$	*YRE* (see Example

14.1)

$^-1.59E^-1$	$1.59E^-1$	$0.00E0$	
$1.59E^-1$	$^-1.53E^-1$	$^-6.28E^-3$	The imaginary part
$0.00E0$	$^-6.28E^-3$	$6.28E^-3$	was produced from

BC and *BL* (see

       ICOMPLEX   Example 14.1)

1

0

0   The a.c. node source

                 current array in

                 column form contain-

0   ing real and imagi-

0   nary parts in two

0   planes

Table 14.4
The function *ACSOL*

∇ *ACSOL* ; *WW* ; *KMAX* ; *K*

This function carries out the actual circuit solutions, one for each entry in *F*. The results are placed in the array *VREIM*. Plane 1 of this result contains the real node voltages, and plane 2 contains the imaginary node voltages

[1]   *T* ← □*AI*[2]

Record the starting value of the CPU time clock

[2]   *K* ← 0

Initialize the frequency index counter *K* to zero

[3]   *ICOMPLEX* ← *IREIM*

Compute the a.c. node source current array and store in *ICOMPLEX* with real and imaginary terms in planes 1 and 2 respectively. The shape is (2, *NR*, 1)

[4]   *VREIM* ← (ρ *ICOMPLEX*)ρ 0

Initialize the voltage result array *VREIM* to have the same shape and contain all zeros

[5]   *WW* ← O2 × , *F*

Convert the frequency vector from hertz in *F* to radians/second in *WW*

[6]   *KMAX* ← ρ , *F*

Determine the number of frequencies at which solutions are required

[7]   *L*: *W* ← *WW*[*K* ← *K* +1]

Label this line *L*; add 1 to *K* to choose the next frequency point (i.e. the first for *K*←1), and store its value (in radians per second) in *W*

[8]   *VREIM* ← *VREIM* , *YNET REALSOLV ICOMPLEX*

Compute the complex node voltages at this frequency point using the solution routine *REALSOLV*; laminate these results as an additional column in each plane of *VREIM*

[9]   → *L* × ι (*K* < *KMAX*)

Test for the last frequency point: if not, loop to line 7 and repeat

[10]   *VREIM* ← 0 0 1 ↓ *VREIM*

After the solution at the last frequency point, drop the first column from each plane of *VREIM* to delete the zeros used to set up the initial structure

[11]   (⍕'*CPUSECONDS* = '), ⍕ (□*AI*[2]−*T*)÷1000
        ∇

Compute and print out the elapsed CPU time

14.3 Basic Processing Functions

We now consider the detailed operation of the functions from levels 3 and 4 of the tree structure in Fig. 14.2. Included here are the functions $FORM$, $IREIM$, $IFORM$, $YNET$, BB, and $REALSOLV$. Each of these functions carries out specific tasks associated with either the data preparation phase, or the actual solution process itself. Each is called either directly or indirectly by the higher-level functions $ACPREP$ or $ACSOL$.

14.3.1 $FORM$

The function $FORM$ is used three times in $ACPREP$ to form three different matrices for components which follow the same processing rules. These are capacitive (BC), reciprocal inductive (BL), and conductance (BGL). In each case the matrix entries are placed according to the particular node numbers associated with the components. The BC and BL matrices are equivalent to nodal admittance matrices at unit frequency, while BGL is a nodal conductance matrix. Each result matrix produced by $FORM$ has the shape (NH, NH). A similar function, also called $FORM$, was introduced in Section 1.7.

    $FORM$ is implemented with a single main loop as the flow chart in Fig. 14.5 illustrates. Each pass through the loop processes one component obtained from one row of the corresponding storage table. This loop is repeated for the number of times required to process all components in the table. A conditional branch before the main processing loop is used to ensure that the function will accept an empty table for a given component type. In this case the result matrix contains all zeros. A detailed line-by-line description of $FORM$ is given in the Table 14.5.

14.3.2 $IREIM$ and $IFORM$

These two functions are used together to process the a.c. voltage and current source tables. The result is a single array ($ICOMPLEX$) which is provided to $ACSOL$, and which contains the composite a.c. node source current data for the circuit. Real and imaginary parts are in planes one and two respectively. $IFORM$ is called by $IREIM$ to process each type of source table in turn.

    As noted in Section 11.2, both a.c. sources are specified by magnitude and phase angle in degrees. Internal processing ($IFORM[8\ 9]$) converts the phase angles to radians in the course of generating the real and imaginary components. Phase angles may be specified over the range from 0 to $\pm360°$. A further consideration which should be noted here is that both sources are required to have one node grounded (the second node number entry in $VACL$ and $IACL$ must be 0). This is seldom a severe restriction but it permits considerably simpler processing in $IREIM$ and $IFORM$.

    A flow chart for the function $IREIM$ is given in Fig. 14.6. A simple linear processing sequence is used with conditional branches to bypass the processing

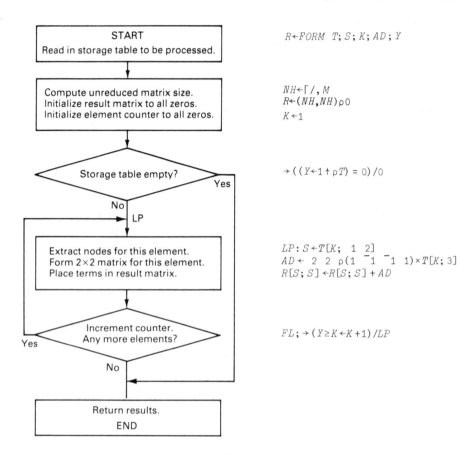

Fig. 14.5
Flowchart for the function $FORM$

for either of the $VACL$ or $IACL$ tables which may be empty. A detailed description of $IREIM$ is given in Table 14.6. Example 14.3 at the end of this subsection illustrates selected processing details within $IREIM$ and $IFORM$.

Table 14.5
The function $FORM$

$\nabla\ R \leftarrow FORM\ T; S; K; AD; Y$

$FORM$ is designed to take as an input argument a storage table containing node numbers and conductance values, and produce the nodal conductance matrix as a result. $FORM$ is similarly useful for other two-terminal component storage tables such as CL

[1] $NH \leftarrow \lceil /, M$
[2] $R \leftarrow (.NH,NH) \rho\ 0$
[3] $K \leftarrow 1$

Compute NH and initialize the result matrix to contain all zeros; start the component count index at 1

[4] $\rightarrow ((Y\leftarrow 1\uparrow\rho\ T) = 0)/0$

If this storage table is empty, return the current result matrix R, and branch out of the function

[5] $LP: S\leftarrow T[K;\ 1\ 2]$

Label this line LP. Extract the two node numbers from the table for this first component

[6] $AD \leftarrow\ 2\ \ 2\ \rho\ (1\ \ ^-1\ \ ^-1\ \ 1)\times T[K; 3]$

Form a 2×2 matrix containing four component entries with appropriate signs, i.e. positive on the leading diagonal

[7] $R[S;\ S]\leftarrow R[S;\ S]+AD$

Add these matrix entries into the result matrix at locations determined by the two node numbers in S

[8] $FL: \rightarrow (Y\geq K\leftarrow K+1)/LP$
  ∇

Label this line FL; increment the component counter K; loop to line 4 if not the last component; otherwise return the result and exit

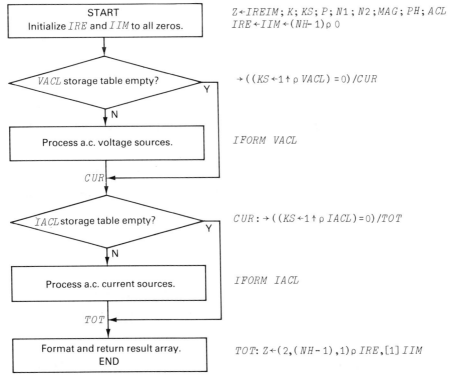

START
Initialize IRE and IIM to all zeros.

$Z\leftarrow IREIM; K; KS; P; N1; N2; MAG; PH; ACL$
$IRE\leftarrow IIM\leftarrow (NH-1)\rho\ 0$

$VACL$ storage table empty?

$\rightarrow ((KS\leftarrow 1\uparrow\rho\ VACL) = 0)/CUR$

Y

N

Process a.c. voltage sources.

$IFORM\ VACL$

CUR

$IACL$ storage table empty?

$CUR: \rightarrow ((KS\leftarrow 1\uparrow\rho\ IACL) = 0)/TOT$

Y

N

Process a.c. current sources.

$IFORM\ IACL$

TOT

Format and return result array.
END

$TOT: Z\leftarrow (2,(NH-1),1)\rho\ IRE,[1]\ IIM$

Fig. 14.6
Flowchart for the function $IREIM$

Table 14.6
The function *IREIM*

$\nabla\ Z \leftarrow IREIM\ ;\ K\ ;\ KS\ ;\ P\ ;\ N1\ ;\ N2\ ;\ MAG\ ;\ PH\ ;\ ACL$

The purpose of *IREIM* is to compute and return as an explicit result, the complex a.c. node source current array in the standard *BASECAP* format with real parts in plane 1 and imaginary parts in plane 2. The major global variables accessed by *IREIM* are the a.c. source storage tables *VACL* and *IACL*

[1] $IRE \leftarrow IIM \leftarrow (NH-1)\rho\ 0$

Initialize the real (*IRE*) and imaginary (*IIM*)current component vectors to all zeros

[2] $\rightarrow ((KS \leftarrow 1 \uparrow \rho\ VACL) = 0)/CUR$

Skip the immediately following line if no a.c. voltage sources exist in the circuit

[3] *IFORM VACL*

Call *IFORM* to make the necessary entries in *IRE* and *IIM* for each a.c. voltage source in *VACL*

[4] $CUR: \rightarrow ((KS \leftarrow 1 \uparrow \rho IACL) = 0)/TOT$

Skip the immediately following line if no a.c. current sources exist in the circuit

[5] *IFORM IACL*

Call *IFORM* to make the necessary entries in the *IRE* and *IIM* vectors for each a.c. current source in *IACL*

[6] $TOT: Z \leftarrow (2, (NH-1), 1)\rho\ IRE, [1]\ IIM$
∇

Format *IRE* and *IIM* into one array with two planes of shape (2, *NR*, 1)

As shown in Fig. 14.7, *IFORM* contains a single loop, used to process one entry (i.e. row) from the given source table at a time. By means of the conditional branch in the last line, the processing within the loop is repeated for each subsequent entry as required. A detailed description of this function is given in Table 14.7.

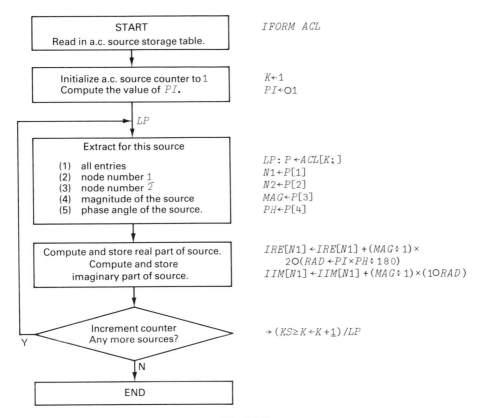

Fig. 14.7
Flowchart for the function $IFORM$

Example 14.3 $IREIM$ **and** $IFORM$ **(Tables 14.6, 14.7)**

AC sources are processed by $IREIM$ and $IFORM$ as illustrated in this
example using the test circuit LPF from Section 11.6

LPF	Set up LPF
$DCSOLVE$	Obtain the d.c. solution

$CPUSECONDS = 1.24$
$ITERATIONS = 3$

$\downarrow\downarrow DC\ NODE\ VOLTAGES\ 1\ TO\ 3\downarrow\downarrow$
0   0   0

$IACL$	Confirm that $IACL$ is empty
$VACL$	The a.c. voltage source at node 1 is
1 0 1 0	$1+J0$ volts
NH	There are four nodes counting
4	the reference node
$T\triangle IREIM \leftarrow 1$	Set a trace on $IREIM$ line 1
$T\triangle IFORM \leftarrow 8\ \ 9$	Set a trace on $IFORM$ lines 8,9

```
          F←1000              Set F to 1000 Hz
          ACSOLVE             Obtain the a.c. solution
 SOLVING FOR 1 FREQUENCY POINTS
 IREIM[1] 0 0 0              IRE and IIM initial values
 IFORM[8] 1                     IRE[1]
 IFORM[9] 0                    'IIM[1]
 CPUSECONDS = 0.336
 DONE
```

```
          IRE              Vector of real currents
 1  0  0
          IIM              Vector of imaginary currents
 0  0  0
```

```
          (2,(NH-1),1)ρ IRE,[1] IIM
 1
 0                          The processing used on line 6 of IREIM
 0                          to format the current array

 0
 0
 0

          ICOMPLEX          ICOMPLEX contains the above result.
 1
 0
 0

 0
 0
 0
```

Table 14.7
The function *IFORM*

```
 ▽ IFORM ACL
```
The purpose of *IFORM* is to extract the magnitude and phase angle components
of a.c. source currents from *IACL*, and the magnitude and phase angle components of equivalent source currents from *VACL*, convert these to real and
imaginary form, and add appropriate entries into the global *IRE* and *IIM* result
vectors

```
[1]    K←1
[2]    PI←○1
```

Set the source counter K to 1 to select the first source; for convenience, PI
equals π

```
[3]    LP: P←ACL[K;]
```

Label this line LP for convenience in looping; extract all of the entries corresponding to a given source from one row (initially the first source when K equals
1)

[4] $N1 \leftarrow P[1]$
[5] $N2 \leftarrow P[2]$
[6] $MAG \leftarrow P[3]$
[7] $PH \leftarrow P[4]$

For convenience in the subsequent processing, separately extract each node and the specified current magnitude and phase angle. Note that this applies equally to a.c. voltage sources since a unit conductance is used in the representation of each a.c. voltage in the function VAC

[8] $IRE[N1] \leftarrow IRE[N1] + (MAG \div 1) \times 2 \circ (RAD \leftarrow PI \times PH \div 180)$

Compute the real part of the node source current and add this into the IRE vector corresponding to the first node number entry $N1$. Note that, for simplicity, $N2$ is always 0

[9] $IIM[N1] \leftarrow IIM[N1] + (MAG \div 1) \times (1 \circ RAD)$

Similarly compute the imaginary part and add this into the IIM result vector

[10] $\rightarrow (KS \geq K \leftarrow K + 1)/LP$
        ∇

Increment the source current counter K and branch to line 3 if other sources remain, otherwise exit from the function

14.3.3 $YNET$ **and** BB

The functions $YNET$ and BB form a dedicated pair. These are used in $ACSOL$ for each new solution frequency to produce the corresponding complex nodal admittance array $(YCOMP)$. BB specifically processes the BC and BL matrices within YRE at each frequency to produce the imaginary part of the overall nodal admittance array $(YIMAG)$, which is then combined with the real part (YRE).

Table 14.8
The function $YNET$

        $\nabla \ Y \leftarrow YNET$

The purpose of $YNET$ is to produce the complex admittance array for the circuit at a given frequency. The explicit result has two planes, one real, and one imaginary. Input arguments are the global variables used in the function BB, plus YRE

[1] $YIMAG \leftarrow \ ^{-}1 \ ^{-}1 \ \downarrow YREACT \leftarrow BB$

Compute the reduced form of the imaginary part of the nodal admittance array and assign to $YIMAG$

[2] $Y \leftarrow YCOMP \leftarrow YRE, [0.5] \ YIMAG$
        ∇

Laminate YRE and $YIMAG$ into two separate planes; assign to $YCOMP$ and return this as an explicit result argument

Table 14.9
The function BB

   ∇ $R \leftarrow BB$

   BB is used in the function $YNET$ to update the imaginary part of the admittance matrix at each new frequency, and to return this matrix as an explicit result. Global variables used are BC, BL and W

[1] $R \leftarrow (W \times BC) - BL \div W$
    ∇

Compute capacitive and inductive reactance terms for the corresponding W (radians/second), and subtract on an element-by-element basis, to yield the imaginary part of the admittance matrix

---

14.3.4 *REALSOLV*

REALSOLV constitutes the heart of the a.c. solution process in *BASECAP*. It is called once for each frequency by *ACSOL* to produce the complex a.c. node voltage solutions, given the complex array of a.c. source currents and the circuit's nodal admittance array valid at the frequency concerned. The methods used in this function are explained in Section 4.4, and a similar function, also called *REALSOLV*, was introduced in Fig. 4.10.

    In the function presented here, the current array I and the voltage array V both have the shape $(2, NR, 1)$, where NR is the number of circuit nodes not counting the reference node. Y is the reduced complex nodal admittance array of shape $(2, NR, NR)$. The first plane of all three arrays contains the real terms, and the second plane contains the imaginary terms.

    In the context of these variables, *REALSOLV* forms and solves the following matrix equation, which is equivalent to equation (4.6):

$$\begin{bmatrix} ,I \end{bmatrix} \leftarrow \begin{bmatrix} \begin{bmatrix} Y[1;;] \end{bmatrix} & \begin{bmatrix} {}^{-}1 \times Y[2;;] \end{bmatrix} \\ \begin{bmatrix} Y[2;;] \end{bmatrix} & \begin{bmatrix} Y[1;;] \end{bmatrix} \end{bmatrix} +.\times \begin{bmatrix} ,V \end{bmatrix} \qquad (14.1)$$

Here $(,I)$ and $(,V)$ each represents a real-number vector containing real- followed-by-imaginary terms. The description of *REALSOLV* in Table 14.10 contains further details, and Example 14.4 illustrates *REALSOLV* in operation.

Table 14.10
The function *REALSOLV*

   ∇ $V \leftarrow Y$ *REALSOLV* I; YT; YB

   REALSOLV formulates a $(2N \times 2N)$ set of real equations given a set of N complex equations, carries out their solution using the APL quad divide operator, and formats and returns the complex node voltages as an explicit result in a form compatible with *VREIM*. Input arguments are the complex a.c. node source current array I, and the complex nodal admittance array Y

```
[1]    YT←Y[1;;],(‾1×Y[2;;])
[2]    YB←Y[2;;],Y[1;;]
```

Separately formulate the top and bottom halves of the intended $2N \times 2N$ real-only matrix prior to solution

```
[3]    V←(ρI)ρ(,I)⊞YT,[1] YB
       ∇
```

Form one $2N \times 2N$ matrix by laminating YT and YB; execute the matrix solution with I configured as a vector (this dictates that the resultant voltages will also have the same vector shape); then format these voltages into the original shape of $I(2, NR, 1)$, return the result, and exit from the function

Example 14.4   *Tracing the operation of* REALSOLV

In this example we use the APL system trace and stop facilities with the function *REALSOLV* to recreate and follow in detail the main processing steps used to produce the solution for the test circuit *LPF* from Section 11.6.

Valid d.c. and a.c. solutions were first obtained for the circuit *LPF* at 1000 Hz (see Example 14.3) before proceeding as follows:

            VREIM   The voltage solution array for
    1   *LPF* at 1000 Hz
    0.5
    0.5

    ‾0.0005
    ‾0.5
    ‾0.5
          YNET   The complex admittance array
    1   ‾0.001 0   for *LPF* at 1000 Hz
  ‾0.001   0.001 0
    0   ‾0.001 0.001

    0   0   0
    0   0.001 0
    0   0   0
        IREIM   The current source array
    1
    0
    0

    0
    0
    0

$T \Delta REALSOLV \leftarrow \iota 9$ Trace *REALSOLV* lines 1 to 9
$S \Delta REALSOLV \leftarrow 3$ Stop *REALSOLV* before line 3

     YNET REALSOLV IREIM
REALSOLV[1]
```
  1        ¯0.001  0        0        0        0
¯0.001     0.001  0        0      ¯0.001     0
  0       ¯0.001  0.001     0        0        0
```

The upper half matrix *YT*

REALSOLV[2]
```
  0        0        0        1      ¯0.001     0
  0        0.001   0      ¯0.001     0.001     0
  0        0        0        0      ¯0.001     0.001
```

The lower half matrix *YB*

REALSOLV[3]   Execution stops before line 3

     □←*YTEMP*←*YT*,[1] *YB*   Compute and display *YTEMP*
```
  1        ¯0.001  0        0        0        0
¯0.001     0.001  0        0      ¯0.001     0
  0       ¯0.001  0.001     0        0        0
  0        0        0        1      ¯0.001     0
  0        0.001   0      ¯0.001     0.001     0
  0        0        0        0      ¯0.001     0.001
```

     I
```
1
0
0
```
The internal currents are the
same as given by *IREIM* above
```
0
0
0
```

     ,*I*   Various processing details are
1 0 0 0 0 0   illustrated manually
     (,*I*) ⊞*YTEMP*
1 0.5 0.5 ¯0.0005 ¯0.5 ¯0.5
     ρ*I*
2 3 1
     (2 3 1)ρ(,*I*)⊞*YTEMP*   The results calculated here are
```
  1
  0.5
  0.5
```
seen to be the same as in *VREIM*
displayed above
```
¯0.0005
¯0.5
¯0.5
```

14.4 Supplementary Functions

We now describe three supplementary functions that are not directly included in the tree structure of Fig. 14.2. These are $FLIN$, $FSTEP$ and $FLOG$, which may be called directly by the user to facilitate setting the global frequency vector F.

14.4.1 $FLIN$ and $FSTEP$

These functions are essentially alternatives to each other, and have a very similar structure, as the detailed descriptions in the subsequent tables reveal. In both cases, a linear sequence of equal steps is chosen for F. $FLIN$ allows the user to specify the desired number of frequency points to be used, whereas $FSTEP$ allows the user to specify the step size between frequency points.

Table 14.11
The function $FLIN$

∇ $FLIN$ P; FH; FL; NP

$FLIN$ is designed to facilitate setting the frequency vector F by specifying an argument vector consisting only of the lower frequency limit, the upper frequency limit, and the number of frequency points desired. Equal frequency intervals are chosen between the two limits

[1] $FL \leftarrow P[1]$
[2] $FH \leftarrow P[2]$
[3] $NP \leftarrow P[3]$

Extract the two limits and the number of frequency points from the input parameter list

[4] $F \leftarrow FL , FL + (\iota NP) \times (FH - FL) \div NP$
        ∇

Compute the desired frequency interval and generate the desired frequency vector F accordingly

Table 14.12
The function $FSTEP$

∇ $FSTEP$ P; FL; FH; ΔF; NP

This function is identical in purpose to the function $FLIN$ except the user specifies a desired frequency increment size instead of the number of points. Again, equal intervals are chosen between the two limits

[1] $FL \leftarrow P[1]$
[2] $FH \leftarrow P[2]$
[3] $\Delta F \leftarrow P[3]$

Extract the two limits and the frequency increment size

[4] $NP \leftarrow \lceil (FH - FL) \div \Delta F$

Compute the (rounded) number of frequency intervals between the two limits

[5] $F \leftarrow FL + 0, (\iota NP) \times \Delta F$
        ∇

Compute the vector of frequency increments, add this to the desired low-frequency value, and assign the result to F

14.4.2 *FLOG*

The processing in this function is a little more complex. Log and antilog operators are used to produce frequency points with logarithmic spacing at a specified number of frequency points per decade.

Table 14.13
The function *FLOG*

        ∇ $FLOG\ P; LF; HF; PPD; NF; FV$
        FLOG enables the user to easily set up the frequency vector F to contain logarithmically spaced increments. Input arguments are the lower and upper frequency limits, and the desired number of frequency points per decade

[1] $LF \leftarrow 10 \circledast P[1]$
[2] $HF \leftarrow 10 \circledast P[2]$
[3] $PPD \leftarrow P[3]$

Extract the two frequency limits from the input parameter list and take the logarithm of each; also extract the number of points per decade (PPD)

[4] $NF \leftarrow \lfloor (HF - LF) \times PPD$

Compute the number of frequency points (rounded down)

[5] $FV \leftarrow LF + 0, (\iota NF) \times \div PPD$

Compute the vector of log frequencies

[6] $F \leftarrow 10 \ast FV$
        ∇

Compute the antilog of each of the entries in the above vector, giving frequencies having equal spacing on a logarithmic basis, and assign to the frequency vector F

Exercises

14.1 Compute *VREIM* with the *BASECAP* functions for the circuit *ACTEST* (Fig. E14.1) below, and confirm that the voltage at node 2 agrees with expectations. Use a convenient test frequency such as 1000 Hz.

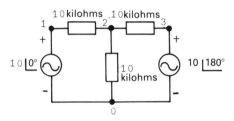

Fig. E14.1 *ACTEST*

14.2 Repeat exercise 14.1 using the following a.c. sources:

```
VAC 1 0 10 0
VAC 3 0 10 0
```

14.3 The function *FLOG* in Section 14.4 requires the following three arguments to set up the frequency vector *F* with logarithmically spaced frequencies: the lower frequency limit, the upper frequency limit, and the number of points per decade (*PPD*). Produce a modified function called *FLOGTP* which does the same job but where the third input argument is the total number of frequency points desired rather than the number of points per decade.

14.4 *BASECAP* currently represents an a.c. voltage source by means of a 1-ohm resistor in parallel with an ideal a.c. current source. In some circuit applications this source resistance may be large enough to affect solution accuracy. Make modifications to the necessary *BASECAP* functions to reduce this source resistance to 0.001 ohm. Note that this will require modifications to *VAC* as well as one of the functions in Section 14.3.2.

14.5 *BASECAP* automatically places a large resistor in the circuit in parallel with each capacitor (*C* line 4). These resistors are required to maintain circuit continuity under d.c. conditions, but they may cause unacceptable errors in some circuits under certain a.c. conditions.

Design and test the necessary modifications and additional functions to be incorporated into the *BASECAP* package to eliminate the effects of these resistors from the nodal conductance matrix *SLOPE* after the d.c. solution but prior to generating the final *YRE* matrix in the function *ACPREP* .

Solutions

14.1   The circuit description function and $VREIM$ at 1000 Hz follow:

```
            ∇ ACTEST
[1]         STARTCCT
[2]         R 1 2 10000
[3]         R 2 3 10000
[4]         R 2 0 10000
[5]         VAC 1 0 10 0
[6]         VAC 3 0 10 180
[7]         END
            ∇

            VREIM
      9.999E0
     ¯1.807E¯16
     ¯9.999E0

      5.812E¯20
      5.813E¯16
      1.744E¯15
```

Essentially zero volts at node 2 (within computational accuracy)

14.2   Following is the solution with the new a.c. sources:

```
            ∇ ACTEST1
[1]         STARTCCT
[2]         R 1 2 10000
[3]         R 2 3 10000
[4]         R 2 0 10000
[5]         VAC 1 0 10 0
[6]         VAC 3 0 10 0
[7]         END
            ∇

            VREIM

     10
      6.666
     10

      0
      0
      0
```

The expected voltage results at node 2

14.3   A possible $FLOGTP$ follows:

```
            ∇ FLOGTP P; LF; HF; PPD; NF: FV
[1]         LF←10⊛P[1]
[2]         HF←10⊛P[2]
[3]         NF←P[3] - 1
[4]         PPD ← NF÷HF- LF'
[5]         FV←LF+0,(⍳NF) ×÷ PPD
[6]         F←10*FV
            ∇
```

```
        FLOGTP 1 1E4 9              Testing FLOGTP
        ρ F
9
        F
1 3.162 10 31.62 100 316.2 1000 3162 1E4

        FLOGTP 1.75 1.63E5 11
        ρ F
11
        F
1.75 5.495 17.25 54.17 170.1 534.1 1677 5266 1.653E4 5.191E4 1.63E5
```

14.4   Make the following replacements and additions

(a)   to *VAC*

[4]   *GIN←GIN*, 1000

(b)   and to *IREIM*

[2.5]   *VACLT←VACL*

[2.6]   *VACLT*[;3] ←*VACLT*[;3] × 1000

[3]   *IFORM VACLT*

CHAPTER

15

Results Processing and Display

The functions in the *OUT* group enable the user to access, process, and display selected a.c. and d.c. node voltage results. The composition of this group is given in Table 15.1. The principal functions here are *PLOT* and *PRINT.* The remaining ten functions provide the processing support for these two display functions. Most functions in Table 15.1 are designed to be executed directly by the user to produce the desired displays. Appendix 6 includes detailed listings for most of these functions.

The first section in this chapter provides additional material on the structure of the *OUT* group, and detailed descriptions of the functions follow in Sections 15.2–15.4. In Section 15.5, we introduce the topic of postprocessing and show how additional types of results can be produced by means of some relatively simple additional APL programming.

Table 15.1
Composition of the OUT Group

OUT	)*GRPS*					
	)*GRP OUT*					
ACVOUT	*ALL*	*AND*	*ANGLE*	*LOG*	*LOGF*	*MAG*
MAGDB	*PLOT*	*PRINT*	*VS*	*ZATN*		
	)*VARS*					
CCT	*F*	*PAR*	*PVAL*	*VALL*	*VANS* *VBN*	*VREIM*

15.1 *OUT* group structure

This section contains an overview of the data and the functions used in the *OUT* group. Clearly the functions in this group must interact directly with the data produced by the d.c. and a.c. solution functions in Chapters 13 and 14. Figures 13.1 and 14.1 show these data items.

The structure of each of the principal items of 'raw' data is summarized in Table 15.2. In the formal sense, it is this data which is the subject of the selection and processing operations, the results of which are displayed by $PRINT$ or $PLOT$. In addition, it is appropriate to recall that APL readily permits the user to display any of the global variables produced throughout the solution process, simply by entering the variable name. Examples of selected data items from Table 15.2 are shown below for the circuit LPF (see Section 11.6).

Table 15.2
Principal data structures
used by the OUT group

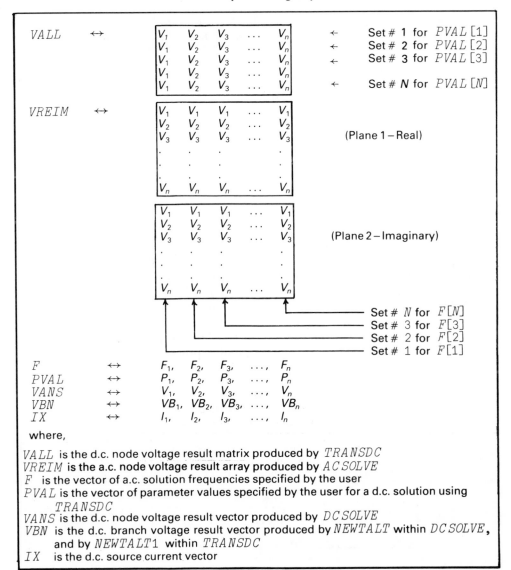

where,

$VALL$ is the d.c. node voltage result matrix produced by $TRANSDC$

$VREIM$ is the a.c. node voltage result array produced by $ACSOLVE$

F is the vector of a.c. solution frequencies specified by the user

$PVAL$ is the vector of parameter values specified by the user for a d.c. solution using $TRANSDC$

$VANS$ is the d.c. node voltage result vector produced by $DCSOLVE$

VBN is the d.c. branch voltage result vector produced by $NEWTALT$ within $DCSOLVE$, and by $NEWTALT1$ within $TRANSDC$

IX is the d.c. source current vector

```
        VALL                    VALL does not exist since no d.c. transfer
VALUE ERROR                     function solution was executed
        VALL
        ∧
        IX                      No d.c. source currents exist in this circuit
0  0  0
        VBN
8.03E‾34  1.6E‾30  1.6E‾30  3.21E‾27  3.21E‾27
        VANS
8.03E‾34  1.6E‾30  3.21E‾27
        F                       F contains five frequencies
100  1E3  1E4  1E5  1E6
        ρVREIM                  Each column of VREIM contains the a.c.
2  3  5                         node voltages for one frequency
        VREIM
1.00E0     1.00E0     9.99E‾1    9.99E‾1    9.99E‾1
9.90E‾1    5.00E‾1    9.88E‾3    9.98E‾5    9.98E‾7
9.90E‾1    5.00E‾1    9.88E‾3    9.98E‾5    9.98E‾7

‾9.90E‾5   ‾5.00E‾4   ‾9.88E‾5   ‾9.98E‾6   ‾9.98E‾7
‾9.91E‾2   ‾5.00E‾1   ‾9.89E‾2   ‾9.99E‾3   ‾9.99E‾4
‾9.91E‾2   ‾5.00E‾1   ‾9.89E‾2   ‾9.99E‾3   ‾9.99E‾4
```

The selection and processing operations specified by the user are carried out on the 'raw' data to meet the input requirements of *PLOT* or *PRINT*. Both of these functions accept data in matrix form, one column per item. *PLOT* requires the independent data for the *X* axis to be in column 1. Up to six columns of dependent data for the *Y* axis may be used, and the columns may have any number of entries so long as all columns are of equal length.

The *OUT* group has virtually no tree structure, since most of the functions are invoked directly by the user. However, it is essential for these functions to be used in the prescribed sequence, as processed data is passed directly from one function to another. The sequence required to process and display a.c. results was specified in Chapter 11, Section 11.5, and is summarized in Table 15.3.

Table 15.3
OUT group function call sequence
for a.c. node voltage displays

Display function	Optional *F* modifier function	Node voltage modifier function	Node selection function	Node number arguments
PRINT or *PLOT*	*F,* or *LOGF,*	*MAG* or *MAGDB* or *ANGLE*	*ACVOUT*	*NOUT*

◄────────────────── Data transfer direction ──────────────────

(Exercise 15.1 elaborates on these concepts.) In the d.c. case, most of the necessary formatting is carried out automatically in the function $TRANSDC$. The supplementary functions VS and AND are also useful for formatting data as discussed in Section 15.4.

15.2 Selection and Display Functions

The functions discussed in this section relate to node selection and results presentation. As specified in Table 15.3, the normal input sequence requires that the function $PRINT$ or $PLOT$ be specified first, while the desired node or nodes are input to $ACVOUT$ last. Recalling, however, that APL processes each input line from right to left, it will be recognized that $ACVOUT$ is actually executed first, so the selected results are fed back to $PRINT$ or $PLOT$ (that is from right to left in Table 15.3) for presentation to the user.

Between the node selection and the display processes, it is necessary to specify one of the node voltage modifier functions and, optionally, the frequency modifier function. These functions are discussed later in Section 15.3.

15.2.1 $PRINT$ and $PLOT$

Of these two display functions, only $PRINT$ will be presented in detail. The $BASECAP$ $PLOT$ function is typical of those available in the APL function libraries on most large APL installations. A detailed study of the internal design and operation of $PLOT$ is beyond the scope of this text. Instead, the discussion here will be limited to its functional capabilities and use. A simple plot function is also included in Section 4.5.

The input data for either $PRINT$ or $PLOT$ is normally a matrix, with the independent (X) data in column 1, and one or more columns of dependent (Y) data. $PLOT$ will display up to six curves on the same set of axes, with each curve having a different symbol. $PRINT$ will accommodate as many columns of data as will fit conveniently on the paper or display screen.

The $PLOT$ function used in $BASECAP$ is auto-scaling, which means that the scales on the X and Y axes are set automatically to accommodate the largest

Table 15.4
The function $PRINT$

   ∇ $PRINT$ P

      This function causes the contents of the input argument P to be printed or displayed at the user's terminal. P may be a scalar, a vector, or a matrix

[1]   ☐ ← P
     ∇

      The contents of P are output to the terminal by means of the APL system quad or window operator

---

numbers encountered. The size of the resulting plot is determined by the two-number vector making up the left argument for $PLOT$. These entries specify the approximate number of character positions vertically and horizontally to be used in laying out the plot. No scaling of any kind is applied to the display produced by $PRINT$.

15.2.2 $ACVOUT$ **and** ALL

The function $ACVOUT$ is used to select, and pass on, a subset of the a.c. node voltage results stored in the array $VREIM$. (Exercise 15.1 clearly demonstrates this processing.) ALL is a function which produces a vector result containing all valid node numbers (not counting the reference node) for the active circuit. ALL can be used, for example, as an argument for $ACVOUT$ or $TRANSDC$, when it is desired to access or display all node results. For the circuit LPF, for example, ALL gives the following result:

```
        LPF
        IX
   0  0  0
        ALL
   1  2  3
```

Table 15.5
The function $ACVOUT$

∇ $Z{\leftarrow}ACVOUT$ NNO
   This function accesses the a.c. node voltage data in the solution result array $VREIM$. It produces an explicit array result for the one or more node numbers specified in the input argument list

[1] $Z{\leftarrow}VREIM[;,NNO;]$
   ∇

Select one or more rows from $VREIM$ by indexing, and return those node voltages for all frequency points in two planes with real and imaginary parts respectively

---

Table 15.6
The function ALL

∇ $Z{\leftarrow}ALL$
   ALL selects all node numbers in a given circuit from 1 to NH and returns these as an explicit result

[1] $Z{\leftarrow}\iota\rho IX$
   ∇

Compute the shape of the node source current vector IX, and convert this to a vector of numbers from 1 up to ρIX, and return this result

15.3 Modifier Functions

All of the functions described in this section, except for the support function *LOG*, are normally called directly by the user as desired. The first three functions *MAG*, *MAGDB*, and *ANGLE* are used with a.c. node voltages, while *LOGF* is used with the frequency vector *F*.

15.3.1 *MAG*, *MAGDB*, and *ANGLE*

The functions *MAG*, *MAGDB*, and *ANGLE* are essentially alternatives, and each is designed to accept, as an input argument, the array of a.c. node voltages produced by *ACVOUT*. The real and imaginary node voltage pairs in this argument are processed to produce respectively their magnitude, their magnitude in dB, or their phase angle in degrees. In each case, the results are formatted in column form and passed on for printing or plotting as described further in the accompanying tables. (Exercise 15.1 also deals with this processing.)

Table 15.7
The function *MAG*

∇ *Z←MAG V*

> The purpose of *MAG* is to produce a result matrix of voltage magnitudes from an input array containing one or more sets of a.c. node voltages

[1]   *Z←⍉(+/[1] V*2)*0.5*
∇

> First compute the square of each term in the input argument array *V*; then add corresponding squared terms on an element-by-element basis along the first dimension, i.e. real with corresponding imaginary terms; then take the square root of each resultant term to obtain the vector magnitude; finally, transpose rows and columns to hold one set of node voltage magnitudes per column

---

Table 15.8
The function *MAGDB*

∇ *Z←MAGDB V*

> *MAGDB* accepts the same input argument as the function *MAG*. It produces a matrix result containing 20 times the logarithm of the voltage magnitudes in dB

[1]   *Z←MAG V*

> Call *MAG* to produce the node voltage magnitudes

[2]   *Z←20×10⍟Z*
∇

> Convert these magnitudes to dB and return them as an explicit result

---

15.3.2 *LOGF* and *LOG*

The function *LOGF* may be called by the user to produce the logarithm of each element in the frequency vector *F*. This is typically used to generate X-axis data for a standard log–log plot as illustrated in Chapter 11. All of the internal processing in *LOGF* is actually carried out by the function *LOG*. Following are sample results produced by *LOGF*:

```
        F←1E2 1E4 1E6 1E8
        LOGF
   2  4  6  8
```

In accordance with the sequence specified in Section 11.5 and Table 15.3, *LOGF* (or *F*) must appear immediately to the right of *PRINT* or *PLOT* in the display argument list. It is also necessary to include the symbol ',' immediately to the right of *LOGF* (or *F*). This is the lamination operator which arranges the frequency data in *F* (or that produced by *LOGF*) as a column in front of the

Table 15.9
The function *ANGLE*

∇ *ANGLV* ←*ANGLE V*; *RE*; *IM*; *K*; *KMAX*; *ANG*

The purpose of *ANGLE* is to produce a matrix having one or more sets of a.c. node voltage phase angles, arranged in column format by selected node number. The input argument is the same as for the function *MAG*

[1]   *RE* ←,*V*[1;;]
[2]   *IM* ←,*V*[2;;]

For convenience in the subsequent processing, separately extract the real (plane 1) and imaginary (plane 2) parts of the a.c. node voltages in the input argument. Convert each to vector form

[3]   *KMAX* ←ρ*RE*

Compute the shape of the *RE* vector in order to ascertain how many times to proceed through the subsequent calculation loop

[4]   *ANG*←ι0
[5]   *K*←1

Set up *ANG* to be an empty vector to permit subsequent catenation of computed results: initialize loop counter *K* to 1

[6]   *LP*: *ANG*←*ANG*, *ZATN RE*[*K*],*IM*[*K*]

For the current value of *K*, select corresponding real and imaginary node voltage parts and compute the vector angle in radians per second with the function *ZATN*; store this angle in *ANG* by catenation

[7]   →*LP*×ι *KMAX*≥(*K*←*K*+1)

> Check if all real and imaginary node voltage pairs have been processed, and if not, increment the counter and loop back to the previous line to repeat the angle calculation with the next pair; otherwise, execute the next line

[8] $ANG \leftarrow ANG \times 180 \div O1$

> Convert all of the angles in the vector ANG to degrees (in the range ± 180)

[9] $ANGLV \leftarrow \lozenge ((1 \uparrow \rho V[1;;]), \bar{\ }1 \uparrow \rho V[1;;]) \rho ANG$
> ∇

> Convert the vector ANG into a matrix having the same number of rows and columns as either plane of the input array argument V; then transpose rows and columns so that all of the phase angles for the first selected node appear in column one, and occupy one row per frequency point used in the a.c. solution

column matrix generated by MAG, or $MAGDB$, or $ANGLE$. Thus, the resultant matrix passed on to $PRINT$ or $PLOT$ has the required column format, with the (optionally modified) independent frequency data in column 1, and the modified node voltage data in columns 2 to 7 as selected. Further details are contained in the accompanying boxes. (See Exercise 15.1 for an illustration of these concepts.)

Table 15.10
The function $LOGF$

> $\nabla Z \leftarrow LOGF$
> The purpose of $LOGF$ is to compute the logarithm of each of the frequencies in the frequency vector F, and return them as an explicit result. The frequency vector F is used as a global input variable

[1] $Z \leftarrow LOG \ F$
> ∇

> The function LOG produces the desired logarithmic result vector using F as the input argument

Table 15.11
The function LOG

> $\nabla Z \leftarrow LOG \ F$
> This function computes the logarithm to the base 10 of one or more entries in the input argument. In this function the input argument F is a local variable

[1] $Z \leftarrow 10 \circledast F$
> ∇

> Use the in-built log operator to compute the desired logarithmic result

15.4 Supplementary Library Functions

The functions described in this section are support routines typically provided by a central APL function library. The discussion here will be limited to user-oriented features, and will not deal with the details of their internal design and operation.

15.4.1 *ZATN*

An example of the use of *ZATN* is contained on line 6 of the function *ANGLE* presented in Section 15.3. *ZATN* accepts one set of real and imaginary numbers as the input argument and returns the corresponding vector angle in radians as an explicit result. The APL arctangent operator is utilized for the processing in *ZATN*.

15.4.2 *VS* and *AND*

The functions *VS* and *AND* may be used to format data into column form as required by *PRINT* and *PLOT*. Both functions require two arguments which may be either equal length vectors or a vector and a column matrix. Example 15.1 illustrates various ways of using these functions with convenient test data, where the data in *X* must reside in column 1 after invoking the function *VS*. See also line 17 of *TRANSDC* in Section 13.2 for another example of the use of *VS*.

Example 15.1   Illustrating uses of *VS* and *AND*

```
          X ←11   12  13  14  15    Define convenient test data;
          Y1←21   22  23  24  25    each is defined as a vector in
          Y2←31   32  33  34  35    row format
          Y3←41   42  43  44  45

          Y1  VS  X                  VS causes the contents of X to be in
    11  21                           column 1, and the contents of Y to be
    12  22                           in column 2, this is compatible with
    13  23                           PRINT and PLOT
    14  24
    15  25

          Y1  AND  Y2                AND produces a column matrix in the
    21  31                           order of the arguments i.e.,  Y1
    22  32                           followed by Y2
    23  33
    24  34
    25  35
```

```
        Y1 AND Y2 VS X
  11   21   31
  12   22   32
  13   23   33
  14   24   34
  15   25   35
```

X data is in column 1, and the Y data in columns 2 and 3 is in sequential order

```
        Y1 AND Y2 AND Y3 VS X
  11   21   31   41
  12   22   32   42
  13   23   33   43
  14   24   34   44
  15   25   35   45
```

X data is still in column 1, and Y data is in columns 2, 3 and 4 in sequential order

The same result is given by the following arrangement.

```
        Y1 AND (Y2 AND (Y3 VS X))

  11   21   31   41
  12   22   32   42
  13   23   33   43
  14   24   34   44
  15   25   35   45
```

These results are identical, but the form of use gives greater insight into how the arguments are interpreted

15.5 Postprocessing

In this section we introduce the subject of postprocessing and illustrate several useful postprocessing applications in the context of *BASECAP*. Postprocessing normally requires access to one or more of the data variables generated by a regular circuit solution. In some cases, postprocessing might involve simply displaying selected data in a special format for convenience. In other cases, more extensive processing of the data is required, for example, to yield new information on circuit behavior not directly provided by the standard processing facilities in the original package.

The development of new postprocessing capabilities normally requires additional programming to achieve the intended purpose: one or more special APL functions in the context of *BASECAP*. To do this effectively, one needs a moderately detailed understanding of the internal package design and operation, and a good knowledge of the data structures used. Such postprocessing is often highly advantageous, however, as it enables the knowledgeable user to tailor a package to accommodate special preferences or requirements. It will be recalled that this was an important consideration in the design and development of *BASECAP* (see Section 10.2.3).

In the subsequent material, we present a number of *BASECAP* post-processing examples of varying complexity, to illustrate the principles involved. In each case, the postprocessing is embodied in an APL function, resulting in a convenient and useful processing tool. For each function, we begin by describing its purpose and the nature of the processing involved. We then illustrate what each function does and how it is used, with sample test data from an actual *BASECAP* circuit solution.

The following postprocessing examples are discussed and illustrated in the remainder of this section:

(1)   a function *IRDC* which computes and displays the d.c. current in a resistor;

(2)   a function *VGAINMAG* which computes and displays the magnitude of the a.c. voltage gain;

(3)   a function *VGAINANG* which computes and displays the phase angle of the a.c. voltage gain;

(4)   a function *IDIODE* which computes and displays the d.c. current in a diode;

(5)   a function *BJTBIAS* which presents a summary of the d.c. bias conditions for all bipolar transistors in the circuit.

For convenience, detailed listings of these functions are included in Appendix 6.

15.5.1   Resistor d.c. Current: *IRDC*

Following a regular d.c. node voltage solution of a circuit with *DCSOLVE*, it is a very simple matter to compute the magnitude and direction of the d.c. current flowing in any given resistor using Ohm's law (in conventional notation)

$$I = \frac{V_1 - V_2}{R}$$

where V_1 and V_2 are the resistor node voltages and *R* is the resistance value. The function *IRDC* was designed to carry out this calculation by accessing the d.c. node voltage result vector *VANS.* Three parameters must be specified in the input argument list in the form:

   IRDC   Node 1   Node 2   Resistance (ohms)

The result gives the d.c. current in amperes, with appropriate sign, according to the actual voltages at the two nodes.

The indexing process in this function is slightly complicated by the fact that the circuit reference node is designated as 0, while the normal indexing of a vector in APL begins with 1. We overcome this by placing 0 ahead of the voltage values in the vector *VANS*, and adding 1 to the node number indices. The actual computation is carried out using 'minus reduction'.

$$\nabla IRDC[\Box]\nabla$$
$$\nabla\ \ Z\leftarrow IRDC\ P$$
[1] $Z\leftarrow(-/(0,VANS)[1+P[1\ 2]])\div P[3]$
$$\nabla$$

Example 15.2   **Interacting with** $RDTEST$

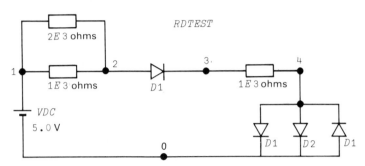

Fig. 15.1

```
      ∇RDTEST[□] ∇
      ∇ RDTEST              Listing for the test circuit RDTEST
[1]      STARTCCT
[2]      VDC 1 0 5
[3]      R 1 2 2000
[4]      R 1 2 1000
[5]      D 2 3 ,D1
[6]      R 3 4 1000
[7]      D 4 0 ,D2
[8]      D 4 0 ,D1
[9]      D 0 4 ,D1
[10]     END
      ∇
      D1                    Diode parameters
1E⁻9 3E⁻12
      D2
1E⁻11 1E⁻11


      RDTEST               Execute the circuit description function
      DCSOLVE                RDTEST and obtain a d.c. solution
CPUSECONDS=2.04
ITERATIONS=4


↓↓DC NODE VOLTAGES 1 TO 4 ↓↓

5 3.44 2.89 0.555          The d.c. node voltage results

      IRDC 1 2 1E3         Using IRDC to check the d.c. currents
0.00155
```

```
        IRDC 1 2 2E3
0.000777
        IRDC 3 4 1E3
0.00233
        IRDC 4 3 1E3
‾0.00233
```

```
        .00155+.000777  The sum of these currents agrees with
0.00233                        the above result
```

15.5.2   A.c. voltage gain: *VGAINMAG* **and** *VGAINANG*

The a.c. node voltage results obtained from an a.c. solution with *ACSOLVE* are stored in real and imaginary form in the array *VREIM*. The functions *VGAINMAG* and *VGAINANG* have been written to access *VREIM*, and to compute and display respectively the magnitude and phase angle of the voltage gain as a function of frequency. One or more sets of input or output nodes may be specified in the left and right input arguments to give one or more sets of gain values. The results are in column form with one row per frequency to be compatible with *PRINT* and *PLOT*. Output units are either volts or degrees. The following statements are typical:

```
30 40 PLOT F,Innode(s)  VGAINMAG  Outnode(s)
30 40 PLOT F,Innode(s)  VGAINANG  Outnode(s)
```

The internal design of these two functions is very similar as their listings show. We make use of the output processing functions *ACVOUT*, *MAG* and *ANGLE* as should be expected, and compute the desired magnitude ratio or phase angle difference accordingly. The reshape operator is used to force the number of input node entries to correspond with the number of output node entries:

```
        ∇VGAINMAG[□] ∇
    ∇ Z←NI VGAINMAG NO; ZO; ZI
[1]   ZO←MAG ACVOUT NO
[2]   NI←(ρNO)ρNI
[3]   ZI←MAG ACVOUT NI
[4]   Z←ZO÷ZI
    ∇
```

```
        ∇VGAINANG[□] ∇
    ∇ Z←NI VGAINANG NO; ZO; ZI
[1]   ZO←ANGLE ACVOUT NO
[2]   NI←(ρNO)ρNI
[3]   ZI←ANGLE ACVOUT NI
[4]   Z←ZO-ZI
    ∇
```

Example 15.3   *Interacting with*   *VGAINMAG* **and** *VGAINANG*

`        R1,R2,RI`	Display parameters to be used for
`2E4 8.2E3 1E3`	*FBTRIPLE*. Refer to Section 11.6 for
`        C1,C2`	the circuit diagram and circuit listing
`0.0001 1E⁻12`	
`        D1`	
`1E⁻9 3E⁻12`	
`        Q1`	
`100 0.1 1E⁻9 1E⁻9 8E⁻11 3E⁻12`	
`        QN`	
`50 0.1 1E⁻9 1E⁻9 8E⁻11 3E⁻12`	
`        ETA`	
`1.5`	
`        TEMP`	
`20`	

`        FBTRIPLE`	Set up the circuit and obtain the d.c.
`        DC SOLVE`	solution
`CPUSECONDS = 7.88`	
`ITERATIONS = 5`	

`↓↓DC NODE VOLTAGES 1 TO 11↓↓`

`0.533 2.09 1.07 0.533 1.58 0.526 1.05 3.64 20 5.33E⁻13`
`        5.34E⁻10`

`        F←10 1E4 1E8`	
`        AC SOLVE`	
`SOLVING FOR 3 FREQUENCY POINTS`	Obtain the a.c. solution for
`CPUSECONDS = 1.87`	three frequencies
`DONE`	

`        PRINT F,MAG ACVOUT 1 3 5 8 10` Display node voltages
`1.00E1  6.77E⁻5 6.20E⁻3 4.02E⁻1 3.95E1  9.99E⁻1`
`1.00E4  6.91E⁻5 6.30E⁻3 4.08E⁻1 4.00E1  9.99E⁻1`
`1.00E8  1.87E⁻2 2.53E⁻2 2.68E⁻2 1.04E⁻1 9.99E⁻1`

Display gains with
respect to node 1
`        PRINT F,1 VGAINMAG  1 3 5 8 10`
`1.00E1 1.00E0 9.16E1 5.94E3 5.83E5 1.48E4`
`1.00E4 1.00E0 9.12E1 5.90E3 5.78E5 1.45E4`
`1.00E8 1.00E0 1.36E0 1.44E0 5.56E0 5.36E1`

`        PRINT F,1 3 5 8 10 VGAINMAG 1 3 5 8 10`
`1E1 1E0 1E0 1E0 1E0 1E0` Multiple gain magnitude results,
`1E4 1E0 1E0 1E0 1E0 1E0` each with the same output and input
`1E8 1E0 1E0 1E0 1E0 1E0` nodes respectively, produce unity
ratios

```
        PRINT F,1 3 VGAINMAG 1 3 5 8
1.00E1 1.00E0 1.00E0 5.94E3 6.36E3
1.00E4 1.00E0 1.00E0 5.90E3 6.34E3
1.00E8 1.00E0 1.00E0 1.44E0 4.10E0
```
Input   nodes   are
repeated in sequence

```
        PRINT F, 8 8 8 8 VGAINMAG 8
1E1  1E0
1E4  1E0
1E8  1E0
```
One result is pro-
duced   for   each
output node

```
     PRINT F,1 VGAINANG 1 3 5 8 10
1.00E1    0.00E0  ⁻1.80E2  ⁻7.41E⁻3  ⁻1.80E2  ⁻9.05E0
1.00E4    0.00E0  ⁻1.83E2  ⁻7.37E0   ⁻1.88E2  ⁻7.55E0
1.00E8    0.00E0   1.20E2   2.36E2   ⁻8.76E1   8.60E1
```

```
     PRINT F,1 3 VGAINANG 1 3 5 8
1.00E1    0.00E0   0.00E0  ⁻7.41E⁻3  ⁻4.80E⁻3
1.00E4    0.00E0   0.00E0  ⁻7.37E0   ⁻4.79E0
1.00E8    0.00E0   0.00E0   2.36E2   ⁻2.07E2
```
These   gain
angle results
follow   the
same pattern
as the gain
magnitude

15.5.3   Diode d.c. current: *IDIODE*

The function *IDIODE* may be used to compute and display the d.c. current in a diode following a d.c. circuit solution with *DCSOLVE*. To obtain the current value in amperes for a given diode, it is only necessary to specify the anode and cathode node numbers in sequence as indicated.

IDIODE NA NC

If two or more diodes exist in parallel with identical anode and cathode node numbers, each such diode current is displayed in its sequential branch number order. A diode current is positive if it flows from NA to NC as shown above. Specifying a diode with the sequence NC NA produces no result.

The majority of the processing in *IDIODE* is used to check for a node number match in the branch connection matrix *M*. This can only be done after each reference node number in *M* is changed to a '0' from the value *NH*. This version of *M* is assigned to *MO*. Once the branch number is obtained, it is used to extract the desired current from the branch current vector obtained with the function *DIODE*, using the branch voltage vector *VBN* as an argument:

```
        ∇IDIODE[□] ∇
     ∇  Z←IDIODE NAC
[1]     M0←M ≠(ρM)ρNH
[2]     M0←M×M0
[3]     H1 ←NAC[1] =M0[1; ]
[4]     H2 ←NAC[2] =M0[2; ]
[5]     H←H1∧H2
[6]     BRI ← ((H≠0)∧(IS ≠0))/ι ρIS
[7]     Z←(DIODE VBN)[BRI]
     ∇
```

Example 15.4 **Interacting with** *IDIODE*

```
         D1
1E⁻9 3E⁻12
         D2
1E⁻11 1E⁻11
         ETA
1.5
         TEMP
20
         RDTEST
         DCSOLVE
CPUSECONDS = 2.07
ITERATIONS = 4
```
Parameters to be used for the circuit *RDTEST* from Example 15.2

```
↓↓DC NODE VOLTAGES 1 TO 4 ↓↓
```

```
5 3.44 2.89 0.555
         IRDC 3 4 1E3
0.00233
```
The d.c. node voltages for *RDTEST*

The current in the 1 kilohm resistor is equal to the diode current as expected

```
         IDIODE 2 3
0.00233
         IDIODE 4 0
0.0000231 0.00231
```
Only two diode currents are produced with the 4-0 node sequence

```
         IDIODE 0 4
⁻1E⁻9
```
This is just the reverse saturation current for the third parallel diode

15.5.4 **Transistor biassing conditions:** *BJTBIAS*

It is important to check the d.c. biassing currents and voltages that result for each transistor in the circuit. Such a summary is conveniently produced by the postprocessing implemented in the niladic function *BJTBIAS*. This function is quite comprehensive, and the results it produces are labelled and neatly formatted. For each transistor *BJTBIAS* produces:

the transistor type and sequence number;
the emitter, base, and collector node numbers;
the base-to-emitter voltage and the collector-to-emitter voltage, each
   in volts;
the emitter and collector currents in milliamperes.

Both the emitter and collector currents are considered to be positive if they actually flow into the transistor. Thus for an NPN transistor in the active region. I_E will be negative and I_C will be positive. For a PNP transistor in the active region, I_E will be positive, and I_C negative.

The key processing steps in *BJTBIAS* are based directly on the simplified Ebers–Moll model presented earlier in Fig. 13.7, where for an NPN transistor (in conventional notation)

$$I_C = \alpha_F I_E' - I_C'$$

and

$$I_E = \alpha_R I_C' - I_E'$$

In this function, the primed terms are shown as *IEP* and *ICP* while the alpha terms are designated *AF* and *AR*. The various current components are calculated with the diode function *DIODE*, and extensive use is made of the node numbers and alpha terms in the transistor storage table *Q*. The use of *TYPE* is a convenient way to adjust signs for NPN or PNP transistors. We again use 'minus reduction' to compute *VBE* and *VCE* from the node voltage vector *VANS* (lines 18 and 19):

```
        ∇BJTBIAS[□] ∇
     ∇  Z←BJTBIAS;IEP;ICP;AF;AR;IE;IC;VBE;VCB;PP;T;AA;TT
[1]     ' '
[2]     → 0×ι (1↑ρQ) =0
[3]     ' '
[4]     'TYPE QNO ENODE  BNODE  CNODE VBE VCE IE(MA.)IC(MA.)'
[5]     QNO←1
[6]     T←' '
[7]     Z←' '
[8]   L:IEP←(DIODE VBN)[Q[QNO;2]]
[9]     ICP←(DIODE VBN)[Q[QNO;2]+1]
[10]    AF←Q[QNO;6]
[11]    AR←Q[QNO;7]
[12]    IE←TYPE[QNO]×(-IEP)+AR×ICP
[13]    IC←TYPE[QNO]×(-ICP)+AF×IEP
[14]    AA ←'NPN'
```

```
[15]    →LL×ι(TYPE[QNO]=1)
[16]    AA←'PNP'
[17]    LL:T←T,AA
[18]    VBE←-/(VANS,0)[Q[QNO;  4  3]]
[19]    VCE←-/(VANS,0)[Q[QNO;  5  3]]
[20]    Z←Z,QNO,((ιNH-1),0)[Q[QNO;3  4  5]],VBE,VCE,(1000×IE),(1000×IC)
[21]    →L×ι(QNO←QNO+1)≤1↑ρQ
[22]    TT←((1↑ρQ),3)ρT
[23]    PP←((1↑ρQ),8)ρZ
[24]    Z←5  0  8  0  7  0  7  0  12  3  8  3  8  3  8  3  ⍕PP
[25]    Z←TT,Z
        ∇
```

Example 15.5   Interacting with *BJTBIAS*

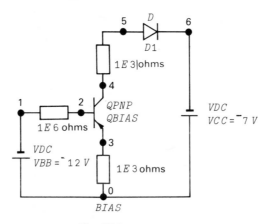

Fig. 15.2
The test circuit $BIAS$

```
        ∇BIAS[□]∇
    ∇   BIAS                        Listing  for  the  test  circuit
[1]     STARTCCT                    (Fig. 15.2) BIAS
[2]     VDC  1  0  ,VBB
[3]     R  1  2  1000000
[4]     QPNP  3  2  4  ,QBIAS
[5]     R  3  0  1000
[6]     R  4  5  1000
[7]     D  5  6  ,D1
[8]     VDC  6  0  ,VCC
[9]     END
    ∇
        D1
1E⁻9  3E⁻12                         Device and circuit parameters used for
        ETA                         the d.c. solution
1.5
        TEMP
20
```

```
        QBIAS
100  1  1E⁻11  1E⁻11  1E⁻12  1E⁻12
        VBB
⁻12
        VCC
⁻7
```

```
        BIAS
        DCSOLVE
CPUSECONDS = 0.867
ITERATIONS = 6
```
Set up *BIAS* and obtain the d.c. solution

```
↓↓DC NODE VOLTAGES 1 TO 6↓↓
```
The d.c. node voltages

```
⁻12 ⁻1.736 ⁻1.037 ⁻5.448 ⁻6.474 ⁻6.999
```

```
        BJTBIAS
```
Call *BJTBIAS* to display the PNP transistor bias conditions

TYPE	QNO	ENODE	BNODE	CNODE	VBE	VCE	IE(MA.)	IC(MA.)
PNP	1	3	2	4	⁻0.700	⁻4.411	1.037	⁻1.026

```
        IRDC  0  3  1E3
0.001037
        IRDC  4  5  1E3
0.001026
        IDIODE  5  6
0.001026
        IDIODE  6  5
```
Check the transistor d.c. currents with *IRDC* and *IDIODE*

Next we utilize the circuit *FBTRIPLE*

```
        FBTRIPLE
        DCSOLVE
CPUSECONDS = 7.9
ITERATIONS = 5
```
Testing *FBTRIPLE* again. See Section 11.6 for circuit details

```
↓↓DC NODE VOLTAGES 1 TO 11↓↓
```

```
0.533 2.09 1.07 0.533 1.58 0.526 1.05 3.64 20 5.33E⁻13 5.34E⁻10
```

```
        BJTBIAS
```
Display the transistor d.c. bias conditions with *BJTBIAS*

TYPE	QNO	ENODE	BNODE	CNODE	VBE	VCE	IE(MA.)	IC(MA.)
NPN	1	0	1	3	0.533	1.065	⁻1.285	1.272
NPN	2	4	3	5	0.533	1.045	⁻1.265	1.241
NPN	3	7	5	8	0.526	2.591	⁻1.063	1.042

```
        IDIODE  7  6
0.00106
        IDIODE  6  0
0.00106
```
The emitter current for *Q*3 agrees with the current in the series diode (within the printing precision, ie. 1.063 versus 1.06 mA.)

Exercises

15.1   Display and check the resultant data produced after each main processing step in the execution sequence specified by the following line taken from Table 15.3:

$$F, MAG\ ACVOUT\ NOUT$$

Use the circuit LPF from Section 11.6 for illustration purposes, assuming

$$F \leftarrow 1E2\ 1E3\ 1E4\ 1E5\ 1E6\ \text{and}\ NOUT \leftarrow 1$$

15.2   Refer to the postprocessing function $IDIODE$ described in Section 15.5.3. Investigate how this function behaves with the circuit $FBTRIPLE$ from Section 11.6, if

   (a)   an invalid node number is used as an input argument e.g. the number 50

   and

   (b)   an odd number of nodes is specified as inputs.

15.3   Modify the postprocessing function $BJTBIAS$ to include the transistor current gain (Beta forward) in the data displayed.

15.4   Develop and test the necessary postprocessing algorithm(s) which, when used following a regular d.c. circuit solution with $BASECAP$ will compute and display the total d.c. power supplied to the circuit from all d.c. sources.

15.5   Write an APL function $ZATN$ to carry out the processing specified in Section 15.4.1. Demonstrate that your function works correctly using appropriate test data, including actual solution results from a circuit example in Section 11.6. $ZATN$ should have the following syntax:

$$R \leftarrow ZATN\ P$$

15.6   Write your own function VS as specified in Section 15.4.2. Demonstrate the correct operation of VS with the test data from Example 15.1. Use the following syntax for VS

$$R \leftarrow P\ VS\ Q$$

Solutions

15.1 We first display the contents of *VREIM* and *F*, and then display the required processing results.

```
    VREIM
1.00E0     1.00E0     9.99E¯1     9.99E¯1     9.99E¯1  The a.c. node voltage
9.90E¯1    5.00E¯1    9.88E¯3     9.98E¯5     9.98E¯7  result array
9.90E¯1    5.00E¯1    9.88E¯3     9.98E¯5     9.98E¯7

¯9.90E¯5   ¯5.00E¯4   ¯9.88E¯5    ¯9.98E¯6    ¯9.98E¯7
¯9.91E¯2   ¯5.00E¯1   ¯9.89E¯2    ¯9.99E¯3    ¯9.99E¯4
¯9.91E¯2   ¯5.00E¯1   ¯9.89E¯2    ¯9.99E¯3    ¯9.99E¯4
    F
100 1E3 1E4 1E5 1E6                       The solution frequencies

    ACVOUT 1
1.00E0     1.00E0     9.99E¯1     9.99E¯1     9.99E¯1  Specifying      node    1
                                                      extracts  row  one  from
¯9.90E¯5   ¯5.00E¯4   ¯9.88E¯5    ¯9.98E¯6    ¯9.98E¯7  each plane of VREIM

    MAG ACVOUT 1
1                                        The required voltage magnitudes for node
1                                        1  at all five frequencies
0.999
0.999
0.999

    F,MAG ACVOUT 1
1.00E2     1.00E0                        The resultant column matrix has the fre-
1.00E3     1.00E0                        quencies in column 1
1.00E4     9.99E¯1
1.00E5     9.99E¯1
1.00E6     9.99E¯1
```

15.2 *DIODE* is tested below:

```
    IDIODE 4  5  6  7  6  0                      IDIODE 4 50

    IDIODE 3                                     IDIODE 50 100
RANK ERROR
IDIODE[3]  H1 ←NAC[1] =M0[1;]                        IDIODE 50
           ∧                             RANK ERROR
    IDIODE 4  0  5                       IDIODE[3]  H1 ← NAC[1] =M0[1;]
0.001294                                            ∧
    IDIODE 4  0
0.001294
```

The single argument case cannot be accommodated without error.

16

Summary and Additional Projects

CACD techniques have become widely accepted and are now being used extensively in industry and in universities. Various examples and exercises (particularly in Chapter 11) have shown that a CACD package such as *BASECAP* is a powerful and useful circuit solution tool. It can give the circuit designer a more detailed understanding of the performance of a proposed circuit by testing its operation over a wide range of operational and environmental conditions. Some packages also help the designer optimize circuit performance or increase manufacturing yields when using toleranced components. Thus, the appropriate use of CACD techniques can, at relatively low cost, give the designer a very high degree of confidence that his circuits will meet specified requirements when actually fabricated.

In this text, the authors have endeavored to provide readers with an interesting and meaningful introduction to this exciting field. An understanding of the material on circuit theory and algorithms in Part 1, and on package design in Part 2, combined with the experience and insight gained from using the CACD package *BASECA* constitutes an excellent foundation for further CACD work. Examples include the evaluation and use of CACD packages as tools for the analysis and design of new electronic circuits, and the design and development of improved CACD algorithms, methods, and packages.

In this chapter, we first review *BASECAP*'s main features and limitations, and then introduce a number of projects which can be undertaken by interested readers. These projects are intended to help readers become more familiar with recent developments and current research activities in CACD, and to enable them to extend their learning in practical and useful ways. The suggested projects will also greatly enhance the capabilities and usefulness of *BASECAP* as a more comprehensive circuit analysis and design package.

16.1 Features and Limitations

The major characteristics of *BASECAP* obviously reflect the particular bias of the authors, and relate directly to the major overall design objectives that have guided its development. To a very large degree, the use of the APL language has had a significant impact on the character of *BASECAP* and has contributed to many of its strengths. As a general-purpose circuit solution tool, *BASECAP* is relatively powerful, highly interactive, and easy to use. The utilization of APL for its implementation has resulted in a package which is relatively simple, and

remarkably free from certain complexities and restrictions found in many other
CACD packages (see the discussions in Sections 10.2 and 11.7.3). The APL system provides extensive support for *BASECAP* users with such facilities as error
messages, diagnostic controls, libraries, etc. The use of a simple, modular, open
structure allows easy internal access to the package for instructional purposes,
and greatly facilitates the incorporation of enhancements and new features.

The requirement to have some knowledge of APL in order to use *BASECAP*
will constitute a limitation for some users. Of course, the open program structure could be replaced by a closed structure, with the advantage that virtually
no knowledge of APL would then be required. In this case the user would
communicate with the CACD solution routines and the APL system through a
user-friendly command-dialogue interface tailored for the circuit analysis tasks
to be performed. Various editing and help features could also be incorporated
for additional convenience, and for the benefit of novice users in particular.

In practice, some users may prefer the comfort and support provided by a
closed system. Others, such as experienced users and those familiar with APL,
may prefer the present type of open system, finding that it is far less restrictive,
and that it readily gives them access to the full power and flexibility of the APL
system to supplement the *BASECAP* solutions. Cost is another important factor,
since closed systems are much more time-consuming to produce, and also more
expensive in terms of computing resources to use. Ideally, both types of systems
should be available to accommodate the preferences of various users.

One of the most serious limitations in *BASECAP* is the relatively simplified
models used for devices such as the bipolar transistor and the operational
amplifier. Clearly, more accurate models are required for these devices, and
models for additional devices should be included. Examples of other important
features and capabilities which are not available in the *BASECAP* package are
time-domain analysis, subcircuit handling, tolerance analysis and design, noise
analysis, and sensitivity analysis. The compact, modular, open structure of
BASECAP makes it entirely feasible for serious users to incorporate one or more
of these features into the present package as suggested in the projects in the next
section.

16.2 Suggested Student Projects

In this section we present a number of possible modifications and extensions
which may be added to the *BASECAP* package. In part, the material in this
section constitutes a 'wish list' of additional features and capabilities which the
authors believe are desirable or essential for a circuit analysis package to
have.[†] This material should be of interest to instructors and students as a source

---

[†] Readers may be interested to learn of a related APL-based commercial circuit analysis and design
package developed by the authors. This package is designed to emphasize features such as
(a) sensitivity analysis and (b) tolerance analysis and design, in order to provide more effective
support for the electronic circuit design process as well as for analysis. Called SCADS, it will be widely
available on the international APL time-shared computer network of I.P. Sharp Associates, Toronto,
Canada.

of ideas for useful 'hands-on' learning projects which are compatible with the study objectives of this text.

(1)   Improved bipolar transistor model

The main emphasis in this project is to develop a more accurate representation of the bipolar transistor to account for a number of secondary effects. It may be convenient to utilize the existing model and add such effects as base-width modulation (Early effect); parasitic resistances in the base, emitter, and collector regions; nonlinear small-signal junction capacitances, and their dependence on d.c. bias currents and voltages; provision for a separate temperature and eta for each transistor (specified as individual device parameters); the dependence of beta on d.c. current levels and temperature; the dependence of junction saturation leakage currents on temperature; and substrate parasitics. Additional device parameters will be required for the new model. However, default values can be utilized so the user does not have to specify all of them unless different values are required.

An important concern in this and other advanced model implementations is the requirement for additional internal nodes and branches, for example, to represent parasitic base and emitter contact resistances. A means should be developed in *BASECAP* for handling such internal nodes and branches in a way that makes them transparent to the user. This is a nontrivial task, and two possibilities are suggested.

One is to use separate additional data storage variables, and incorporate these nodes and branches into the solution matrices (for example *SLOPE*) as required. Temporary numbers can be assigned initially to internal nodes, with final node numbers being assigned automatically in sequence following the user-assigned node numbers after all the elements have been processed (see the function *END*). The d.c. and a.c. node voltages are normally computed for all nodes, both internal and user-assigned. If desired, the normal display of results may be restricted to only the user-assigned nodes and branches, but results for internal nodes and branches should be obtainable by special request for diagnostic purposes.

A second approach to handling model internal nodes and branches automatically is to use a form of preprocessing to prepare data for the existing *BASECAP* solution functions. This is part of a more general subcircuit and model processing capability discussed later in item (8).

(2)   Improved diode model

Produce a more accurate diode model for use in *BASECAP*. Include various second-order effects as discussed in (1) above.

(3)   Models for the JFET and MOSFET

Implement a full-range, nonlinear d.c. and small-signal a.c. model for each of these devices in *BASECAP*. Nonlinear d.c. equations will be required, with appropriate testing to ensure that the correct d.c. relationship is used in each different range of operation depending on terminal conditions. These equations must be incorporated into the Newton–Raphson d.c. solution process in much the same way as the equations for the diode and bipolar transistor (see Chapter 13). Small-signal junction capacitances and parasitic source and drain resistances should be included in the model. The user should be able to specify parameters for these effects, along with other parameters such as pinch-off voltage or threshold voltage, transconductance, channel length modulation parameter, junction contact potentials, and junction saturation currents.

(4)   *Improved op amp model*
Many variations are possible here, but a simple and useful representation of the op amp for use in *BASECAP* is one that incorporates internal frequency compensation, and allows the user to specify the d.c. gain and break frequency. A minimum of one additional node and one or more internal branches are required to implement the frequency-dependent gain (see the discussion under (1) above). Other possible improvements include d.c. offset effects, a multiple break-frequency gain characteristic, d.c. power supply connections, output signal voltage limiting, and output current limiting.

(5)   *Models for other devices*
Many other devices are very useful in various electrical and electronic circuits. The development of circuit simulation models for *BASECAP* would follow approaches similar to those described above. Possibilities include the following: zener diode, transformer, crystal filter, SCR, UJT, and various nonlinear controlled sources.

(6)   *Transient analysis*
Provide *BASECAP* with the capability for analyzing circuits in the time domain. Use one of several methods of numerical integration such as backward Euler or trapezoidal (see Chapter 8). Automatic time-step control is a very useful feature, and several new time-dependent current and voltage sources should be implemented. Examples are sine, pulse, step and ramp. Appropriate data storage and display facilities will be required to generate plots of one or more node voltages versus time over a user-specified interval.

(7)   *Complex arithmetic*
Recent versions of the APL language can now accommodate numbers in complex form and carry out the necessary complex arithmetic operations automatically. See for example, APL2 from IBM, and Sharp APL (and Sharp APL/PC for the IBM personal computer) from I.P. Sharp Associates Ltd.

Redesign the a.c.-related data structures in *BASECAP* to accommodate complex variables, and modify the processing (for example in *REALSOLV*) to take advantage of the new complex arithmetic capabilities. An important example is the new matrix divide operator which directly produces a complex matrix of node voltages, given complex arrays for both the currents and the admittances. The use of complex arithmetic can be expected to have several advantages, including simplicity, enhanced operating speed, and more effective space utilization in the workspace.

(8) *Subcircuit and generalized modelling facility*

A subcircuit is a circuit containing *BASECAP* elements and other subcircuits, which can be separately defined and tested, and is referenced in a manner similar to a regular device model. With this facility, all elements in a subcircuit are effectively inserted into another circuit by means of a single subcircuit statement which specifies the node interconnections. An example of a useful subcircuit is a nand gate consisting of a number of transistors. Nand gate subcircuits can be interconnected to form a two-bit adder, and two-bit adders, in turn, can be connected to form a four-bit adder subcircuit, and so on.

Implement such a subcircuit facility as a front-end processor for the existing *BASECAP* package. It is feasible to permit subcircuit nesting to any depth, and to keep track of all the internal node numbering assignments using a recursive approach which effectively expands each subcircuit into its most fundamental components. The final result of the subcircuit preprocessing should be an expanded circuit description function which contains only element description function statements acceptable to *BASECAP*.

A complex model of a device can also be defined as a subcircuit by the user (though only existing elements can be used). A generalized set of model parameters can be defined in such cases to allow the user to specify different internal parameter values each time the model or subcircuit is used. Subcircuit interconnections can become very complex, but such a comprehensive subcircuit and modelling facility is very powerful and simple to use.

(9) *Text editor*

Write a simple line-based text editor, or adapt an existing one, to facilitate the creation and modification of circuit data for use with *BASECAP*. It may be convenient to process and store such data in the form of character strings, and then create the necessary function format to describe the circuit to *BASECAP*, using APL operators provided for this purpose. To manipulate the data, various editing commands must be supported, such as list, delete, input, change, and find.

(10) *User-friendly interface*

Implement a comprehensive command-based, or menu-based, user-friendly control interface for *BASECAP*. By means of simple commands or menu selections, the user should be able to input and modify circuit and model data, select or set solution control and display parameters, obtain d.c. and a.c. solutions, display results, obtain various levels of on-line help, etc. This description is for a closed type of interface structure which will require an extensive amount of programming to implement fully. Partial implementations are also useful, however.

A menu-based user-interface is very convenient for use with video display terminals. To be effective, the data communication path between the terminal and the computer should operate at 1200 baud or faster. Menus are also appropriate with microcomputers, such as the IBM/PC running Sharp APL/PC, or APL*PLUS/PC from STSC Inc. Note that menu-based arrangements which control the display screen may be somewhat system dependent. Command-based dialogues, on the other hand, are generally system independent, and are usable at relatively slow data communication rates.

(11) Improved plot facilities

Many APL computing systems have libraries with advanced plotting or graphics functions. These can be used to produce high-quality line plots on graphics terminals or plotting devices. If such facilities are available, it is relatively easy to interface them to *BASECAP.* The IBM product called GRAPHPAK is an example.

(12) Sparse matrix solutions

Implement a sparse matrix solution algorithm in *BASECAP* to achieve greater solution efficiency with very large circuits. Indications are that the APL matrix divide operator will be more efficient for solutions involving less than about 30 simultaneous equations. A test should be used to select the most efficient solution method according to matrix size. See Chapter 6 for additional information.

(13) Sensitivity analysis

Sensitivity calculations can provide valuable information to the circuit designer. They can show, for example, which components most significantly affect a critical node voltage. Refer to Chapter 9 for detailed information, and implement a differential sensitivity algorithm for use in *BASECAP.*

(14) Noise analysis

Investigate the use of the adjoint method, and implement a noise calculation facility in *BASECAP.* Include representations for all important sources of noise in a circuit, including resistors and semiconductor devices. Provide facilities for the computation and display of meaningful measures of circuit noise performance, such as plots of equivalent input noise voltage versus frequency, and noise figure versus frequency.

(15) AC tolerance analysis

Design and implement a system with *BASECAP* to estimate the manufacturing yield for a circuit by means of a Monte Carlo analysis technique. Tolerance limits must be specified for selected component values, and pass/fail specifications are required for a designated performance parameter such as voltage gain. The user should be provided with an estimate of the accuracy of the yield calculation, and the opportunity should exist for the user to select the number of samples, and to add to this number if he wishes to refine the yield estimate with additional calculations. Also include facilities for producing histograms of passing and failing responses for a toleranced parameter (Spence, 1984).

(16) AC tolerance design

This project extends the methods implemented above for tolerance analysis. Additional processing is required to compute new nominal values for designated components based on a knowledge of the distribution of passing and failing samples from the Monte Carlo solutions. In this process known as design centering, the new nominal component values are chosen to increase the manufacturing yield. An approach such as the centers-of-gravity method can be used to implement the required processing. Dramatic improvements in yield are possible with a relatively small number of design centering iterations using these techniques (Soin & Spence, 1980 and Spence 1984).

┌─────────┐
│ PART   │
│   │
│ **3** │
│   │
└─────────┘

APPENDICES

───────────

APPENDIX

1

A Sample APL Terminal Session with Annotation

Perusal of this session – or, better still, its repetition at an APL interactive terminal – may help the newcomer to gain sufficient understanding of the APL notation to be able to follow the foregoing text. This session should not be regarded as concise and extensive reference material on the APL notation: that is provided by Appendix 2.

What the user types is automatically indented. Results returned by the computer begin at the left margin. For the first two entries, activation of the carriage return key (**cr**) is indicated explicitly. Thereafter it should be understood that the key is activated at the end of each line typed by the user.

Actual terminal copy	Annotation
	Scalars
$\quad$ 4.5+6.3 **cr**	Scalar addition, multiplication,
10.8	division and subtraction.
$\quad$ 2×.5 **cr**	
1	
$\quad$ 8÷⁻16	The symbol for subtraction (-) is
⁻0.5	different from the negative sign (⁻)
$\quad$ 5-11	which denotes a negative value.
⁻6	
$\quad$ 4=5	
0	
$\quad$ 3≥1	A logical proposition (=, ≥, <, etc.)
1	is either true (1) or false (0).
$\quad$ 16<.1	
0	
$\quad$ 1∧1	The logic functions AND, OR and
1	NOT.
$\quad$ 0∨1	
1	
$\quad$ ∼0	
1	

```
      3⌈ 6
6
      2*3
8
      *1
2.718281828
      |‾4.2
4.2

      X← 4



      X
4
      □←X←7

7
      X+23
30
      2×5+1
12
      (2×5)+1
11

      2÷3
0.6666666667
      )DIGITS 3
WAS 10
      2÷3
0.667
```

Maximum of two numbers.

Raise to a power.

Exponential.

Magnitude.

Assignment of a value to a named variable (computer returns no result; a definition is simply remembered).

Interrogation of the value of X

An alternative means of defining and then displaying the value of a variable.

Variables having values can be used as the arguments of APL functions.

APL expressions are executed from right to left, unless precedence is indicated by parentheses.

Control of the number of significant digits to which all numerical values will subsequently be displayed.

Vectors

```
      2 7 1+5 3 ‾2
7 10 ‾1
      1 2 3×4 5 6
4 10 18
      1 1 0 1∧ 1 0 1 1
1 0 0 1

      X←ι 5
      X
1 2 3 4 5
      □←Y←2×ι 5
2 4 6 8 10
      X+Y
3 6 9 12 15
```

The primitive functions defined above are extended, element by element, to vectors.

One or more spaces serve to separate the elements of a vector.

The 'iota' function generates a vector of integers.

$+/X$ 15	Reduction of a vector to a scalar (the operation to the left of $/$ is placed between each element of the vector appearing to the right of $/$).
$\vee/0\ 1\ 0\ 0\ 0$ 1	
$\lceil/Y$ 10	
$Y[2]$ 4	Indexing a vector.
ρX 5	Dimension of a vector.
$2\uparrow X$ 1 2	Take the first two elements of the vector X.
$S\leftarrow{}^-1\uparrow X$ S 5	Take the last element of X and call it S.
${}^-2\downarrow Y$ 2 4 6	Drop the last two elements of Y

Matrices

$M\leftarrow3\ 3\rho\iota9$ M 1 2 3 4 5 6 7 8 9	A 3-row, 3-column reshape (ρ) of the vector of integers from 1 to 9.
$\square\leftarrow N\leftarrow3\ 3\rho\iota5$ 1 2 3 4 5 1 2 3 4	Vector to right of ρ is repeated sufficiently to form a matrix of the required dimension.

$M+N$
  2 4 6
  8 10 7
  9 11 13
	Primitive functions defined for scalars extend, element by element, to matrices.

$M-N$ 0 0 0 0 0 5 5 5 5	
$M[2;3]$ 6	Indexing a matrix to get an element, row or column.
$M[2;]$ 4 5 6	
$M[;3]$ 3 6 9	
ρM 3 3	The dimension of a matrix.
$1\ {}^-1\downarrow N$ 4 5 2 3	Drop first row and last column of N.

```
        ⍉M
1  4  7
2  5  8
3  6  9
```
Matrix transpose (○ backspace \ )

```
        M+.×N
15  21  17
36  51  41
57  81  65
```
Ordinary matrix product. Operator +.× signifies element-by-element multiplication of rows of M with columns of N, followed by sum reduction over these products.

```
        M⌈.+N
  6   7   7
  9  10  10
 12  13  13
```
The *same* rule applies if other primitive functions replace + and × in the operator +.× ('generalized inner product')

```
        ⌹N
 ‾3.4    ‾.2   2.6
  2.8     .4  ‾2.2
 ‾.4    ‾.2    .6
```
Matrix inversion.

```
        X←10 20 30
        Y←(⌹N)+.×X
        Y
40 ‾30 10
```
Solution of the simultaneous linear equation $X←N+.×Y$

```
        Y←X⌹N
        Y
40 ‾30 10
```
Alternative expression.

```
        1 2 3∘.×1 2 3
1  2  3
2  4  6
3  6  9
```
Outer product of two vectors, i.e., products of all pairs of elements. Vectors can be of unequal length.

```
        1 2 3∘.=1 2 3
1  0  0
0  1  0
0  0  1
```
Same rule applies if another primitive function replaces × in the operator ∘.× ('generalized outer product')

Higher-dimensional arrays

```
        3 2 4ρ⍳24
 1   2   3   4
 5   6   7   8

 9  10  11  12
13  14  15  16

17  18  19  20
21  22  23  24
```
Definition of a three-dimensional array. Vector to the left of ρ defines dimension of array.

Function definition

$\nabla R \leftarrow UNIT\ N$ **cr**

Definition of a function, named $UNIT$, for the generation of a unit matrix of any size.

[1] $R \leftarrow (\iota N) \circ . = \iota N\ \nabla$

After carriage return, computer types [1].

    $UNIT\ 3$

The function $UNIT$ can be used just like any other APL function.

```
1 0 0
0 1 0
0 0 1
```

    $4 \times UNIT\ 2$

```
4 0
0 4
```

$\nabla\ R \leftarrow A\ BOX\ N$

[1] $R \leftarrow A \times (\iota N) \circ . = \iota N\ \nabla$

Defined functions can take one argument, on the right (as above), or two arguments.

    $S \leftarrow 71\ BOX\ 3$
    S

```
71  0   0
 0 71   0
 0  0  71
```

2

Familiarity with APL

It is suggested that you repeat the terminal session shown below. Nevertheless, you are encouraged, if time permits, to carry out your own exploration.

```
        VO←6
        VO
6
        V←VO
        V
6
        X←¯6+ι11
        X
¯5 ¯4 ¯3 ¯2 ¯1 0 1 2 3 4 5
        |X
5 4 3 2 1 0 1 2 3 4 5
        3≥|X
0 0 1 1 1 1 1 1 1 0 0
        ⌈/X
5
        ⌊/X
¯5
        ρX
11
        11ρ0
0 0 0 0 0 0 0 0 0 0 0
        (ρX)ρ4
4 4 4 4 4 4 4 4 4 4 4
        2*3
8
        3*2
9
        X*2
25 16 9 4 1 0 1 4 9 16 25
        ¯2↑X
4 5
        3↑X
¯5 ¯4 ¯3
```

```
      4↓X
¯1 0 1 2 3 4 5
      4×3⋆2                        APL execution is from
36                                right to left
      3⋆2×4
6561
      Y←ι3
      Y
1 2 3
      X,Y
¯5 ¯4 ¯3 ¯2 ¯1 0 1 2 3 4 5 1 2 3
      V←1 2 3
      ⋆V
2.718281828 7.389056099 20.08553692
      ⋆2×V
7.389056099 54.59815003 403.4287935
      2⌊X
¯5 ¯4 ¯3 ¯2 ¯1 0 1 2 2 2 2
      0⌈X-3
0 0 0 0 0 0 0 0 0 1 2
      X[9 10 11]
3 4 5
      A←X
      A
¯5 ¯4 ¯3 ¯2 ¯1 0 1 2 3 4 5
      A[9 10 11]←100 200 300
      A
¯5 ¯4 ¯3 ¯2 ¯1 0 1 2 100 200 300
      A=X
1 1 1 1 1 1 1 1 0 0 0
```

Common errors and associated messages

```
      QQ
VALUE ERROR
      QQ
      ∧                           QQ  has not been assigned a value

      1 2 3+4 5 6 7
LENGTH ERROR
      1 2 3 + 4 5 6 7             The caret ∧ indicates the
            ∧                     point at which the error
                                  was identified
      ⋆6
403.4287935
      ⋆2000
DOMAIN ERROR
      ⋆2000
      ∧
```

```
      QQ-MM×3+2×A=X
VALUE ERROR                        MM  has not been given a value
      QQ-MM×3+2×A=X
         ∧
```

Defined functions

```
      ∇PRINT                       A niladic function
[1]   X∇                           (takes no arguments)
      PRINT
¯5 ¯4 ¯3 ¯2 ¯1 0 1 2 3 4 5
      2×PRINT                      The result is only printed:
¯5 ¯4 ¯3 ¯2 ¯1 0 1 2 3 4 5         it is not stored and thus
VALUE ERROR                        cannot be multiplied by 2
      2×PRINT
         ∧
```

```
      ∇R←PRINTA                    A niladic function returning a result
[1]   R←X∇
      PRINTA
¯5 ¯4 ¯3 ¯2 ¯1 0 1 2 3 4 5
      2×PRINTA
¯10 ¯8 ¯6 ¯4 ¯2 0 2 4 6 8 10
```

```
      ∇EXP X                       A monadic function
[1]   *X∇                          (one argument, on the right)
      EXP 1 2 3
2.718281828 7.389056099 20.08553692
      4×EXP 1 2 3
2.718281828 7.389056099 20.08553692
VALUE ERROR                        Result is printed, but cannot be
      4×EXP 1 2 3                  multiplied by 4 because no result
         ∧                         is stored
```

```
      ∇R←EXP X                     An attempt to redefine the function
DEFN ERROR                         EXP. Not allowed, in case previous
      ∇R←EXP X                     function is accidentally lost.
         ∧
```

```
      ∇R←EXPO X                    Monadic function returning a result
[1]   R←*X∇
      EXPO 1 2 3
2.718281828 7.389056099 20.08553692
      ¯2+EXPO 1 2 3
0.7182818285 5.389056099 18.08553692
```

```
      ∇A PLUS B                    A dyadic function (two arguments)
[1]   A+B∇
      1 3 5 PLUS 5 3 1
```

```
6  6  6
        100×1  3  5  PLUS  5  3  1
6  6  6
VALUE  ERROR                              No result was stored
        100×  1  3  5  PLUS  5  3  1
               ∧
```

```
        ∇R←A  ADD  B       New dyadic function, with returned result
[1]     R←A+B∇
        300×1  2  ADD  3  4
1200 1800
        R                  R cannot be accessed by name: it is
VALUE  ERROR               a local variable
        R
        ∧
```

```
        ∇R←A  FUNCTION  B   A new dyadic function, with
[1]     H←A - B            a returned result
[2]     R←H*2∇
```

```
        2  FUNCTION  3
1
        H                  H is a global variable
⁻1
        R                  R is a local variable
VALUE  ERROR
        R
        ∧
```

```
        )ERASE  FUNCTION    Both the function FUNCTION
        )ERASE  H           and the variable H are erased from
                            the workspace
        ∇R←A  FUNCTION  B;H  In defining FUNCTION again, the
[1]     H←A - B            variable H is localized by including
[2]     R←H*2∇             it, preceded by a semicolon, in the
                            function header.
        Q←4  FUNCTION  5
        Q                  After the execution of FUNCTION we find
1                           that H is a local variable.
        H
VALUE  ERROR
        H
        ∧
```

APL Primitive Scalar and Mixed Functions

Scalar Functions

Monadic form $f\omega$		dyadic form $\alpha\,f\,\omega$		
ω	Conjugate	$+$	Plus	$4+6.4 \leftrightarrow 10.4$
$0-\omega$	Negative	$-$	Minus	$4-6.4 \leftrightarrow \bar{2}.4$
$(\omega>0)-\omega<0$	Signum	$\times$	Times	$4\times6.4 \leftrightarrow 25.6$
$1\div\omega$	Reciprocal	$\div$	Divide	$4\div6.4 \leftrightarrow .625$
$\omega\lceil-\omega$	Magnitude	$\|$	Residue	$\omega-\alpha\times\lfloor\omega\div\alpha+\alpha=0$
Integer part	Floor	$\lfloor$	Minimum	$(\omega\times\omega<\alpha)+\alpha\times\omega\geq\alpha$
$-\lfloor-\omega$	Ceiling	$\lceil$	Maximum	$-(-\alpha)\lfloor-\omega$
Choice of $\iota\omega$	Roll	$?$	Deal *See under* Mixed Functions *below*	
$2.71828\ldots*\omega$	Exponential	$*$	Power	$3*4 \leftrightarrow 81$
$*\omega$ inverse	Natural log	$\circledast$	Log	Log ω base α
$\omega\times3.14159\ldots$	Pi times	$\circ$	Circular, hyperbolic, Pythagorean (*below left*)	
$\times/\iota\omega$ or $\lceil\omega+1$	Factorial	$!$	Binomial	$(!\omega)\div(!\alpha)\times!\omega-\alpha$
$1-\omega$ $[\omega=0$ or $1]$	Not	$\sim$	Not defined	

					α	ω	$\wedge$	$\vee$	$\tilde{\wedge}$	$\tilde{\vee}$
$(-\alpha)\circ\omega$	α	$\alpha\circ\omega$		And	0	0	0	0	1	1
			$\vee$	Or	0	1	0	1	1	0
$(1-\omega*2)*.5$	0	$(1-\omega*2)*.5$	$\tilde{\wedge}$	Not and	1	0	0	1	1	0
Arcsin ω	1	Sine ω	$\tilde{\vee}$	Not or	1	1	1	1	0	0
Arccos ω	2	Cosine ω	$<$	Less						
Arctan ω	3	Tangent ω	$\leq$	Not greater			Result is 1			
$(\bar{1}+\omega*2)*.5$	4	$(1+\omega*2)*.5$	$=$	Equal			if relation			
Arcsinh ω	5	Sinh ω	$\geq$	Not less			holds, and 0			
Arccosh ω	6	Cosh ω	$>$	Greater			if not; e.g.			
Arctanh ω	7	Tanh ω	$\neq$	Not equal			$5\leq8 \leftrightarrow 1$			

Relations:

System variables

$\square CT$	Comparison tolerance
$\square IO$	Index origin
$\square LX$	Latent expression
$\square PP$	Print precision

System functions

$\square CR$	Canonical rep. of function
$\square FX$	Fix function definition
$\square EX$	Expunge
$\square NL$	Name list: 1 2 3 are

□PW	Print width
□RL	Random link　(?)
□AI	Account information
□AV	Atomic vector
□LC	Line counter
□TS	Time stamp
□TT	Terminal type
□UL	User load
□WA	Work area

	label, variable, function;
	limited to initial α if given
□NC	Name class: 0 1 2 3 4
	are free, label, variable,
	function, other
□DL	Delay (in seconds)
□SVO	Share offer　　　　inquiry
□SVC	Set access　　　　if no α
□SVR	Retract offer
□SVQ	User numbers of in-offers

© 1983 by APL PRESS, Palo Alto 94306-1683

Mixed functions

Examples:	$P \leftrightarrow$ 2　3　5	$E \leftrightarrow$ 1　2　3 　　4　5　6	$X \leftrightarrow ABC$ 　　DEF
Shape	ρA	ρ$P \leftrightarrow$ 3　　ρ$E \leftrightarrow$ 2 3　　ρ5 $\leftrightarrow$ ι0	
Reshape	VρA	Reshape A to dimension V 　2 3ρι6 $\leftrightarrow E$　　0ρ$E \leftrightarrow$ ι0	
Ravel	,A	,$A \leftrightarrow (\times/\rho A)\rho A$　　　,$E \leftrightarrow$ ι6	
Reverse	φA	φ$X \leftrightarrow CBA$　　φ[1]$X \leftrightarrow DEF$ 　　　　FED　　　　　ABC	
Rotate	AφA	¯1φ$P \leftrightarrow$ 5　2　3　　1 ¯1φ$X \leftrightarrow BCA$ 　　　　　　　　　　　　　　FDE	
Catenate	A,A	$P,7 \leftrightarrow$ 2　3　5　7　　$X,\phi X \leftrightarrow ABCCBA$ 　　　　　　　　　　　　　　　$DEFFED$	
Transpose	V⍉A	Axis I of A becomes axis $V[I]$ of result. 3　2 $\leftrightarrow$ ρ2 1⍉E　1　5 $\leftrightarrow$ 1 1⍉E	
	⍉A	Reverse order of axes	
Take	V↑A	Take or drop $|V[I]|$ first ($V[I] \geq 0$)	
Drop	V↓A	or last ($V[I] < 0$) items along axis I.	
Compress	V/A	1　0　1/$P \leftrightarrow$ 2　5　　1　0　1/$X \leftrightarrow AC$ 　　　　　　　　　　　　　　　　DF	
Expand	$V\backslash A$	1　0　1\ι2 $\leftrightarrow$ 1　0　2 1　1　0　1\$X \leftrightarrow AB\ C$ 　　　　　　　　$DE\ F$	
Indexing	$V[A]$ $M[A;A]$ $A[A; \ldots ;A]$	$P[1\ 3] \leftrightarrow$ 2　5　　$X[2\ 1;1\ 3] \leftrightarrow DF$ 　　　　　　　　　　　　　　　　AC 　　　　　　　　$X[2;] \leftrightarrow DEF$	
Index 　generator	ιS	First S integers	
Index of	VιA	Least index of A in V (or 1+ρV) Pι3　0 $\leftrightarrow$ 2　4	
Membership	$A \in A$	ρ$W \in Y \leftrightarrow$ ρW　　$E \in P \leftrightarrow$ 0　1　1 　　　　　　　　　　　　　　　0　1　0	
Grade up	⍋V	Permutation which orders V up	

Grade down	ψV	Permutation which orders V down
Deal	$S?S$	$W?Y \leftrightarrow W$ items from ιY
Matrix inverse	$\boxminus M$	$M+.\times\boxminus M$ is identity matrix.
Matrix divide	$M\boxminus M$	$A\boxminus B \leftrightarrow (\boxminus B)+.\times A$
Decode	$A\perp A$	$10\perp1\ 8\ 6\ 7 \leftrightarrow 1867$
		$24\ 60\ 60\perp1\ 2\ 3 \leftrightarrow 3723$
Encode	$A\top A$	$24\ 60\ 60\top3723 \leftrightarrow 1\ 2\ 3$
Execute	$\textipa{\textbaro} V$	$\textipa{\textbaro}'1+2' \leftrightarrow 3 \qquad \textipa{\textbaro}'P' \leftrightarrow 2\ 3\ 5$
Format	$\textipa{\textbaro} A$	$\textipa{\textbaro}\ ^-2.4 \leftrightarrow\ '^-2.4' \qquad \rho\,\textipa{\textbaro}E \leftrightarrow 2\ 5$
	$V\textipa{\textbaro} A$	$1\textipa{\textbaro}P \leftrightarrow\ '2.0\ 3.0\ 5.0'$
		$^-1\textipa{\textbaro}1\textipa{\textdownstep}P \leftrightarrow\ '3E0\quad 5E0'$

OPERATORS: Reduction $f/$ Scan $f\setminus$ Axis $[I]$
PRODUCTS: Inner $f.g$ Outer $\circ.g$

Workspace Management and System Commands

If you have painstakingly defined some APL functions and variables, you will certainly need some mechanism for storing (i.e. saving) them so that, at some time in the future, you do not have to repeat the definitions. Additionally, some functions may have been defined by someone else for your use, and you will need to access them in some way.

Upon switching on, or signing onto, an APL service, you are provided with a clear (i.e., empty) active workspace (Fig. 1). It has a name (*CLEAR*) for

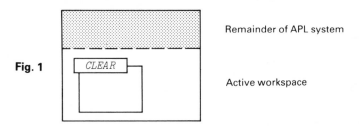

Fig. 1

Remainder of APL system

Active workspace

identification (ID) purposes. Within it you can proceed to define functions and variables in the simple manner illustrated in the book. Suppose that, in this way, you have defined a function *FORM* and a variable *GG,* which now (Fig. 2)

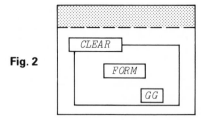

Fig. 2

reside in your active workspace. At this point you may use *system commands* to find

```
      )WSID
CLEAR
```

the workspace's identification (i.e. name),

404

```
    )FNS
FORM
```

the list of functions, and

```
    )VARS
  GG
```

the list of variables. You may wish to save the function *FORM* and the variable *GG* to work on tomorrow. So you give the workspace a distinctive name (Fig. 3):

Fig. 3

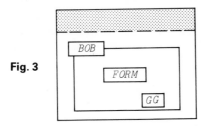

```
    )WSID BOB
WAS CLEAR WS
```

and then save an image of it within the APL system (Fig. 4):

Fig. 4

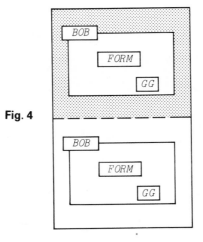

```
    ) SAVE
14.02.24 01/03/84 BOB
```

The time-and-date stamp is useful for reference purposes. On terminating your use of the APL service

```
    )OFF
14.04.36 01/03/84 STUDENT
```

you can then be assured that (Fig. 5) an image of the workspace *BOB* is now secure.

Fig. 5

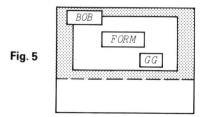

On returning to the APL system, suppose that a new workspace *JOHN* had earlier been added to the *library* of workspaces to which you have access (Fig. 6). How would you know about this addition? By entering the system command

```
        )LIB
BOB
JOHN
```

Suppose also that, in the clear active workspace you were provided with on accessing the APL service, you define a function *INCID* and a variable *M* (Fig. 7).

At this point you may wish to proceed in a number of ways. A few possibilities will now be illustrated.

Copying   Suppose that you want to use the function *FORM*, which is stored in the workspace *BOB*, together with the function *INCID* To do so, you must copy *FORM* into your active workspace (Fig. 8):

```
    )COPY BOB FORM
SAVED 14.02.24 01/03/84
```

Loading   If, however, you want to make use *only* of the workspace *JOHN*, you may *load* a copy of it into your active workspace (Fig. 9):

```
    )LOAD JOHN
SAVED 9.00.08 02/10/83
```

You will find that this action has eliminated your previous clear workspace:

```
    )WSID
JOHN
```

as well as the functions and variables it contained.

Loading   If a workspace called *EXERCISES* has been provided (perhaps by
from a   your instructor) in Library 23 for access by any authorized
library   student, this workspace can be loaded into your active workspace
            (Fig. 10) by the command

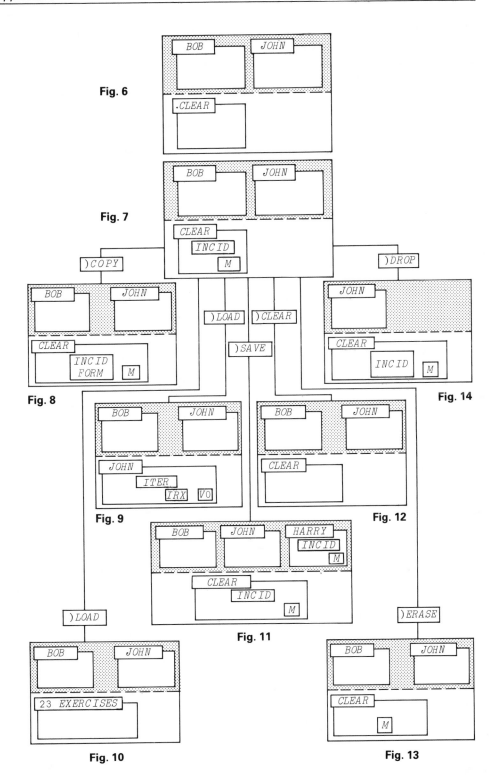

Fig. 6

Fig. 7

)COPY

)DROP

Fig. 8

)LOAD

)CLEAR

)SAVE

Fig. 14

Fig. 9

Fig. 12

Fig. 11

)LOAD

)ERASE

Fig. 10

Fig. 13

)LOAD 23 EXERCISES
SAVED 18.04.11 08/07/83

Saving If you wish to save the variable *M* and the function *INCID*, though not in an existing workspace, you can easily do so by indicating the name you wish to give to the new saved workspace (Fig. 11).

)SAVE HARRY
15.12.05 02/04/84

Clearing If *INCID* and *M* are no longer required, and you wish to begin again with a clear workspace (Fig. 12), the appropriate system command is

)CLEAR
CLEAR WS

Erase If, for some reason, you wish to remove the function *INCID* from your active workspace (Fig. 13) the correct command is:

)ERASE INCID

There is now no way of retrieving *INCID* as a function: it would have to be redefined.

Drop If you wish to discard one of your saved workspaces (say *BOB*), the command

)DROP BOB
11.06.15 02/10/84

should be used, though with care because (Fig. 14) there is then no way of retrieving it.

State indicator Entry of the command

)SI
UPDATE[5]*
PRESENT[2]

where *SI* denotes 'State Indicator', provides a response which shows which functions were executing when an *ERROR* (of any kind) occurred and halted execution. In the above example, the *ERROR* occurred on line 5 of *UPDATE* which was, at the time, being called on line 2 of *PRESENT*. The state indicator offers a valuable diagnostic aid. However, the problems which can occur if the state indicator becomes too long can be avoided by restoring the state indicator to its initial (empty) condition by means of the command

>)*RESET*

This short introduction covers the system commands necessary to start making effective use of an APL service. It is strongly recommended that, to gain familiarity, you explore the commands discussed above. When needed, a more comprehensive reference should be consulted (Gilman and Rose, 1984).

Function Listings for Part 1

(Note: some functions may exist in several near-identical versions, only one of which is listed below. The page number is where the function is first defined.)

AA page 47

```
        ∇   YB ←X AA Y; XX; YY; I
[1]         XX←(¯2↑1 1,ρX)ρX
[2]         YY←(¯2↑1 1,ρY)ρY
[3]         I ←(ρXX)+ρYY
[4]         YB ←(I↑XX)+(-I)↑YY∇
```

ACSOLVE, 83

```
        ∇   ACSOLVE; W; N; NR; IY; RY; K; YR
[1]         W←,2×○ F
[2]         N←⌈/(,G[; 1 2]),(,C[; 1 2]),,IL[; 1 2]
[3]         NR←N- 1
[4]         RY←(N,N)ρ0
[5]         VREIM←(2,NR,(ρW))ρ0
[6]         RY←RY FORM G
[7]         K←1
[8]         LOOP: IY←(N,N)ρ0
[9]         IY←IY FORM C[; 1 2],C[; 3]×W[K]
[10]        IY←IY FORM IL[; 1 2],IL[; 3]÷-W[K]
[11]        YR← 0 ¯1 ¯1 ↓RY,[0.5]IY
[12]        VREIM[;;K]←YR REALSOLV IR
[13]        →LOOP ×(ρW)≥K←K+1      ∇
```

ADDGM , 65

```
        ∇   R←Y ADDGM  GM ; C; VNDS; INDS; ∆Y
[1]         R←Y
[2]         C←1
[3]      L : VNDS←GM[C; 1 2]
[4]         INDS←GM[C; 3 4]
[5]         ∆Y← 2 2 ρ 1 ¯1 ¯1 1 ×GM[C; 5]
[6]         R[INDS; VNDS]←R[INDS; VNDS]+∆Y
[7]         →L ×(1↑ρ GM )≥C←C+1
        ∇
```

BACKEULER, 158

```
    ∇  VC←X BACKEULER I;H;K
[1]     VC←X[1],(⁻1+ρI)ρ0
[2]     H←X[2]
[3]     K←1
[4]  LOOP: VC[K+1]←(I[K+1]+VC[K]×CT[3]÷H)÷GT[3]+CT[3]÷H
[5]     →LOOP×(ρI)>K←K+1
    ∇
```

BACKEULERNL, 167

```
    ∇  VC←X BACKEULERNL I;H;K;ICS
[1]     VC←X[1],(⁻1+ρI)ρ0
[2]     H←X[2]
[3]     K←1
[4]  LOOP: ICS←VC[K]×CT[3]÷H
[5]     VC[K+1]←(I[K+1]+ICS)ITER VC[K]
[6]     →LOOP×(ρI)>K←K+1
    ∇
```

BKSUB (Chapter 5), 99

```
    ∇  VR←YR BKSUB IR
[1]     VR←(K←ρIR)ρ0
[2]  L1: VR[K]←IR[K]-YR[K;]+.×VR
[3]     →(0<K←K-1)/L1
    ∇
```

BKSUB (Chapter 5 solutions), 114

```
    ∇  V←Y BKSUB I
[1]     V←(ρI)ρ0
[2]     C←0
[3]  L: S←(1↑ρI)-C
[4]     V[S;]←(÷Y[S;S])×I[S;]-Y[S;]+.×V
[5]     →((1↑ρI)>C←C+1)/L
    ∇
```

BRANCHADD, 193

```
    ∇  ZRN←ZR BRANCHADD NNG
[1]     KM←NNG[1 2]
[2]     G←NNG[3]
[3]     A←-/ZR[;KM]
[4]     B←÷(÷G)+-/-/[1]ZR[KM;KM]
[5]     C←-/[1]ZR[KM;]
[6]     ZRN←ZR-A∘.×B×C∇
```

BRANCHTOREFADD, *194*

```
        ∇  ZRN←ZR BRANCHTOREFADD NG
[1]        K ←NG[1]
[2]        G ←NG[2]
[3]        A←ZR[;K]
[4]        B←÷(÷G)+ZR[K;K]
[5]        C←ZR[K;]
[6]        ZRN ←ZR - A∘.×B ×C∇
```

CAPCURRENT, *178*

```
        ∇  ICC←CAPCURRENT; VBC
[1]        ICC←Nρ0
[2]        →(~((1↑ρCT)>0))/0
[3]        VRC←-/(V[;K])[CT[; 1 2]]
[4]        IC←CT[;3]×VBC÷H
[5]        ICC←IFORM CT[; 1 2],IC
        ∇
```

COM, *26*

```
        ∇  IR←COM VR
[1]        IR←VR+0.1×VR⋆3
        ∇
```

COMLIN, *39*

```
        ∇  IR←COMLIN VR
[1]        IR←(DIODE VR⌊VT)+(DIFDIODE VT)×0⌈(VR-VT)
        ∇
```

COMPANCURRENTS, *178*

```
        ∇  ICX←COMPANCURRENTS
[1]        ICC←CAPCURRENT
[2]        ICL←INDCURRENT
[3]        ICX←ICC+ICL
        ∇
```

COMT, *167*

```
        ∇  IR←COMT VR
[1]        IR←(DIODE VR⌊VT)+(DIFDIODE VT)×0⌈(VR-VT)
        ∇
```

COND, *13*

```
        ∇  GB←COND G
[1]        B ←ρG
[2]        GB ←((B,B)ρG)×UNIT B
        ∇
```

DIA, *151*

```
        ∇ R←DIA X
[1]        R←((B,B←ρX)ρX)×UNIT ρX
        ∇
```

DIFCOM, *26*

```
        ∇ S←DIFCOM VR
[1]        S←1+0.3×VR⋆2
        ∇
```

DIFCOMLIN, *40*

```
        ∇ S←DIFCOMLIN VR
[1]        S←DIFDIODE VR⌊VT
        ∇
```

DIFDIODE, *38*

```
        ∇ S←DIFDIODE VR
[1]        S←1E¯11×40×⋆40×VR
        ∇
```

DIFF, *39*

```
        ∇ R←DIFF X
[1]        R←(¯1↓X)-1↓X
        ∇
```

DIFNET, *151*

```
        ∇ S←DIFNET VR
[1]        VR←(⍉A)+.×VR,0
[2]        X←VB>0
[3]        S1←DIFDIODE VT⌊X×VB
[4]        S2←Y\((Y/IS)×¯1+⋆40×VBR)÷VBR←(Y←~X)/VB
[5]        S←¯1 ¯1 ↓A+.×(DIA S1+S2+G)+.×⍉A
        ∇
```

DIODE, *38*

```
        ∇ IR←DIODE VR
[1]        IR←1E¯11×¯1+⋆40×VR
        ∇
```

FIND, *132*

```
        ∇ R←J FIND K
[1]        X← 0 ¯1 +RP[J+ 0 1]
[2]        CIINDICES←X[1]+0,⍳X[2]-X[1]
[3]        Z←(K=CI[CIINDICES])/CIINDICES
[4]        R←+/0,EV[Z]
        ∇
```

FORM (Chapter 1), 17

```
       ∇   Y←FORM  GT
 [1]       N←⌈/,GT[;1 2]
 [2]       B←1↑ρGT
 [3]       Y←(N,N)ρ0
 [4]       K←1
 [5]    L:ΔY←2 2 ρ 1 ¯1 ¯1 1×GT[K;3]
 [6]       NDS←GT[K;1 2]
 [7]       Y[NDS;NDS]←Y[NDS;NDS]+ΔY
 [8]       →L×(K←K+1)≤B
       ∇
```

FORM (Chapter 4), 73

```
       ∇   R←Y FORM  G;K;I
 [1]       R←Y
 [2]       K←0
 [3]    S:→L×(1↑ρG)≥K←K+1
 [4]    L:I←G[K; 1 2]
 [5]       R[I;I]←R[I;I]+ 2 2 ρ 1 ¯1 ¯1 1 ×G[K;3]
 [6]       →S
       ∇
```

FORM (Chapter 8), 161

```
       ∇   Y←FORM  GT;B;K;ΔY;NDS
 [1]       B←1↑ρGT
 [2]       Y←(N,N)ρ0
 [3]       K←1
 [4]    L:ΔY← 2 2 ρ 1 ¯1 ¯1 1 ×GT[K;3]
 [5]       NDS←GT[K; 1 2]
 [6]       Y[NDS;NDS]←Y[NDS;NDS]+ΔY
 [7]       →L×(K←K+1)≤B
       ∇
```

FREDUC, 98

```
       ∇   YRI←YR FREDUC IR
 [1]       YRI←YR,IR
 [2]       K←0
 [3]       NR←1↑ρYR
 [4]    L1:RR←(K←K+1)↓ιNR
 [5]       YRI[K;]←YRI[K;]÷YRI[K;K]
 [6]       →L2×NR>K
 [7]    L2:YRI[RR;]←YRI[RR;]- YRI[RR;K]∘.×YRI[K;]
 [8]       →L1
       ∇
```

FREDUC1, 113

```
      ∇  A←Y FREDUC1 I
[1]      A←Y,I
[2]      C←1
[3]    L:A←A PIV C
[4]      →L×(C←C+1)≤1↑ρA
      ∇
```

FREDUCPIV, 103

```
      ∇  YRI←YR FREDUCPIV IR
[1]      YRI←YR,IR
[2]      K←0
[3]      NR←1↑ρYR
[4]    L1:RR←(K←K+1)↓ιNR
[5]      YRI←YRI PIVCH K
[6]      YRI[K;]←YRI[K;]÷YRI[K;K]
[7]      →L2×NR>K
[8]    L2:YRI[RR;]←YRI[RR;]-YRI[RR;K]∘.×YRI[K;]
[9]      →L1
      ∇
```

FSUB, 119

```
      ∇  R←Y FSUB I
[1]      R←ϕ(ϕ⊖Y)LUBKSUB ϕI
      ∇
```

GAUSS, 100

```
      ∇  VR←YR GAUSS IR
[1]      YRI←YR FREDUC IR
[2]      VR←(0 ¯1↓YRI) BKSUB,YRI[;1+¯1↑ρYR]
      ∇
```

IFORM, 163

```
      ∇  I←IFORM X;K;NN
[1]      I←Nρ0
[2]      K←1
[3]    L:NN←X[K; 1 2]
[4]      I[NN]←I[NN]+ 1 ¯1 ×X[K;3]
[5]      →L×ι(1↑ρX)≥K←K+1
      ∇
```

INCID, 13

```
      ∇  A←INCID M
[1]      N←⌈/,M
[2]      AP←(ιN)∘.=M[1;]
[3]      AN←(ιN)∘.=M[2;]
[4]      A←AP-AN
      ∇
```

INDCURRENT, *178*

```
        ∇   ICL←INDCURRENT
[1]         ICL←Nρ0
[2]         →(~((1↑ρILT)>0))/0
[3]         ICL←IFORM ILT[; 1 2],−IL
        ∇
```

INITIALIZE, *178*

```
        ∇   INITIALIZE
[1]         N←⌈/,(GT,[1]CT,[1]ILT)[; 1 2]
[2]         V←(ρI)ρ0
[3]         V[; 1]←V0
[4]         IL←IL
        ∇
```

INV, 115

```
        ∇   R←INV M;N;C;V;S
[1]         R←M,UNIT N←1↑ρ M
[2]         C←0
[3]         V←ιN
[4]     L : C←C+1
[5]         S←1↓(‾1+C)⌽V
[6]         R[C;]←R[C;]÷R[C; C]
[7]         R[S;]←R[S;]−R[S; C]∘.×R[C;]
[8]         →(C≠N)/L
[9]         R←(−ρM)↑R
        ∇
```

ITER, 26

```
        ∇   VR←IRX ITER VRO;ERROR;SLOPE;ΔVR
[1]         VR←VRO
[2]         L1:→L2×1E‾6<| ERROR←IRX−COM VR
[3]         L2:SLOPE←DIFCOM VR
[4]         VR←VR+ΔVR←ERROR÷SLOPE
[5]         →L1 ∇
```

ITER1, 29

```
        ∇   VR←IRX ITER1 VRO;ERROR;IR;SLOPE;IEQ
[1]         VR←VRO
[2]         L1:→L2×1E‾6<|ERROR←IRX−IR←COM VR
[3]         L2:SLOPE←DIFCOM VR
[4]         IEQ←IR−VR×SLOPE
[5]         VR←(IRX−IEQ)÷SLOPE
[6]         →L1 ∇
```

ITERDIODE, 38

```
        ∇  VR←IRX ITERDIODE VRO
[1]        VR←VRO
[2]       L1:→L2×1E¯6<|ERROR←IRX-DIODE VR
[3]       L2:SLOPE←DIFDIODE VR
[4]          VR←VR+∆VR←ERROR÷SLOPE
[5]          →L1
        ∇
```

ITERNET, 143

```
        ∇  VR←IRX ITERNET VRO
[1]        VR←VRO
[2]       L1:→L2×1E¯6<⌈/|ERROR←IRX-NET VR
[3]       L2:SLOPE←DIFNET VR
[4]          VR←VR+∆VR←ERROR⌹SLOPE
[5]          →L1
        ∇
```

ITERSPLINE, 40

```
        ∇  VR←IRX ITERSPLINE VRO
[1]        W←0
[2]        VR←VRO
[3]       L1:→L2×1E¯6<|ERROR←IRX-COMLIN VR
[4]       L2:SLOPE←DIFCOMLIN VR
[5]          VR←VR+∆VR←ERROR÷SLOPE
[6]          W←W+1
[7]          →L1
        ∇
```

LO, 135

```
        ∇  R←LO X
[1]        R←(((ιN)∘.≥ιN←1↑ρX)×X∇
```

LUBKSUB, 119

```
        ∇  V←Y LUBKSUB I;S
[1]        V←(S←ρI)ρ0
[2]       L1:V[S]←(÷Y[S;S])×I[S]-Y[S;]+.×V
[3]          →(0<S←S-1)/L1
        ∇
```

LUFAC , 121

```
        ∇  LU←LUFAC Y;C;S
[1]        C←1
[2]        LU←Y
[3]       L1:S←C↓ι1↑ρLU
[4]          LU[C;S]←LU[C;S]÷LU[C;C]
[5]          LU[S;S]←LU[S;S]-LU[S;C]∘.×LU[C;S]
[6]          →L1×(1↑ρLU)>C←C+1  ∇
```

LUFACTOR, 176

```
        ∇   LUFACTOR GG; LU
[1]         LU←LUFAC GG
[2]         L←LO LU
[3]         U←UP LU
        ∇
```

LUSOLV, 176

```
        ∇   VR←II LUSOLV GG
[1]         VR←U LUBKSUB L FSUB II
        ∇
```

MAG, 84

```
        ∇   Z←MAG X
[1]         Z←(+/[1] X* 2)*0.5
        ∇
```

MINFILLS, 131

```
        ∇   NF←MINFILLS G
[1]         NF←(1↑ρG)ρ0
[2]         K←1
[3]     L: W←G[; 1]∘.∧G[1;]
[4]         LF←W∧~G
[5]         NF[K]←+/1 =,LF
[6]         G←1⊖1⌽G
[7]         →L×ι(ρNF)≥K←K+1
        ∇
```

NET, 151

```
        ∇   IR←NET VR
[1]         VB←(⍉A)+.×VR,0
[2]         IBD←COMLIN VB
[3]         IB←IBD+G×VB
[4]         IR←¯1↓A+.×IB
        ∇
```

NEWDIFCOMLIN, 41

```
        ∇   S←NEWDIFCOMLIN VR
[1]         →(VR>0)/L 1
[2]         S←(COMLIN VR)÷VR
[3]         →0
[4]     L 1: S←DIFCOMLIN VR
        ∇
```

NODCON, 178

```
        ∇   GG←NODCON
[1]         G←FORM GT
[2]         GC←FORM CT[; 1 2],CT[; 3]÷H
[3]         GL←FORM ILT[; 1 2],ILT[; 3]×H
[4]         GG←¯1 ¯1 ↓G+GC+GL
        ∇
```

NOFILLS, 129

```
        ∇  NG←NOFILLS G;B;Q
[1]        NG←G
[2]        →L×0≠+/B←1 = ¯1←+/G
[3]     L:Q←+/0=∨\B
[4]        NG←Q RENUMBER G
        ∇
```

PCOND, 108

```
        ∇  YN←PCOND Y
[1]        YN← 1 1 ↓Y-Y[;1]∘.×Y[1;]÷Y[1;1]
        ∇
```

PIV, 110

```
        ∇  R←Y PIV N
[1]        D←1↑ρY
[2]        S←N↓ιD
[3]        Y[N;]←Y[N;]÷Y[N;N]
[4]        Y[S;]←Y[S;]-Y[S;N]∘.×Y[N;]
[5]        R←Y
        ∇
```

PIVCH, 103

```
        ∇  YI←YR PIVCH R
[1]        YI←YR
[2]        ROS←(R-1)↓ι NR←1↑ρYR
[3]        M←(R-1)+1↑ ⍒ |YI[ROS;R]
[4]        YI[R,M;]←YI[M,R;]
        ∇
```

PLOT, 84

```
        ∇  X PLOT Y;SS;SX;SY
[1]        SS← 40 14 ,(L/X),(⌈/X),(L/Y),(⌈/Y)
[2]        SX←⌊0.5+(X-SS[3])×SS[1]÷SS[4]-SS[3]
[3]        SY←⌊0.5+(Y-SS[5])×SS[2]÷SS[6]-SS[5]
[4]     LP:'|','  *'[1+(0,ιSS[1])∈.(SY∈SS[2])/SX]
[5]        →(0≤SS[2]←SS[2]-1)/LP
[6]        '∘',(SS[1]ρ'-'),'→'
        ∇
```

REALSOLV, 79

```
        ∇  VR←YR REALSOLV IR;YT;YB;Y;V
[1]        YT←YR[1;;],¯1×YR[2;;]
[2]        YB←YR[2;;],YR[1;;]
[3]        Y←YT,[1]YB
[4]        V←(,IR)⌹Y
[5]        VR←(ρIR)ρV
        ∇
```

RENUMBER, *129*

```
        ∇  NG←Q RENUMBER G;NN
[1]        NN←Qφ ι 1↑ρG
[2]        NG← 1  1  ↓G[NN;NN]
        ∇
```

SP, *139*

```
        ∇  R←SP M
[1]        R←∼0=M
        ∇
```

SPV, *139*

```
        ∇  R←SPV M
[1]        R←(+/+/∼ 0=M ),×/ρM
        ∇
```

STOREIL, *178*

```
        ∇  R←STOREIL V;VL
[1]        VL←-/(V[;K+1])[ILT[; 1 2]]
[2]        R←IL+VL×H×ILT[; 3]
        ∇
```

TIMEDOMAIN, *163*

```
        ∇  V←VO TIMEDOMAIN NT
[1]        N←⌈/,GT[; 1 2],[1]CT[; 1 2]
[2]        V←((N-1),NT)ρ0
[3]        V[; 1]←VO
[4]        G←FORM GT
[5]        GC←FORM CT[; 1 2],CT[; 3]÷H
[6]        GG← ¯1 ¯1 ↓G+GC
[7]        K←1
[8]      L:VBC←-/(V[;K])[CT[; 1 2]]
[9]        IC←CT[; 3]×VBC÷H
[10]       IN←IFORM CT[; 1 2],IC
[11]       II←¯1↓I+IN
[12]       V[;K+1]←II⌹GG
[13]       →L×NT>K←K+1
        ∇
```

TIME△RLC△BE, *177*

```
        ∇  V←VO TIME△RLC△BE I;N;GG;K;ICX;II;IL
[1]        INITIALIZE
[2]        GG←NODCON
[3]        K←1
[4]      LOOP:ICX←COMPANCURRENTS
[5]        II←¯1↓I[;K]+ICX
[6]        V[;K+1]←(II⌹GG),0
[7]        IL←STOREIL V
[8]        →LOOP×(¯1↑ρI)>K←K+1
        ∇
```

TRAP, *172*

```
        ∇  VC←X TRAP I;H;ICS;K;IC
[1]        VC←X[1],(¯1+ρI)ρ0
[2]        H←X[3]
[3]        ICS←X[2]+(2×CT[3]÷H)×VC[1]
[4]        K←2
[5]     LOOP: VC[K]←(I[K]+ ICS)÷GT[3]+CT[3]÷H÷2
[6]        IC←I[K]-VC[K]×GT[3]
[7]        ICS←IC+(2×CT[3]÷H)×VC[K]
[8]        →LOOP×(ρI)≥K←K+1
        ∇
```

TRUE, *177*

```
        ∇  R←TRUE H
[1]        R←1-*-(H×0,ιNT)
        ∇
```

UNIT, *13*

```
        ∇  R←UNIT N
[1]        R←(ιN)∘.=ιN∇
```

UP, *135*

```
        ∇  R←UP X
[1]        R←((ιN)∘.=ιN)+((ιN)∘.<ιN←1↑ρX)×X∇
```

VOLTAGE, *84*

```
        ∇  Z←VOLTAGE NN
[1]        Z←VREIM[;NN;]
        ∇
```

Function Listings for Part 2 *(BASECAP)*

* *ACBJT*, *266*

```
      ∇  ACBJT P
[1]      ⍝ PARAMETER SEQUENCE IS
[2]      ⍝ NE NB NC NBP NCP BF RPI
[3]      ⍝ RO RX RC RU [CPI CU CCS]
[4]      ⍝ []MEANS OPTIONAL IF IN SEQUENCE
[5]      ⍝
[6]       P ←P, 0 0 0
[7]       R P[2 4 9]
[8]       R P[4 1 7]
[9]       R P[4 5 11]
[10]      C P[4 5 13]
[11]      R P[5 1 8]
[12]      R P[5 3 10]
[13]      C P[5],0,P[14]
[14]      C P[4 1 12]
[15]      VCIS P[1 5 4 1],P[6]÷P[7]
      ∇
```

ACPREP, pages *345-346*

```
      ∇  ACPREP
[1]       BC←FORM CL
[2]       BL←FORM LL[; 1 2],÷LL[;3]
[3]       BGL←FORM RL[; 1 2],÷RL[;3]
[4]       YRE←SLOPE- ‾1 ‾1 ↓BGL
      ∇
```

ACSOL, *348*

```
      ∇  ACSOL;WW;KMAX;K
[1]       T←⎕AI[2]
[2]       K←0
[3]       ICOMPLEX←IREIM
[4]       VREIM←(⍴ICOMPLEX)⍴0
[5]       WW←○2×,F
[6]       KMAX←⍴,F
[7]    L:W←WW[K←K+1]
[8]       VREIM←VREIM, YNET REALSOLV ICOMPLEX
[9]       →L×⍳(K<KMAX)
[10]      VREIM← 0 0 1 ↓VREIM
[11]      (⍕'CPUSECONDS= '),⍕(⎕AI[2]-T)÷1000
      ∇
```

---

*Functions that are not part of the *BASECAP* package per se .

ACSOLVE, 343-344

```
        ∇  ACSOLVE
[1]        ACPREP
[2]        (⍕'SOLVING FOR '),(⍕⍴,F),⍕'FREQUENCY POINTS'
[3]        ACSOL
[4]        'DONE'
        ∇
```

* *ACTEST, 362*

```
        ∇  ACTEST
[1]        STARTCCT
[2]        R 1 2 10000
[3]        R 2 3 10000
[4]        R 2 0 10000
[5]        VAC 1 0 10 0
[6]        VAC 3 0 10 180
[7]        END
        ∇
```

* *ACTEST1, 362*

```
        ∇  ACTEST1
[1]        STARTCCT
[2]        R 1 2 10000
[3]        R 2 3 10000
[4]        R 2 0 10000
[5]        VAC 1 0 10 0
[6]        VAC 3 0 10 0
[7]        END
        ∇
```

ACVOUT, 368

```
        ∇  Z←ACVOUT NNO
[1]        Z←VREIM[;,NNO;]
        ∇
```

ALL, 368

```
        ∇  Z←ALL
[1]        Z←⍳⍴IX
        ∇
```

ALPHA, 281

```
        ∇  Z←ALPHA B
[1]        Z←B ÷ 1+B
        ∇
```

$ALTBASE1$, *314-315*

```
       ∇   V←ΔVB ALTBASE1 VB
[1]        V←VB + ΔVB
[2]        →NZ×ι (⌈/ΔVB[DBI])≤ 0
[3]        DPI←((IS≠0)∧(ΔVB≥0)∧(V≥ALTREF))/ιρIS
[4]        IDNEW←(DIODE VB)+ΔVB×DIFDIODE VB
[5]        V[DPI]←(÷JF)×⊛(1+(IDNEW[DPI])÷IS[DPI])
[6]      NZ:V[DBI]←V[DBI]⌊VBMAX
       ∇
```

† $A\mathring{N}D$

$ANGLE$, *370-371*

```
       ∇   ANGLV←ANGLE V;RE;IM;K;KMAX;ANG
[1]        RE ←,V[1; ;]
[2]        IM ←,V[2; ;]
[3]        KMAX←ρ RE
[4]        ANG←ι0
[5]        K←1
[6]      LP:ANG←ANG,ZATN RE[K],IM[K]
[7]        →LP×ιKMAX≥(K←K+1)
[8]        ANG←ANG×18÷○1
[9]        ANGLV←⍉((1↑ρV[1; ;]),¯1↑ρV[1; ;])ρANG
       ∇
```

BB, *356*

```
       ∇   R←BB
[1]        R←(W×BC)−BL ÷ W
       ∇
```

* $BIAS$, *381*

```
       ∇   BIAS
[1]        STARTCCT
[2]        VDC 1 0 ,VBB
[3]        R 1 2 1000000
[4]        QPNP 3 2 4 ,QBIAS
[5]        R 3 0 1000
[6]        R 4 5 1000
[7]        D 5 6 ,D1
[8]        VDC 6 0 ,VCC
[9]        END
       ∇
```

†The functions AND, $PLOT$, VS, and $ZATN$ are found in the APL system library.

BJTBIAS, 380-381

```
      ∇ Z←BJTBIAS;IEP;ICP;AF;AR;IE;IC;VBE;VCB;PP;T;AA;TT
[1]      ' '
[2]      →0×ι(1↑ρQ)=0
[3]      ' '
[4]      'TYPE QNO ENODE BNODE CNODE VBE VCE IE(MA.)IC(MA.)'
[5]      QNO←1
[6]      T←' '
[7]      Z←' '
[8]   L:IEP←(DIODE VBN)[Q[QNO;2]]
[9]      ICP←(DIODE VBN)[Q[QNO;2]+1]
[10]     AF←Q[QNO;6]
[11]     AR←Q[QNO;7]
[12]     IE←TYPE[QNO]×(-IEP)+AP×ICP
[13]     IC←TYPE[QNO]×(-ICP)+AF×IEP
[14]     AA←'NPN'
[15]     →LL×ι(TYPE[QNO]=1)
[16]     AA←'PNP'
[17]  LL:T←T,AA
[18]     VBE←-/(VANS,0)[Q[QNO; 4 3]]
[19]     VCE←-/(VANS,0)[Q[QNO; 5 3]]
[20]     Z←Z,QNO,((ιNH-1),0)[Q[QNO; 3 4 5]],VBE,VCE,(1000×IE),
         (1000×IC)
[21]     →L×ι(QNO←QNO+1)≤1↑ρQ
[22]     TT←((1↑ρQ),3)ρT
[23]     PP←((1↑ρQ),8)ρ Z
[24]     Z←5 0 8 0 7 0 7 0 12 3 8 3 8 3 8 3 ⊤PP
[25]     Z←TT,Z
      ∇
```

* *BRIDGE, 263-264*

```
      ∇ BRIDGE
[1]     STARTCCT
[2]     VDC  1 0  ,F
[3]     R  1 2 5000
[4]     OA 0 2 3
[5]     R  2 3 10000
[6]     D  2 5  ,D1←1E¯14 0
[7]     D  4 2  ,D1
[8]     D  4 3  ,D1
[9]     D  3 5  ,D1
[10]    R  4 6 37500
[11]    VDC 6 0 15
[12]    R  5 7 25000
[13]    VDC 7 0 ¯15
[14]    END
      ∇
```

C, 277

```
        ∇   C P
[1]         MIN←MIN,P[1 2]
[2]         IBIN←IBIN,0
[3]         ISIN←ISIN,0
[4]         GIN ←GIN, ÷1000000000000
[5]         CL←CL,[1] P
        ∇
```

* *CACDCCT*, 331

```
        ∇   CACDCCT
[1]         STARTCCT
[2]         R 1 0 8200
[3]         R 1 5 33000
[4]         R 2 0 1800
[5]         R 5 3 5600
[6]         QNPN 2 1 3 125 0.1 4E⁻9 4E⁻9
[7]         R 4 0 4700
[8]         QNPN 4 3 5 75 0.1 4E⁻9 4E⁻9
[9]         VDC 5 0 15
[10]        END
        ∇
```

* *CCTNAME*, 222

```
        ∇   CCTNAME
[1]         STARTCCT
[2]         VDC 1 0 10
[3]         R 1 0 1000
[4]         ⍝THIS IS A COMMENT LINE
[5]         END
        ∇
```

* *CECBPAIR*, 265

```
        ∇   CECBPAIR
[1]         STARTCCT
[2]         VAC 1 0 1 0
[3]         R 1 2 1000
[4]         R 3 0 75
[5]         R 5 0 1000
[6]         ACBJT 3 2 4 6 7 ,PARS
[7]         ACBJT 4 0 5 8 9 ,PARS
[8]         END
        ∇
```

D, *281*

```
     ∇  D  P
[1]      MIN←MIN,P[1 2]
[2]      IBIN←IBIN,0
[3]      ISIN←ISIN,P[3]
[4]      GIN←GIN,0.000000000001
[5]      P←P,0
[6]      CL←CL,[1] P[1 2 4]
     ∇
```

*DCDIFAMP, *264-265*

```
     ∇  DCDIFAMP
[1]      STARTCCT
[2]      VDC 1 12 ,VONE
[3]      VDC 2 13 ,VTWO
[4]      QNPN 3 1 6 ,QAMP← 100 0.1 1E¯14 1E¯14 5E¯11 1E¯12
[5]      QNPN 3 2 7 ,QAMP
[6]      QNPN 11 9 3 ,QAMP
[7]      QNPN 10 9 8 ,QAMP
[8]      R 8 9 ,RCBS←0.001
[9]      R 5 8 ,RONE
[10]     R 10 11 ,REMIT
[11]     VDC 10 0 ¯15
[12]     VDC 5 0 15
[13]     QPNP 5 4 6 ,QAMP
[14]     R 6 4 ,RCBS
[15]     QPNP 5 4 7 ,QAMP
[16]     R 7 5 ,RBW
[17]     VAC 12 0 2 0
[18]     VAC 13 0 1 0
[19]     C 7 14 1
[20]     R 14 0 1000
[21]     END
     ∇
```

DCSOLVE, *304*

```
     ∇  DCSOLVE
[1]      VANS←IX NEWTALT VBSTART2
[2]      ' '
[3]      ' '
[4]      '↓↓DC NODE VOLTAGES 1 TO  ';(NH-1);'↓↓'
[5]      ' '
[6]      VANS
     ∇
```

‡ *DIA*

   ‡ The functions *DIA* and *'UNIT'* are found in Appendix 5.

DIFDIODE, *319*

```
      ∇  S←DIFDIODE VB
[1]       S←(ρIS)ρ0
[2]       S[DBI]←SMIN⌈IS[DBI]×JF×∗JF×VB[DBI]⌈(-VBMAX)
      ∇
```

DIFNETL, *317-318*

```
      ∇  S←DIFNETL VB;QNO
[1]       S1←DIFDIODE VB
[2]       S←GMAT+ ¯1 ¯1 ↓A+.×(DIA S1+G)+.×⍉A
[3]       QNO←1
[4]    L1: S←S+GT←GTRANS
[5]       →L1×ι(QNO←QNO+1)≤1↑ρQ
      ∇
```

DISPLAY

```
      ∇  DISPLAY CMD;VAR;M1;CONT;RC
[1]       ' '
[2]       'ITEM       CONTENTS'
[3]    L:VAR←ΔSCAN
[4]       →0×ι0=ρVAR
[5]       →V×ι0=1↑ρ⍎VAR
[6]       →V×ι1=ρρ⍎VAR
[7]     ISM:
[8]       RC←ρ CONT←⍎VAR
[9]       →0×ι(ρRC)>3
[10]      →PRC×ι3=ρRC
[11]      M1←((1↑RC),8)ρ' '
[12]      M1[1;]←8↑VAR,8ρ' '
[13]      M1[1; 6 7 8]←'↔ '
[14]      M1,⍕CONT
[15]      →L
[16]    PRC:
[17]      M1←((2↑RC),8)ρ' '
[18]      M1[1;1;]←8↑VAR,8ρ' '
[19]      M1[1;1; 6 7 8]←'↔ '
[20]      M1⍕CONT
[21]      →L
[22]    E:
[23]      VAR←8↑VAR,8ρ' '
[24]      VAR[6 7 8]←'↔ '
[25]      VAR
[26]      →L
[27]    V:
[48]      CONT←⍎VAR
[29]      VAR←9↑VAR,9ρ' '
[30]      VAR[6 7 8]←'↔ '
[31]      VAR,⍕CONT
[32]      →L
      ∇
```

DIODE, *318-319*

```
        ∇   I ←DIODE V
[1]         I ←IS×0
[2]         I[DBI]←IS[DBI]×⁻1+⋆JF×V[DBI]⌈(-VBMAX)
        ∇
```

*DO, 215

```
        ∇   DO
[1]         EXAMPLE
[2]         DCSOLVE
[3]         FLOG 1 1000000 12
[4]         ACSOLVE
[5]         20 40 PLOT LOGF,MAGDB ACVOUT 3
        ∇
```

*DTEST ,316

```
        ∇   IEQ←DTEST VB
[1]         ID←DIODE VB
[2]         GD←DIFDIODE VB
[3]         IG←VB×GD
[4]         IEQ←ID-IG
        ∇
```

END, *289-290*

```
        ∇   END;NQ;MINO
[1]         NH←(⌈/,MIN)+1
[2]         MINO←REFNODE MIN
[3]         M←⍉((ρGIN),2)ρMINO
[4]         A← NH INCID M
[5]         IX←⁻1↓A+.×IBIN
[6]         NQ←(ρQIN)÷7
[7]         Q←QCALC QIN
[8]         G←GIN
[9]         IS←ISIN
[10]        VCISL[; ι4]←((1↑ρVCISL),4)ρREFNODE,VCISL[; ι4]
[11]        CL[; 1 2]←((1↑ρCL),2)ρREFNODE,CL[; 1 2]
[12]        LL[; 1 2]←((1↑ρLL),2)ρREFNODE,LL[; 1 2]
[13]        RL[; 1 2]←LL[; 1 2]
        ∇
```

*EXAMPLE, 208

```
          ∇   EXAMPLE
[1]           STARTCCT
[2]           VAC 1 0 1 0
[3]           R 5 3 500
[4]           C 1 5 100
[5]           VDC 2 0 1
[6]           R 2 3 10000
[7]           R 3 0 5000
[8]           C 3 4 1E⁻6
[9]           R 3 4 5000
[10]          R 4 0 100
[11]          END
          ∇
```

* EXGTRANS, 319

```
          ∇   EXTRANS
[1]           STARTCCT
[2]           R 3 0 10000
[3]           R 4 3 70000
[4]           R 1 0 2000
[5]           R 4 5 10000
[6]           QNPN 1 3 5  ,QMOD13
[7]           R 4 2 10
[8]           VDC 2 0 20
[9]           END
          ∇
```

* FBTRIPLE, 233

```
          ∇   FBTRIPLE
[1]           STARTCCT
[2]           R 1 0  ,R2
[3]           QNPN 0 1 3  ,Q1
[4]           R 9 3 14600
[5]           R 9 5 14600
[6]           QNPN 4 3 5  ,QN
[7]           D 4 0  ,D1
[8]           R 9 8 14600
[9]           QNPN 7 5 8  ,QN
[10]          D 7 6  ,D1
[11]          D 6 0  ,D1
[12]          VDC 9 0 20
[13]          R 8 2  ,R1
[14]          R 2 1  ,R1
[15]          R 10 11  ,RI
[16]          C 11 1  ,C1
[17]          C 2 0  ,C2
[18]          VAC 10 0 1 0
[19]          END
          ∇
```

FLIN, 359

```
      ∇  FLIN P; FH; FL; NP
[1]      FL←P[1]
[2]      FH←P[2]
[3]      NP←P[3]
[4]      F←FL,FL+(ιNP)×(FH-FL)÷NP
      ∇
```

FLOG, 360

```
      ∇  FLOG P; LF; HF; PPD; NF; FV
[1]      LF←10⊛P[1]
[2]      HF←10⊛P[2]
[3]      PPD←P[3]
[4]      NF←⌊(HF-LF)×PPD
[5]      FV←LF+0,(ιNF)×÷PPD
[6]      F←10*FV
      ∇
```

* FLOGTP, 362

```
      ∇  FLOGTP P; LF; HF; PPD; NF; FV
[1]      LF←10⊛P[1]
[2]      HF←10⊛P[2]
[3]      NF←P[3]-1
[4]      PPD←NF÷HF-LF
[5]      FV←LF+0,(ιNF)×÷PPD
[6]      F←10*FV
      ∇
```

FORM, 350-351

```
      ∇  R←FORM T; S; K; AD; Y
[1]      NH←⌈/,M
[2]      R←(NH,NH)ρ0
[3]      K←1
[4]      →((Y←1↑ρT)=0)/0
[5]   LP: S←T[K; 1 2]
[6]      AD← 2 2 ρ(1 ¯1 ¯1 1)×T[K; 3]
[7]      R[S; S]←R[S; S]+AD
[8]   FL: →(Y≥K←K+1)/LP
      ∇
```

```
        ∇   FRANK
[1]         STARTCCT
[2]         R  6  0  76000
[3]         R  6  7  10000
[4]         R  2  7  1000
[5]         R  5  0  5.5
[6]         L  4  5  0.069
[7]         C  6  1  1E⁻6
[8]         VDC  0  7  15
[9]         VAC  1  0  1  0
[10]        OA  4  3  3
[11]        QNPN  2  6  4  ,Q1
[12]        END
        ∇
```

```
        ∇   FRANKOSC
[1]       ⍝ BASED ON ELECTRONICS II PROJECTS
[2]       ⍝ BY MICHAEL POTHIER AND BERNARD PLOURDE
[3]         STARTCCT
[4]         R  6  0  76000
[5]         R  6  7  10000
[6]         R  10  7  1000
[7]         R  8  9  22000
[8]         R  1  2  ,RF
[9]         R  3  1  20000
[10]        R  5  0  5.5
[11]        R  0  4  ,RP
[12]        L  4  5  0.069
[13]        C  4  0  ,C1
[14]        C  6  8  1E⁻6
[15]        VDC  0  7  15
[16]        VAC  9  0  1  0
[17]        OA  4  3  3
[18]        OA  0  1  2
[19]        QNPN  10  6  4  ,Q1
[20]        END
        ∇
```

```
        ∇   FSTEP P; FL; FH; ΔF; NP
[1]         FL←P[1]
[2]         FH←P[2]
[3]         ΔF←P[3]
[4]         NP←⌈(FH-FL)÷ΔF
[5]         F←FL+0,(⍳NP)×ΔF
        ∇
```

* *GICBP*, *263*

```
        ∇   GICBP
[1]         STARTCCT
[2]         VAC  1  0  1  0
[3]         R  1  2,RNOW
[4]         C  2  0,CNOW
[5]         R  2  3  10E3
[6]         C  3  4  1.59E¯9
[7]         R  4  5  10E3
[8]         R  5  6  10E3
[9]         R  6  0  10E3
[10]        OA1  2  4  5  1E6
[11]        OA1  6  4  3  1E6
[12]        END
        ∇
```

GTRANS, *324-325*

```
        ∇   GT ←GTRANS; R; ECB; F
[1]         GT ←  ¯1  ¯1  ↓GTI
[2]         →((1↑ρ Q)=0)/0
[3]         ECB ←Q[QNO;  3  5  4]
[4]         F ←Q[QNO; 6]×S1[BEBRI[QNO]]
[5]         R ←Q[QNO; 7]×S1[BEBRI[QNO]+1]
[6]         GT ←GTI
[7]         GT[ECB; ECB]←  3  3  ρ0,(-R),R,(-F),0,F,F,R,-R+F
[8]         GT ←  ¯1  ¯1  ↓GT
        ∇
```

* *HO*, *265*

```
        ∇   HO
[1]         DCDIFAMP
[2]         DCSOLVE
[3]         BJTBIAS
[4]         ACSOLVE
[5]         MAG ACVOUT 14
        ∇
```

IAC, *279*

```
        ∇   IAC P
[1]         MIN ←MIN ,P[1  2]
[2]         IBIN ←IBIN ,0
[3]         ISIN ←ISIN ,0
[4]         GIN ←GIN ,0
[5]         IACL ←IACL ,[1] P
        ∇
```

IDC, *274*

```
      ∇   IDC P
[1]       MIN←MIN,P[1 2]
[2]       IBIN←IBIN,P[3]
[3]       ISIN←ISIN,0
[4]       GIN←GIN,0
      ∇
```

IDIODE, *379*

```
      ∇   Z←IDIODE NAC
[1]       M0←M ≠(ρM )ρNH
[2]       M0←M ×M0
[3]       H1←NAC[1]=M0[1;]
[4]       H2←NAC[2]=M0[2;]
[5]       H←H1∧H2
[6]       BRI←((H≠0)∧(IS≠0))/ιρIS
[7]       Z←(DIODE VBN)[BRI]
      ∇
```

IEQUIV, *315-316*

```
      ∇   ICOM←IEQUIV VB;IEQ;QNO
[1]       IEQ←(DIODE VB)-VB ×DIFDIODE VB
[2]       ICOM←IX -⁻1↓A+.×IEQ
[3]       QNO←1
[4]     L1:ICOM←ICOM +IQ←ITRANS
[5]       →L1×ι(QNO←QNO+1)≤ 1↑ρQ
      ∇
```

IFORM, *354-355*

```
      ∇   IFORM ACL
[1]       K←1
[2]       PI←○1
[3]     LP:P←ACL[K;]
[4]       N1←P[1]
[5]       N2←P[2]
[6]       MAG←P[3]
[7]       PH←P[4]
[8]       IRE[N1]←IRE[N1]+(MAG ÷1)×2○(RAD ←PI×PH ÷180)
[9]       IIM[N1]←IIM[N1]+(MAG ÷1)×(1○RAD )
[10]      →(KS≥K←K+1)/LP
      ∇
```

INCID (Different from *INCID* of Part 1)

```
      ∇   A←N INCID M
[1]       A←((ιN)∘.=M[1;])-(ιN)∘.=M[2;]
      ∇
```

```
        ∇   INITIALIZE
[1]        NH←(⌈/,M)
[2]        A←NH INCID M
[3]        DBI←(IS≠0)/ιρIS
[4]        ΔVB←(ρIS)ρIT←0
[5]        VB ←(ρ ΔVB)ρ VB
[6]        TYPE ←Q[; 1]
[7]        IQI←(1+ρIX)ρ0
[8]        GTI←((1+ρIX),(1+ρIX))ρ0
[9]        GMAT←(NH,NH)ρ0
[10]       BEBRI←Q[; 2]
[11]       JF←11594.2÷ETA×TEMP+273
[12]       VCFORM VCISL
        ∇
```

```
        ∇   Z←IRDC P
[1]        Z←(-/(0,VANS)[1+P[1 2]])÷P[3]
        ∇
```

```
        ∇   Z←IREIM ; K; KS; P; N1; N2; MAG; PH; ACL
[1]        IRE ←IIM←(NH-1)ρ0
[2]        →((KS←1↑ρ VACL)=0)/CUR
[3]        IFORM VACL
[4]        CUR; →((KS←1↑ρ IACL)=0)/TOT
[5]        IFORM IACL
[6]     TOT: Z←(2,(NH-1),1)ρ IRE,[1]IIM
        ∇
```

```
        ∇   IQ←ITRANS; ECB; FE; RC
[1]        IQ← ¯1↓IQI
[2]        →((1↑ρ Q)=0)/0
[3]        ECB ←Q[QNO; 3 5 4]
[4]        FE ←Q[QNO; 6]×IEQ[BEBRI[QNO]]
[5]        RC←Q[QNO; 7]×IEQ[BEBRI[QNO]+1]
[6]        IQ←IQI
[7]        IQ[ECB]←(-RC),(-FE),RC+FE
[8]        IQ← ¯1↓IQ×TYPE[QNO]
        ∇
```

```
        ∇   L P
[1]        MIN ←MIN,P[1 2]
[2]        IBIN ←IBIN,0
[3]        ISIN ←ISIN,0
[4]        GIN ←GIN,÷0.00001
[5]        LL←LL,[1]P
[6]        RL ←RL,[1]P[1 2],0.00001
        ∇
```

LOG, 371

```
        ∇   Z←LOG F
  [1]       Z←10⊛F
        ∇
```

LOGF, 371

```
        ∇   Z←LOGF
  [1]       Z←LOG F
        ∇
```

* *LPF, 230*

```
        ∇   LPF
  [1]       STARTCCT
  [2]       VAC 1 0 1 0
  [3]       R 1 2 1000
  [4]       C 2 0 ,5E¯7÷○1
  [5]       VCIS 3 0 2 0 0.001
  [6]       R 3 0 1000
  [7]       END
        ∇
```

* *LTEST, 248*

```
        ∇   LTEST
  [1]       STARTCCT
  [2]       IAC 1 0 1 0
  [3]       L 1 2 0.03
  [4]       R 2 0 1000
  [5]       END
        ∇
```

MAG, 369

```
        ∇   Z←MAG V
  [1]       Z←⌹(+/[1]V*2)*0.5
        ∇
```

MAGDB, 369

```
        ∇   Z←MAGDB V
  [1]       Z←MAG V
  [2]       Z←20×10⊛Z
        ∇
```

* *MH133, 292*

```
        ∇   MH133
  [1]       G← 1 5E¯6 0 0 0.0005 0.000333 1
  [2]       IS← 0 0 0.00002 0.00002 0 0 0
  [3]       IX← 5 0 0 10 0
  [4]       M←⌹ 7 2 ρ 1 6 1 2 2 5 2 3 5 6 4 3 4 6
  [5]       Q← 1 7 ρ 1 3 5 2 3 0.99 0.001
        ∇
```

* $MH133IF$, *291*

```
        ∇  MH133IF
[1]        STARTCCT
[2]        VDC 1 0 5
[3]        R 1 2 200000
[4]        QNPN 5 2 3 100 0.1 0.00002 0.00002
[5]        R 5 0 2000
[6]        R 4 3 3000
[7]        VDC 4 0 10
[8]        END
        ∇
```

NEWTALT, 309-310

```
        ∇  V←IX NEWTALT VB;T
[1]        T←⎕AI[2]
[2]        INITIALIZE
[3]     L1:VB←ΔVB ALTBASE1 VB
[4]        ICOM←IEQUIV VB
[5]        SLOPE←DIFNETL VB
[6]        V←ICOM⌹SLOPE
[7]        ΔVB←(VBN←(⍉A)+.×V,0)-VB
[8]        IT←IT+1
[9]        →L1×⍳⌈/(|ΔVB)>(|(TOL÷100)×VBN)+0.000000000001
[10]       (⍕'CPUSECONDS= '),⍕(⎕AI[2]-T)÷1000
[11]    OUT:(⍕'ITERATIONS= ')⍕IT
        ∇
```

NEWTALT1 , 330

```
        ∇  V←IX NEWTALT1 VB;T
[1]        INITIALIZE
[2]     L1:VB←ΔVB ALTBASE1 VB
[3]        ICOM←IEQUIV VB
[4]        SLOPE←DIFNETL VB
[5]        V←ICOM⌹SLOPE
[6]        ΔVB←(VBN←(⍉A)+.×V,0)-VB
[7]        IT←IT+1
[8]        →L1×⍳⌈/(|ΔVB)>(|(TOL÷100)×VBN)+1E‾12
[9]        (⍕'IT= '),⍕IT
        ∇
```

NEWTDC , 330

```
        ∇   V←IX NEWTDC VB;T
[1]         T←□AI[2]
[2]         INITIALIZE
[3]       L1:VB←ΔVB VBCHKCAN VB
[4]         ICOM←IEQUIV VB
[5]         SLOPE←DIFNETL VB
[6]         V←ICOM⊟SLOPE
[7]         ΔVB←(VBN←(⍉A)+.×V,0)-VB
[8]         IT←IT+1
[9]         →L1×ι⌈/(|ΔVB)>(|(TOL÷100)×VBN)+1E‾12
[10]        (⍕'CPUSECONDS= '),⍕(□AI[2]-T)÷1000
[11]      OUT:(⍕'ITERATIONS= '),⍕IT
        ∇
```

NEWTDC1 , 330

```
        ∇   V←IX NEWTDC1 VB;T
[1]         INITIALIZE
[2]       L1:VB←ΔVB VBCHKCAN VB
[3]         ICOM←IEQUIV VB
[4]         SLOPE←DIFNETL VB
[5]         V←ICOM⊟SLOPE
[6]         ΔVB←(VBN←(⍉A)+.×V,0)-VB
[7]         IT←IT+1
[8]         →L1×ι⌈/(|ΔVB)>(|(TOL÷100)×VBN)+1E‾12
[9]       OUT:(⍕'ITERATIONS= '),⍕IT
        ∇
```

OA, 286

```
        ∇   OA P
[1]         R P[1], 0 100000000
[2]         R P[2], 0 100000000
[3]         R P[3],0,1
[4]         VCIS 0,P[3],P[2],0,100000
[5]         VCIS P[3],0,P[1],0,100000
        ∇
```

*OA1, 261

```
        ∇   OA1 P
[1]         P←P,100000
[2]         VCIS P[3],0,P[1],P[2],100×P[4]
[3]         R P[3],0,0.01
        ∇
```

PARMSET, 224

```
      ∇ PARMSET
[1]     ' '
[2]     'VARIOUS PROGRAM  PARAMETERS ARE'
[3]     'SET TO THEIR REFERENCE VALUES.'
[4]     'TYPE PARMTELL  FOR DETAILS.'
[5]     ' '
[6]     ALTREF←0.26
[7]     ETA←1
[8]     TEMP←20
[9]     M1←2
[10]    M2←10
[11]    SMIN←1E⁻12
[12]    VBMAX←1
[13]    VT←0.026
[14]    TOL←0.1
      ∇
```

PARMTELL

```
      ∇ PARMTELL
[1]     'ALTREF IS A CONTROL  PARAMETER FOR THE METHOD OF'
[2]     '    ALTERNATING BASES USED BY THE FUNCTION ALTBASE1'
[3]     '    WITH THE VERSION OF THE DC NEWTON RAPHSON PROCESS'
[4]     '    USING THE FUNCTIONS NEWTALT AND NEWTALT1'
[5]     ' '
[6]     'ETA VARIES BETWEEN 1 AND 2 DEPENDING ON WHETHER'
[7]     '    THE SEMICONDUCTOR DEVICES ARE GE. OR SI.'
[8]     '    IT IS USED BY FUNCTIONS DIODE AND DIFDIODE'
[9]     ' '
[10]    'TEMP IS THE GLOBAL  TEMPERATURE  IN DEGREES C'
[11]    '    OF THE DEVICE JUNCTIONS. IT IS USED BY'
[12]    '    THE FUNCTIONS DIODE AND DIFDIODE.'
[13]    ' '
[14]    'M1,M2,AND VT ARE USED BY VBCHKCAN TO CONTROL'
[15]    '    THE JUNCTION VOLTAGE INCREMENTS DURING'
[16]    '    THE DC NEWTON RAPHSON PROCESS BY THE'
[17]    '    FUNCTIONS NEWTDC AND NEWTDC1.'
[18]    ' '
[19]    'SMIN AND VBMAX ARE USED BY DIODE AND DIFDIODE'
[20]    '    TO PLACE LIMITS ON THE MINIMUM  SLOPE OF THE'
[21]    '    DIODE JUNCTION I-V RELATIONSHIP, AND THE'
[22]    '    MAXIMUM  FORWARD JUNCTION VOLTAGE, WHICH ARE'
[23]    '    PERMITTED DURING THE DC SOLUTION PROCESS'
[24]    ' '
[25]    'TOL  SETS THE  CONVERGENCE ACCURACY FOR THE DC'
[26]    '    NEWTON RAPHSON SOLUTION IN PERCENT.'
      ∇
```

† *PLOT*

PRINT, 367


```
        ∇  PRINT P
[1]        □ ←P
        ∇
```

QCALC, 288


```
        ∇  Q←QCALC QIN; QNO; EBN; CN; BEBRI
[1]        QNO ←1
[2]        Q←(NQ,7)ρQIN
[3]        BEBRI ←Q[; 2]
[4]        →((1↑ρ QIN)=0)/0
[5]      L: EBN ←M [Q[QNO;  3  4]; BEBRI[QNO]]
[6]        CN ←M [Q[QNO; 5]; BEBRI[QNO]+ 1]
[7]        Q[QNO;  3  4  5]←EBN,CN
[8]        →L×ι (QNO ←QNO +1)≤ NQ
        ∇
```

QNPN, 282-283


```
        ∇  QNPN P
[1]        MIN ←MIN,P[2  1  2  3]
[2]        IBIN ←IBIN,0,0
[3]        ISIN ←ISIN,P[6  7]
[4]        GIN ←GIN,0.000000000001, 0.000000000001
[5]        QIN ←QIN,1,((ρ ISIN)-1),2,1,2,ALPHA  P[4  5]
[6]        P ←P, 0  0
[7]        CL←CL,[1] P[1  2  8]
[8]        CL←CL,[1] P[3  2  9]
        ∇
```

QPNP, 283-284


```
        ∇  QPNP P
[1]        MIN ←MIN,P[1  2  3  2]
[2]        IBIN ←IBIN,0,0
[3]        ISIN ←ISIN,P[6  7]
[4]        GIN ←GIN,0.000000000001, 0.000000000001
[5]        QIN ←QIN, ̄1,((ρ ISIN)-1),1,2,1,ALPHA  P[4  5]
[6]        P ←P, 0  0
[7]        CL←CL,[1] P[1  2  8]
[8]        CL←CL,[1] P[3  2  9]
        ∇
```

R, 275

```
     ∇  R P
[1]     MIN←MIN,P[1 2]
[2]     IBIN←IBIN,0
[3]     ISIN←ISIN,0
[4]     GIN←GIN,÷P[3]
     ∇
```

* *RDTEST*, 375

```
     ∇  RDTEST
[1]     STARTCCT
[2]     VDC 1 0 5
[3]     R 1 2 2000
[4]     R 1 2 1000
[5]     D 2 3 ,D1
[6]     R 3 4 1000
[7]     D 4 0 ,D2
[8]     D 4 0 ,D1
[9]     D 0 4 ,D1
[10]    END
     ∇
```

REALSOLV, 356-357

```
     ∇  V←Y REALSOLV I;YT;YB
[1]     YT←Y[1;;],(⁻1×Y[2;;])
[2]     YB←Y[2;;],Y[1;;]
[3]     V←(ρI)ρ(,I)⌹YT,[1]YB
     ∇
```

REFNODE, 287

```
     ∇  MNEW←REFNODE MZ;GNODEI
[1]     MNEW←MZ
[2]     GNODEI←(MZ=0)/ιρMZ
[3]     MNEW[GNODEI]←NH
     ∇
```

STARTCCT, 273

```
     ∇  STARTCCT
[1]     MIN←IBIN←ISIN←GIN←QIN←ι0
[2]     LL←RL←CL←0 3 ρ0
[3]     IACL←VACL← 0 4 ρ0
[4]     VCISL← 0 5 ρ0
     ∇
```

TRANS

```
      ∇  ANS←TRANS NOUT
[1]      'TYPE CIRCUIT NAME'
[2]      CCT ←Ü
[3]      'TYPE NAME OF PARAMETER TO BE VARIED'
[4]      PAR←Ü
[5]      'TYPE PARAMETER VALUES'
[6]      PVAL ←,□
[7]      I ←1
[8]      ♇ PAR,'←PVAL[I]'
[9]      ♇ CCT
[10]     VALL←((ρ PVAL),(ρ IX))ρ 0
[11]     VBN ←VBSTART2
[12]     →J1
[13]     L1: ♇ PAR,' ←PVAL[I]'
[14]     ♇ CCT
[15]     J1: VALL[I; ]←IX NEWTDC1 VBN
[16]     →L1×ι (ρ PVAL)≥I ←I +1
[17]     ANS←VALL[; NOUT]VS PVALL
      ∇
```

TRANSDC, 306-308

```
      ∇  ANS←TRANSDC NOUT
[1]      'TYPE CIRCUIT NAME'
[2]      CCT ←Ü
[3]      'TYPE NAME OF PARAMETER TO BE VARIED'
[4]      PAR←Ü
[5]      'TYPE PARAMETER VALUES'
[6]      PVAL ←,□
[7]      I ←1
[8]      ♇ PAR,' ←PVAL/[I]'
[9]      ♇ CCT
[10]     VALL←((ρ PVAL),(ρ IX))ρ 0
[11]     VBN ←VBSTART2
[12]     →J1
[13]     L1: ♇ PAR,'←PVAL[I]'
[14]     ♇ CCT
[15]     J1:VALL[I; ]←IX NEWTALT1 VBN
[16]     →L1×ι (ρ PVAL)≥I ←I+1
[17]     ANS←VALL[; NOUT]VS PVAL
      ∇
```

‡*UNIT*

VAC, 280

```
     ∇  VAC P
[1]     MIN ←MIN,P[1 2]
[2]     IBIN ←IBIN,0
[3]     ISIN ←ISIN,0
[4]     GIN ←GIN,1
[5]     VACL ←VACL,[1] P
     ∇
```

VBCHKCAN

```
     ∇  V←△VB VBCHKCAN VB;A;B;C;D;E;P;△V1;△V2;
        △V3;△V4;△V5;△VK;VK1;VK
[1]     VK1 ←(VK ←VB[DBI])+△VK ←△VB[DBI]
[2]     P←VK1>VK
[3]     A←VK1<M2×VT
[4]     B ←(VK1-VK)>M1×VT
[5]     C ←(VK+M1×VT)<M2×VT
[6]     D ←VK<M2×VT
[7]     E ←(VK-VK1)>M1×VT
[8]     △V1 ←△VK×(P∧A)∨P∧(~A)∧~B
[9]     △V2 ←M1×VT×P∧(~A)∧B∧~C
[10]    △V3 ←((-VK)+M2×VT)×P∧(~A)∧B∧C
[11]    △V4 ←△VK×((~P)∧D)∨(~P)∧(~D)∧~E
[12]    △V5 ←-M1×VT×(~P)∧(~D)∧E
[13]    △VB[DBI]←△VK ←△V1+△V2+△V3+△V4+△V5
[14]    V←VB+△VB
[15]    V[DBI]←|V[DBI]⌊VBMAX
     ∇
```

VBSTART, 310

```
     ∇  VB ←VBSTART
[1]     VB ←(ρG)ρ 0.4
     ∇
```

VBSTART2, 310-311

```
     ∇  VB ←VBSTART2
[1]     VB ←(ρG)ρ 0.4
[2]     VB[Q[; 2]+1]← ‾1
     ∇
```

VCA, 261

```
      ∇   VCA D
[1]       QNPN P[9 1 7],QVAL←10000 1E¯6 1E¯14 1E¯14 0 0
[2]       QNPN P[9 2 8],QVAL
[3]       R P[7 5],100
[4]      .R P[8 5],100
[5]       OA1 P[8 7 4], 1
[6]       VCIS P[6 9 3], 0 0.001
      ∇
```

VCFORM, 328-329

```
      ∇   VCFORM VCISL;K;P;NTOI;NFPN;GM
[1]       →((1↑ρ P←VCISL)=0)/LAST
[2]       K←1
[3]   LP:NTOI←P[K; 1 2]
[4]       NFPN←P[K; 3 4]
[5]       GM←P[K;5]
[6]       GMAT[NTOI;NFPN]←GMAT[NTOI;NFPN]+ 2 2 ρ ¯1 1 1
          ¯1×GM
[7]       →((1↑ρ P)≥K←K+1)/LP
[8]   LAST:GMAT← ¯1 ¯1 ↓GMAT
      ∇
```

VCIS, 285

```
      ∇   VCIS P
[1]       MIN←MIN,P[1 2]
[2]       IBIN←IBIN,0
[3]       ISIN←ISIN,0
[4]       GIN←GIN,0
[5]       VCISL←VCISL,[1] P
      ∇
```

VCT, 261-262

```
       ∇  VCT
[1]       STARTCCT
[2]       VDC 1 10  ,VDX÷2
[3]       VDC 2 11  ,-VDX÷2
[4]       VAC 10 0  ,(VAX÷2),0
[5]       VAC 11 0  ,(VAX÷2),180
[6]       R 4 0 10000
[7]       VDC 3 0  ,VCX
[8]       VDC 6 0   15
[9]       VDC 5 0 15
[10]      VCA 1 2 3 4 5 6 7 8 9
[11]      END
       ∇
```

VDC, *276*

```
      ∇   VDC P
[1]       MIN←MIN,P[1 2]
[2]       IBIN←IBIN,P[3]÷1
[3]       ISIN←ISIN,0
[4]       GIN←GIN,÷1
      ∇
```

VGAINANG, *376*

```
      ∇   Z←NI VGAINANG NO;ZO;ZI
[1]       ZO←ANGLE ACVOUT NO
[2]       NI←(ρNO)ρNI
[3]       ZI←ANGLE ACVOUT NI
[4]       Z←ZO-ZI
      ∇
```

VGAINMAG, *376*

```
      ∇   Z←NI VGAINMAG NO;ZO;ZI
[1]       ZO←MAG ACVOUT NO
[2]       NI←(ρNO)ρNI
[3]       ZI←MAG ACVOUT NI
[4]       Z←ZO÷ZI
      ∇
```

† *VS*

YNET, *355*

```
      ∇   Y←YNET
[1]       YIMAG← ⁻1 ⁻1 ↓YREACT←BB
[2]       Y←YCOMP←YRE,[0.5]YIMAG
      ∇
```

† *ZATN*

Δ*SCAN* (Used in *DISPLAY*; *see page 428*)

```
      ∇   R←ΔSCAN
[1]       CMD←(⁻1+(,(CMD=' ')∨(CMD=','))ι0)↓CMD
[2]       R←(⁻1+(,(CMD=' ')∨(CMD=','))ι1)↑CMD
[3]       CMD←(ρR)↓CMD
[4]       CMD←(⁻1+(,(CMD=' ')∨(CMD=','))ι0)↓CMD
      ∇
```

Principal Variables for Part 2 in Alphabetical Order

Entries in the *Shape* column of this listing should be interpreted in the APL sense; i.e. in the order planes, rows, columns for a three-dimensional array, where

B	represents the number of circuit branches;
JCNS	represents the total number of diode and transistor junctions;
NH	represents the total number of nodes in the circuit counting the reference node;
NR	represents the number of nodes in the circuit not counting the reference node: NR = NH−1;
T	represents the number of bipolar junction transistors;
#	represents the number of components, sources, etc. of a given type;
S	represents a scalar.

Name	Description	Shape
A	incidence matrix — indicates whether each current flows out of or into a particular node	NH B
ALTREF	see *PARMSET* and *PARMTELL*	S
BC	capacitance matrix in *ACPREP*	NH NH
BEBRI	base–emitter branch index vector; contains the base–emitter branch number for each transistor	T
BGL	matrix of conductance values substituted in place of inductors in the d.c. solution and subtracted out in *ACPREP*	NH NH
BL	reciprocal inductance matrix in *ACPREP*	NH NH
CCT	character variable holding the circuit name in *TRANSDC*	—

CL	storage table for capacitors	# 3
DBI	diode branch index — a vector of all PN junction branch numbers	JCNS
DPI	vector containing the index number of each PN junction with a branch voltage increment that is positive, and greater than the reference parameter $ALTREF$, at a particular iteration step in the function $ALTBASE1$	—
ECB	emitter–collector–base node number vector for a particular transistor; local to $GTRANS$ and $ITRANS$	3
ETA	see $PARMSET$ and $PARMTELL$	S
F	global frequency vector, Hz	$\rho\,F$
F	used as a local variable in $GTRANS$ for the $\alpha_F G_E$ product	S
FE	linearized forward emitter current product term, $\alpha_F I_{EEQ}$ for a given transistor, local to $ITRANS$	S
G	vector of branch conductances	B
GIN	preliminary branch conductance vector; assigned to G in END	B
$GMAT$	a conductance matrix which is initially set to 0 in $INITIALIZE$ and used to accumulate various conductance components, e.g., from controlled sources in $VCFORM$	NH NH also NR NR
GT	reduced dependent source conductance matrix in $GTRANS$ and $DIFNETL$	NR NR
GTI	initial transistor node conductance matrix; contains zeros	NH NH
$IACL$	storage table for a.c. current sources	# 4
$IBIN$	d.c. branch current source vector used in the input functions, and converted to the node source current vector IX in END	B

Name	Description	Shape
ICOM	nodal current vector, including equivalent current sources from diodes and transistors; result variable in *IEQUIV* and global variable in *NEWTALT*,*NEWTDC*,*NEWTALT1* and *NEWTDC1*	NR
ICOMPLEX	array of real (plane 1) and imaginary (plane 2) a.c. node source currents used in *ACSOL*	2 NR 1
IDNEW	revised diode current vector in *ALTBASE1*	B
IEQ	vector of equivalent linearized branch currents for each diode & transistor junction; local to *IEQUIV*	B
IIM	vector of imaginary parts of a.c. node source currents	NR
IQ	equivalent linearized dependent current contributions for a transistor in *IEQUIV*	NR
IQI	a vector containing an initialized set of values for the transistor's nodal current source components	NH
IRE	vector of real parts of a.c. node source currents	NR
IS	vector of diode branch saturation currents; entry is zero for branches not containing diodes	B
ISIN	preliminary diode saturation current vector; assigned to *IS* in *END*	B
IT	the iteration number of Newton–Raphson solutions in *NEWTDC*, *NEWTALT*, etc.	S
IX	d.c. nodal source current vector	NR
JF	semiconductor junction factor defined in *INITIALIZE*	S
LL	storage table for inductors; see *L*	# 3
M	branch connection matrix	2 B
MIN	preliminary form of *M*; converted to *M* in *END*	2×B

M1 and M2	see *PARMSET* and *PARMTELL*	S
NH	total number of nodes in the circuit including the reference node — see *INITIALIZE*	S
NQ	number of bipolar junction transistors; used in *END* and *QCALC*	S
PAR	character variable holding the name of the independent parameter to be varied in *TRANSDC*	—
PVAL	vector of values of the independent parameter used for a d.c. transfer function solution in *TRANSDC*	# of values used
Q	the transistor data storage matrix	T 7
QIN	preliminary vector of transistor connection and parameter data; converted to *Q* matrix in *QCALC* in *END*	7×T
QNO	the transistor index number; local in *DIFNETL* and *IEQUIV*	S
R	used as a local variable in *GTRANS* for the $\alpha_R G_C$ product	S
RC	linearized reverse collector current product term, $\alpha_R I_{CEQ}$ fir a given transistor; local in *ITRANS*	S
RL	storage table for resistors substituted in place of inductors in the d.c. solution; see function *L*	# 3
SLOPE	the circuit's reduced nodal incremental conductance matrix, including resistor, diode and transistor contributions	NR NR
SMIN	see *PARMSET* and *PARMTELL*	S
S1	diode slope conductance vector in *DIFNETL*	B
TEMP	diode and transistor junction temperature; see *PARMSET* and *PARMTELL*	S
TOL	percent tolerance convergence criterion; see *PARMSET* and *PARMTELL*	S
TYPE	a vector of transistor types ($NPN=1$, $PNP = {}^-1$) used in *GTRANS* and *ITRANS*	T

Name	Description	Shape
V	the d.c. node voltage vector — voltages with respect to reference node; result variable used in several functions such as *NEWTDC* and *NEWTALT*	NR
VACL	storage table for a.c. voltage sources	# 4
VALL	matrix of all node voltages produced by a d.c. transfer function solution; see *TRANSDC*	ρ*PVAL*
VANS	vector of d.c. node voltages produced by *DCSOLVE*	NR
VB	the d.c. branch voltage result vector in *NEWTDC*, *NEWTALT*, etc.	B
VBMAX	see *PARMSET* and *PARMTELL*	S
VBN	the 'new', iterated values of d.c. branch voltages in *NEWTDC*,\ *NEWTALT*, etc.	B
VCISL	storage table for voltage-controlled current sources; see function *VCFORM* and *INITIALIZE*	# 5
VREIM	real (plane 1) and imaginary (plane 2) a.c. node voltages	2 NR NR
VT	see *PARMSET* and *PARMTELL*	S
W	radians/second at a particular frequency	S
WW	vector of all frequencies, radians/second , in *ACSOL*	ρ*F*
YCOMP	reduced composite admittance array with reals in plane 1 and imaginaries in plane 2, evaluated at a particular frequency in *YNET*	2 NR NR
YIMAG	reduced admittance matrix, imaginary part, evaluated at a particular frequency in *YNET*	NR NR
YRE	matrix of resultant conductances produced in *ACPREP*, and used in *YNET* for the a.c. solution	NR NR
Δ*VB*	incremental branch voltage vector (equals *VBN-VB*) in *NEWTDC*, *NEWTALT*, etc.	B

References

Calahan, D. A. (1972), *Computer-aided Network Design*, McGraw-Hill.

Chua, L. O. and Lin, P. M. (1975), *Computer-aided Analysis of Electronic Circuits: Algorithms and Computational Techniques*, Prentice-Hall.

Desoer, C.A. and Kuh, E.S. (1969), *Basic Circuit Theory*, McGraw-Hill.

Ebers, J. J. and Moll, J. L. (1954), Large signal behaviour of junction transistors, *Proc. IRE,* **42**, 1761–72.

Fidler, J. K. and Nightingale, C. (1978), *Computer-aided Circuit Design*, Nelson.

Gibson, L. (1982), Designing user-friendly APL systems, *Proceedings 1982 APL Users' Meeting*, Vol. 2, I.P. Sharp Associates, Toronto.

Gilman, L., and Rose, A. (1984), *APL An Interactive Approach*, Third Edition, Wiley.

Grinich, V. H. and Jackson, H. G. (1975), *Introduction to Integrated Circuits*, McGraw-Hill.

Householder, A. S. (1957), A survey of some closed methods for inverting matrices, *J. SIAM*, **5**, 155–69.

LePage, W. (1978), *Applied APL Programming*, Prentice-Hall.

Nagel, L. W. and Roher, R. A. (1971), Computer analysis of nonlinear circuits excluding radiation (CANCER), *IEEE J. Solid State Circuits,* **SC-6**, 166–82.

Pinel, J. F. and Blostein, M. L. (1967), Computer techniques for the frequency analysis of linear electrical networks, *Proc. IEEE,* **55**(11), 1810–19.

Ralston, A. (1965), *A First Course in Numerical Analysis*, McGraw-Hill.

Spence, R. (1979), *Resistive Circuit Theory*, Palo Alto, APL Press.

Spence, R. (1984), *Tolerance analysis and design of electronic circuits*, Computer-Aided Engineering Journal, 1, 3, pp. 91-99

Soin, R. S. and Spence, R. (1980), *Statistical exploration approach to design centring*, IEE Proc. 129G, **4**, pp. 181-185

Index